IEE ELECTROMAGNETIC WAVES SERIES 14
SERIES EDITORS: PROFESSOR P.J.B. CLARRICOATS,
E.D.R. SHEARMAN
AND J.R. WAIT

Leaky Feeders and Subsurface

Radio Communication

P. Delogne

Previous volumes in this series

Leaky Feeders and Subsurface Radio Communication

P. Delogne

PETER PEREGRINUS LTD.
On behalf of the
Institution of Electrical Engineers

Published by: The Institution of Electrical Engineers, London
               and New York
               Peter Peregrinus Ltd., Stevenage, UK, and New York

Delogne, P.
    Leaky feeders and subsurface radio communications.
    — {Electromagnetic wave series; 14)
    1. Signal theory (Telecommunications)
    I. Title   II. Series
    621.38'043   TK5102.5

ISBN 0-906048-77-X

Printed in England by Short Run Press Ltd., Exeter

Contents

★ see Introduction

Preface

This book has its origin in a letter. In August 1978, Professor James R. Wait wrote to me, asking if I would contribute a book on leaky feeders to the IEE Electromagnetic Waves Series.

Participants in the XVIIIth General Assembly of the International Union of Radio Science, held a few days before in Helsinki, will remember that the B. van der Pol Gold Medal had just been awarded to the requester for his work on this topic. The easiest possible answer from myself to Professor Wait would have been that nobody was better placed than himself to carry out this work.

In spite of other heavy commitments, I gave a positive answer. Indeed, I followed closely the remarkable theoretical work performed by Professor Wait and his coworkers. I had myself carried out some theoretical analyses, and I had gained much field experience from my collaboration with the Belgian Institut National des Industries Extractives (INIEX) and the Régie Autonome des Transports Parisiens (RATP). I thus thought that writing the book would be a light but valuable task. I now know that at least the first of these adjectives was fanciful. I nevertheless hope that the reader will endorse the second.

If he does, it will be owing in great measure to the people with whom I have had the good fortune to be associated for several years. Most of my work in the present field has been carried out as a consultant to INIEX. This fertile association with R. Liegeois, R. De Keyser and L. Deryck benefited from frequent contacts with J. R. Wait and D. A. Hill from the USA, Q. V. Davis and D. J. R. Martin from England, R. Gabillard and P. Degauque from France, and many others. Field experience at UHF was gained from cooperation with MM. Jouan and Sniter, and Mme Malet of RATP.

My colleagues and collaborators at the Catholic University of Louvain have been a constant stimulus and encouragement. The manuscript was carefully typed by Mrs E. Colle and H. Lobelle.

Finally, to my wife and children I express warm thanks for their contribution of evenings and weekends which might have been spent otherwise.

Louvain-la-Neuve
22 April 1982

PAUL DELOGNE

Chapter 1

Introduction

Radio waves do not propagate very well in tunnels. This fact is experienced every day by many car drivers, but what is here merely a discomfort becomes a serious drawback for people working below the earth's surface. This is particularly true in mines, where mobile radio communications are today considered essential for increased safety and productivity. Very surprisingly, the subject of electromagnetic wave propagation in subsurface works was almost unexplored until the end of the 1960s. Since then, substantial research efforts have been made on this topic in several countries and a good knowledge of the involved propagation mechanisms is now available. Radio systems now work satisfactorily in many mines and in road and railway tunnels. The time has come to write the first book on this subject and on the related topic of leaky feeders.

The main objective of this monograph is to gather together many results scattered in the open and the less accessible literature rather than to introduce new matter, although several paragraphs, given in the last three chapters, are based on personal unpublished notes. Electromagnetic wave propagation problems related to subsurface radio communications are extremely complex. When commencing this work the author took into consideration that some potential readers of this book would be concerned with the design of subsurface radio systems rather than with their theory. Consequently it was decided to collect all the difficult theoretical approaches in Chapter 2 and to write the book in such a way that the uninterested reader could skip this chapter. Reading of the sections marked by an asterisk is however recommended. Conversely some theoreticians may not be interested in system design; they will nevertheless find interesting data and facts as well as sources for new investigations in the more practical Chapters 3 to 5. This choice unavoidably results in some duplication. Chapter 1 contains a common introduction for both categories of reader. Chapter 3 is devoted to an analysis of modes and mode conversion processes and techniques which are known to play a fundamental role in subsurface propagation. The methods used there are based on transmission line theory, an approach which is justified in Chapter 2: this is still theory, but it is readable by people who are not specialists in electromagnetic theory and it is necessary for a good understanding of the more practically oriented Chapters 4 and 5.

1.1 Material properties

The question of the electrical parameters of natural rock materials is rather complex and could form the subject of another book. The fine structure of many natural rocks is crystalline. The crystals are always very small compared with the wavelengths used in subsurface radio communications and, in so far as they have no preferential orientation, we may consider that the bulk electrical properties of the material are isotropic. In fact they frequently do have a preferred orientation due to geological stratification and this creates some anisotropy. This effect will not be considered here because it is negligible compared with the wide range and high variability of the values of electrical parameters. These are the dielectric constant κ, the permeability μ, and the conductivity σ.

Excluding rocks with a high concentration of metals of the ferromagnetic group, including iron, nickel and cobalt, the permeability of natural materials is very close to that of a vacuum, $\mu_0 = 4\pi \times 10^{-7}$ H m^{-1}. Unless otherwise mentioned, this value will be adopted for all the calculations contained in this book.

The dielectric constant and the conductivity of rocks, on the other hand, are highly variable. Dielectric constants range from 2 to 70 but more frequently from 4 to 10. Conductivities range from 10^{-6} to 1 S m^{-1} and may be spread over more than two decades for a given type of material at a fixed frequency. In general both κ and σ increase with the water content. As a general rule the conductivity increases and the dielectric constant decreases with frequency, because of dispersion effects; these evolutions are not independent for they are related by Kramers–Kronig relations (Landau and Lifshitz, 1960), which may be used to check the validity of experimental data. Furthermore the conductivity and the dielectric constant may depend on the pressure (Onsager effect). Extremely high pressures exist at large depths, but when a tunnel is bored, decompression occurs in the surrounding rock. As a result the pressure effect may be neglected.

The conductivity and the dielectric constant of samples can be measured with good accuracy in the laboratory. One should be very careful about the validity of such measurements because of the difficulty of taking a sample and bringing it to the laboratory without modifying the electrical parameters. Furthermore values obtained in this way only relate to a small sample and are not necessarily representative if the ground is inhomogeneous. *In situ* measurements and particularly the careful comparison of propagation data with calculations made from adequate models are undoubtedly preferable. They are also more difficult to obtain.

Fig. 1.1 shows the range of values obtained by J. C. Cook (1975) from laboratory measurements on various rock materials and on several types of concrete at four frequencies. We have shown only the global range covered by the measurements on each type of rock: the reader interested in detailed values is referred to the original paper. The distinction between wet and dry material has been made only for those materials where this yields larger differences than the spread between samples of the same type of rock.

Although the values shown in this figure result from an intensive series of

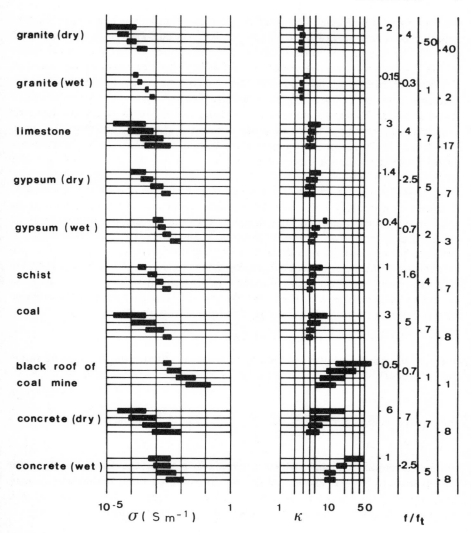

Fig. 1.1 *Electrical parameters of some materials. For each material the heavy bars are from top to bottom for the frequencies of 1, 5, 25 and 100 MHz*

measurements, they should still be considered as indicative. Values lying outside the indicated range may be encountered in practice. We are here particularly interested in the electrical properties of coal and of the surrounding rock. As the figure shows, the conductivity of coal is rather low. A remarkable property used in one projected radio communication system is that the conductivity of the roof in coal mines is much larger than that of the seam. Emslie and Lagace (1978) have reported conductivities of 3×10^{-5} to 4×10^{-3} for coal and $7 \cdot 7 \times 10^{-3}$ to $1 \cdot 09$ S m^{-1} for the roof in US mines at medium frequencies. In the various cases the ratio of the conductivities was between 75 and 5400. Values measured by Balanis *et al.* (1978) for bituminous coal up to 100 MHz are consistent with this, except that the dielectric constant of coal is higher at low frequencies: 16 to 41 at 500 kHz and 10 to 34 at 1 MHz. These authors have observed that the conductivity and the dielectric constant may vary by a factor of about 4 with the direction, but that these parameters are nearly insensitive to temperature up to $370\,^{\circ}$C. They also report measurements made at 9 GHz for bituminous coal: $\kappa = 3 \cdot 4$ to $3 \cdot 9$ and $\sigma = 0 \cdot 12$ to $0 \cdot 73$ S m^{-1}.

A characteristic frequency for a given medium is the transition frequency f_t defined by the equality $2\pi f_t \epsilon = \sigma$. Below this frequency, conduction currents in the material are more important than displacement currents and the material may be regarded as a conductor. Above this frequency, the inverse situation occurs and the material behaves rather like a dielectric. The last column of Fig. 1.1 gives, for each of the four frequencies considered, a mean value of the ratio f/f_t, where f_t has been calculated using the mean values of σ and κ at the frequency f.

For quick reference, Fig. 1.2 shows the real and imaginary parts of the propagation constant for a uniform plane wave in a homogeneous medium

$$\Gamma = \alpha + j\beta = (j\omega\epsilon_0\kappa + \sigma)j\omega\mu_0 \tag{1.1}$$

for $\kappa = 10$ and for various conductivities. Scales are also drawn for the penetration (or skin) depth

$$\delta = 1/\alpha \tag{1.2}$$

and for the wavelength

$$\lambda = 2\pi/\beta \tag{1.3}$$

At the transition frequency, which is marked by a dot on the curves, the argument of Γ is equal to $3\pi/8$. It should however not be overlooked that the material properties, including the transition frequency, vary with frequency. This figure may also be used to obtain the complex refractive index n, which is defined by

$$n = \Gamma/(jk_0) \tag{1.4}$$

where $k_0 = \omega\sqrt{\epsilon_0\mu_0}$ is also shown in the figure.

1.2 Natural propagation in empty tunnels

We start our analysis of subsurface propagation by considering a straight tunnel and we assume that it does not contain any object and in particular any axial conductor. This is a very rare situation, for most tunnels contain conductors such as power cables strung along the lateral walls or under the roof, or rails, water pipes, etc. As will become evident in the following paragraphs, such conductors would change drastically the electromagnetic properties of the structure. In spite of the sometimes mediocre conductivity of the tunnel walls, we can already obtain a good picture of the propagation phenomena by comparing the tunnel with a hollow waveguide with highly conducting walls.

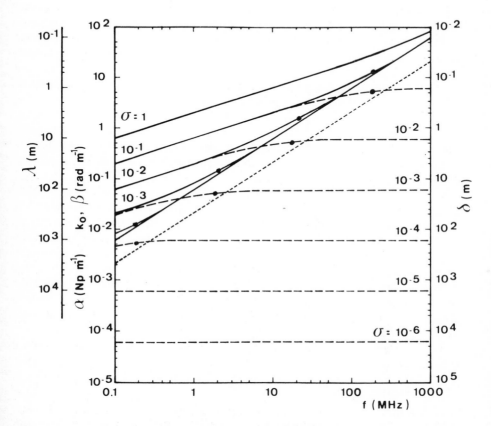

Fig. 1.2   *Propagation parameters of a medium with dielectric constant $\kappa = 10$. Solid lines show β and λ. Broken lines are for α and δ. The dotted line is for the free-space wave number k_0. The dot on the α and β curves shows the transition frequency*

1.2.1 Simplified waveguide model

The theory of waveguides with perfectly conducting walls is available in numerous books, e.g. Marcuvitz (1951), and we will not develop it again here. We will only recall the main results as far as they have implications in the context of electromagnetic wave propagation in tunnels. The analysis is made for the harmonic time dependence $\exp(j\omega t)$. Theory resolves the electromagnetic field into a sum of solutions which are called modes. A mode is a solution with a dependence on the longitudinal coordinate z by a factor $\exp(-\Gamma z)$. The complex constant

$$\Gamma = \alpha + j\beta \tag{1.5}$$

is the propagation constant of the mode. Its real part α is the specific attenuation expressed in Np m^{-1} and its imaginary part β is the specific phase shift or phase constant expressed in rad m^{-1}. The various modes obviously have different propagation constants and different field distributions.

For hollow waveguides with perfectly conducting walls the dependence of the propagation constant of a given mode on the frequency is very simple. Each mode is characterised by a critical frequency f_c which depends on the tunnel shape and size. Below this frequency the mode is evanescent, i.e. it suffers only attenuation and no phase shift: the specific attenuation is given by

$$\alpha = 2\pi f_c/c \; [1 - (f/f_c)^2]^{1/2} \tag{1.6}$$

where $c = 3 \times 10^8$ m s^{-1} is the speed of light. Above its critical frequency the mode propagates without attenuation and with a phase constant given by

$$\beta = 2\pi f_c/c \; [(f/f_c)^{2} - 1]^{1/2} \tag{1.7}$$

The mode having the lowest critical frequency is called the dominant mode and its critical frequency is called the waveguide cutoff frequency. Below this frequency all modes are evanescent and no electromagnetic energy can be conveyed by the waveguide.

The modes of a hollow waveguide with perfectly conducting walls are either transverse electric (TE) or transverse magnetic (TM), which means that either the electric or the magnetic field has no axial component. The dominant mode is always a TE mode. The modes are further labelled by a two-dimensional order number (m, n).

For a rectangular waveguide with cross-sectional dimensions a and b, the critical frequencies are given by

$$f_{cmn} = c/2 \; [(m/a)^2 + (n/b)^2]^{1/2} \tag{1.8}$$

For TE modes m and n may be zero but not simultaneously, while for TM modes both must be strictly positive. Assuming that $a > b$, we see that the cutoff frequency is given by

$$f_c = f_{c10} = c/(2a) \tag{1.9}$$

which means that the free-space wavelength at the cutoff frequency is twice the largest side of the rectangle.

For a circular waveguide with a radius a the critical frequencies are given by

$$f_{cmn} = \begin{cases} (cx_{mn})/2a; & \text{TM modes} \\ (cx'_{mn})/2a; & \text{TE modes} \end{cases} \tag{1.10}$$

where $m \geqslant 0$; x_{mn} and x'_{mn} are the nth zeros of the Bessel function $J_m(x)$ and of its derivative, respectively. The dominant mode is the TE_{11} mode. The critical frequencies of the six lowest-order modes are as follows:

$TE_{11} : 0\cdot293c/a$ $TE_{01} :0\cdot609c/a$
$TM_{01} :0\cdot383c/a$ $TM_{11} :0\cdot609c/a$
$TE_{21} :0\cdot485c/a$ $TE_{31} :0\cdot668c/a$

For tunnels of arbitrary shape, a very rough approximation is that the cutoff frequency is such that the free-space wavelength at this frequency is about equal to the tunnel perimeter and also that, well above the cutoff, the number of propagating modes grows as the square of frequency. The cross-section of most tunnels has dimensions of a few metres and the cutoff frequencies are consequently of a few tens of megahertz. Below the tunnel cutoff the specific attenuation predicted by eqn. 1.6 is extremely high, say 1 dB m^{-1} or more, and no radio communication seems possible over appreciable ranges.

The model of a tunnel with perfectly conducting walls considered up to now is of course highly idealised. The most immediate effect of a finite conductivity is to introduce some attenuation of the modes above their respective critical frequency. In the case of metallic waveguides such as those used at microwave frequencies, the wall conductivity is nevertheless very high and the skin depth in the walls is extremely small when compared with the waveguide cross-section. This allows the calculation of an approximate solution by a perturbation method. For the purpose of this introduction we will recall the formulas obtained by this method for the specific attenuation of the modes above their respective critical frequency in a waveguide with highly conducting walls (Marcuvitz, 1951). For a rectangular waveguide with width a and height b, one has for the TE modes

$$\alpha = \frac{R}{\eta_0 b} \frac{\epsilon_n m^2 b/a + \epsilon_m n^2}{m^2 b/a + n^2 a/b} (1 - f_{cmn}^2/f^2)^{1/2} + \frac{(\epsilon_n + \epsilon_m b/a)f_{cmn}^2/f^2}{(1 - f_{cmn}^2/f^2)^{1/2}}$$

(1.11)

and for the TM modes

$$\alpha = \frac{2R}{\eta_0 a} \frac{m^2 + n^2 a^3/b^3}{m^2 + n^2 a^2/b^2} \frac{1}{(1 - f_{cmn}^2/f^2)^{1/2}}$$

(1.12)

where $\epsilon_m = 1$ if $m = 0$ and $\epsilon_m = 2$ if $m \neq 0$. For a circular waveguide with radius a, one has for TE modes

$$\alpha = \frac{R}{\eta_0 a} \frac{m^2}{x_{mn}'^2 - m^2} + \frac{f_{cmn}^2}{f^2} \frac{1}{(1 - f_{cmn}^2/f^2)^{1/2}}$$

(1.13)

and for TM modes

$$\alpha = \frac{R}{\eta_0 a} \frac{1}{1 - f_{cmn}^2/f^2}$$

(1.14)

In these formulas $\eta_0 = 377 \, \Omega$ is the intrinsic impedance of the air and R is the real part of the intrinsic impedance of the wall, given by

$$R = [\omega\mu_0/(2\sigma)]^{1/2} \tag{1.15}$$

for a good conductor, i.e. well below the transition frequency of this medium. This is a condition for the validity of these formulas; another condition is that the skin depth in the wall should be small compared with the dimensions of the tunnel cross-section.

As the last column of Fig. 1.1 shows, the first assumption will never be justified. Above the cutoff, i.e. at a few tens of megahertz, the ratio f/f_t is larger than unity in spite of the increase of the conductivity with frequency. Consequently the wall material behaves in all cases as a lossy dielectric rather than as a conductor, except perhaps for the most conductive materials ($\sigma > 0\cdot1 \, \text{S m}^{-1}$) and below 100 MHz. It may thus be preferable to use the exact formula

$$R = \text{Re} \, [j\omega\mu_0 \, (j\omega\epsilon_0\kappa + \sigma)]^{1/2}$$

$$= \eta_0 \, \text{Re} \, [\kappa - j\sigma/(\omega\epsilon_0)]^{1/2} \tag{1.16}$$

instead of its approximation 1.15. The use of eqns. 1.11 to 1.14 remains questionable however because, as can be seen on Fig. 1.2, the assumption that the skin depth is small compared with the tunnel cross-section is frequently not fulfilled.

1.2.2. Comparison with experimental results

In spite of the expected weakness of the simple waveguide model, Deryck (1979) has nevertheless obtained fairly good agreement between some experimental results and calculations based on eqns. 1.11 to 1.15. Fig. 1.3 shows his results for the Lanaye tunnel, near Liège in Belgium. This tunnel is dug in calcareous tufa. It is 1600 m long, 5 to 6 m high and 4 to 5 m wide. The rock overburden is about 50 m thick. The value of the wall conductivity was obtained from a measurement of the attenuation of wave propagation through the rock and was found to be 0·01 S m^{-1} at 30 MHz. The tunnel has a flat floor and a round roof: it can be compared with a circular cylinder with a radius of 2·5 m, for which the cutoff frequency is 35 MHz.

The crosses on Fig. 1.3 show the measured values of the specific attenuation. Curve 2 shows the theoretical attenuation of a waveguide with perfectly conducting walls below the cutoff, while line 1 shows the attenuation for through-the-rock propagation. Theoretical attenuation curves above the cutoff obtained from eqns. 1.13 and 1.14 together with the approximation 1.15 are also drawn in solid lines. Deryck's conclusion is that the experimental results are for each frequency in accordance with the theoretical curve which predicts the lowest attenuation. This conclusion should not be considered as general. A counterexample in which the highest predicted attenuation is selected below cutoff will be found below.

The comparison of the measurements with the theoretical curves above the cutoff should also be used carefully. Indeed the steady decrease of the TE$_{01}$ mode attenuation with frequency is a unique property of the circular waveguide and does not generally exist for other cross-sectional shapes. A theoretical calculation based on the high-conductivity assumption and taking into account that the tunnel floor is flat would probably not predict such a decrease.

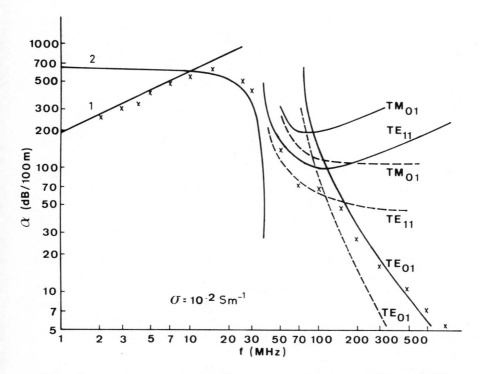

Fig. 1.3   *Measured and calculated specific attenuation of natural propagation in the Lanaye tunnel*

Fig. 1.4 shows results obtained by the same author for a rectangular road tunnel. The width of this tunnel was 17 m and the height 4·9 m. The tunnel walls were made of concrete. The cutoff frequency is 8·8 MHz. The dominant TE_{10} mode has a vertical polarisation. The measurements of the specific attenuation for this polarisation are represented by circles and agree fairly well with the theoretical prediction for the dominant mode up to about 200 MHz. The theoretical curve in solid line was calculated from eqns. 1.11 and 1.15 for a conductivity of 0·1 S m^{-1}. This value seems somewhat high compared with the data of Fig. 1.1, but it may be justified by the existence of concrete reinforcement. The TE_{m0} modes with $m > 1$ have not been taken into consideration although their critical frequencies are relatively low, being multiples of 8·8 MHz. They should indeed not change the conclusions since their specific attenuation is always higher than that of the dominant mode. The waveguide model considered up to now can thus not explain the decrease of the measured attenuation above 200 MHz.

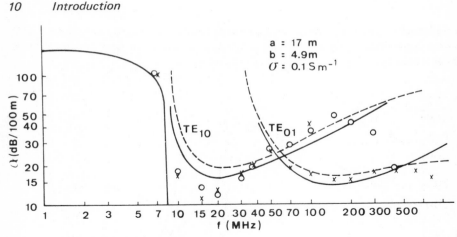

Fig. 1.4   *Measured and calculated specific attenuation of natural propagation in a rectangular road tunnel. Circles are for vertical polarisation and crosses for horizontal polarisation*

Measurements were also made by Deryck for the horizontal polarisation. The results are represented by crosses on Fig. 1.4 and they are compared with the theoretical attenuation curve in solid line for the TE_{01} mode, obtained from eqns. 1.11 and 1.15. This mode is the lowest-order mode with horizontal polarisation in a waveguide with perfectly conducting walls. The measurements agree very well above the critical frequency (30·6 MHz) of this mode but fail completely below it. The discrepancy might be explained if some cross-polarising effect could be invoked, but this seems improbable because the tunnel walls were very smooth. Instead we propose a theoretical explanation that will be examined later. If the wall conductivity is finite the modes are perturbed and their polarisation is slightly changed. The perturbed TE_{m0} modes have a small component with horizontal polarisation and it is not surprising that the measured specific attenuation for the horizontal polarisation is the same as for these modes, since all modes with nominal horizontal polarisation are cutoff. The specific attenuation however does not tell us everything and, to be complete, we also need to consider the coupling of the mode to the antenna, i.e. the absolute level of the field for a given transmitter power. Between 10 and 40 MHz this level should normally be significantly lower for the horizontal polarisation than for the vertical one. Indeed, Deryck confirmed that he observed a difference of 15 dB.

The reader could object that, if this explanation is correct, the vertical polarisation must propagate above 70 MHz with the specific attenuation of the TE_{01} mode rather than with the higher attenuation of the TE_{10} mode, because the former has a component with vertical polarisation. That this is not observed by Deryck can easily be explained. For the measuring distance used by this author the TE_{m0}

modes, which are excited at a higher level by a vertical antenna, remain dominant in the measurement up to about 200 MHz in spite of their higher attenuation. At higher frequencies however the attenuation becomes so high that the vertical component of the TE_{0m} (and other) modes dominates at the end of the measurement path: as a result the measured attenuation of the vertical polarisation tends to that of these modes.

On Figs. 1.3 and 1.4 we have also drawn broken curves showing theoretical results based on eqns. 1.11 and 1.14, but with R given by the exact expression 1.16 instead of its approximation 1.15. We have used the values $\kappa = 10$ for Fig. 1.3 and $\kappa = 5$ for Fig. 1.4. The main characteristic of results obtained in this way is that the theoretical curves have a horizontal asymptote for high frequencies. Indeed, above the transition frequency the value of R calculated by eqn. 1.16 tends to $\eta_0/\kappa^{1/2}$ instead of increasing as $f^{1/2}$ as predicted by eqn. 1.15. However the uncertainty about the actual values of κ and σ, as well as their frequency dependence, are such that we cannot say that eqn. 1.16 gives a better fit to the experimental data than eqn. 1.15.

We may conclude that the simple waveguide model may give relatively good predictions of the attenuation up to a few times the cutoff frequency but that it needs to be refined. Theoretical and experimental investigations that will be reported in a later chapter show that, at higher frequencies, the specific attenuations decrease as f^{-2}. The simple waveguide model used above thus fails completely at these frequencies.

1.2.3 Natural propagation above the cutoff

As the specific attenuation of the waveguide modes takes on relatively low values above the cutoff, it seems possible to communicate in tunnels without supplementary infrastructure. In this process the antennas of the radio sets are coupled to the electromagnetic field of the waveguide modes. Apart from the location and characteristics of the equipment no fundamental difference can be made here between a fixed base station and a mobile radio set. This type of communication based on natural propagation unfortunately has limited application because of several characteristics that we will examine briefly.

The decrease of the specific attenuation with frequency above the cutoff is rather slow. On the other hand, in many applications, it is not desirable to have a marked polarisation effect because the orientation of the mobile antennas may be arbitrary. These factors favour the use of frequencies higher than say two or three times the cutoff. The existence of several propagating modes however has some drawbacks. These modes have different phase constants and their relative phases vary along a path parallel to the tunnel axis. As it is impossible to excite a single mode, the resulting standing waves are unavoidable in natural propagation. They show nodes which may be extremely deep. In cases where any breakdown of the communication link is damaging, as in data transmission, it may be necessary to resort to suitable techniques like frequency or space diversity, error control, etc. For moving radio sets, the phase variations can also create some problems.

In speech transmission on the contrary some loss of communication may be tolerated. As the width of the deep standing waves' nodes is minute, not exceeding a small fraction of the wavelength, this effect is acceptable for mobile communications. It is however necessary to have a safety margin such that good communications are obtained at a large percentage of places. When numerous modes with comparable amplitudes exist, the standing wave pattern looks random and the statistics of the field amplitudes tend to a Rayleigh distribution. The margin, defined as the ratio of the amplitudes exceeded at 1% and x% of the places, amounts for this distribution to 16·4, 19·5 and 26·6 dB for $x = 90, 95$ and 99%, respectively.

Another important drawback in natural propagation is the disastrous effect of obstacles present in the tunnel. This is best understood if the propagation is viewed in terms of geometrical optics rather than a decomposition into modes. It will be shown in a later chapter that both approaches are equivalent well above the tunnel cutoff. Geometrical optics decomposes the transmission between two antennas as a sum of a direct ray and of numerous rays which undergo reflections on the tunnel walls. Obviously an obstacle present in the tunnel will intercept some of the rays. This may suppress rays which have an unfavourable phase for a given receiving location and the standing wave pattern will thus be changed. However the main effect of an obstacle is to absorb a part of the energy, thereby creating a global loss. Indeed, the rays incident on the obstacle are partly reflected and partly refracted into the latter. The energy of the refracted rays is absorbed if the obstacle thickness is important compared with the skin depth. Most of the reflected rays emerge toward the nearby tunnel wall on which they fall with arbitrary incidence angles: remembering that the reflection is quasi-total only for those rays which have a grazing incidence, we see that a large part of the power scattered by the obstacle toward the tunnel walls will be absorbed into them.

This dependence of the reflection coefficient on the incidence angle allows us to predict qualitatively the effect of an obstacle in a straight tunnel well above the cutoff. If the cross-section is clear, the rays contributing effectively to the received signal have a grazing incidence. If we admit that the energy of the rays intercepted by the obstacle is completely lost, we arrive at the conclusion that the power loss is equal to the ratio of the tunnel and obstacle cross-sections.

The effect of a curvature of the tunnel axis may be analysed by similar methods. Experience has shown that, in the UHF and higher bands, no communication can be established with receivers located in the shadow zone beyond a bend if the total distance exceeds about 100 m. It can easily be understood from geometrical optics' considerations that sharp bends are more disastrous.

The limiting case is that of an abrupt corner. No ray from the geometrical approach can reach a receiver located beyond the corner and the only received signal comes from the wedge diffraction and subsequent reflections on the tunnel walls. A complementary and instructive picture is provided by physical optics. The cross-section of the transverse tunnel is illuminated by the sum of the incident wave and of the wave reflected by the corner into the main tunnel. As this sum is a standing wave with a phase change of 180° per half wavelength, the power transmitted to the transverse tunnel shows a global decrease with the ratio of the tunnel width to the wavelength. This evolution is oscillatory, the transmitted power showing deep minima when this ratio is an integer number, separated by maxima at the half integer values. Similar considerations apply to the transmission in the main and transverse tunnels of a crossing or for the penetration of waves from open air into a tunnel.

As an example we will give some results obtained in an iron mine. The tunnels had a width of 7 m and a height of 8 m. The total attenuation between two antennas separated by a distance d with n corners could be written in the form (Delogne, 1976a)

$$a = a_0 + \alpha d + nc + s \qquad dB \qquad (1.17)$$

where a_0 is the standing wave envelope extrapolated to a zero distance, α is the global specific attenuation, c is the attenuation due to a corner and s is a margin for 95% of the places. The measured values of these parameters are shown in Table 1.1 and can be interpreted in the light of the previous discussion.

Table 1.1

f	a_0	s	α	c
MHz	dB	dB	dB/100 m	dB
36	17	5	60	6
68	20	10	40	2
150	21	15	35	15
450	24	22	15	25

From these remarks we may conclude that diffraction effects play an important role in natural propagation in tunnels. Because of the general decrease of the specific attenuations, the highest frequency is undoubtedly the best choice for straight and unobstructed tunnels, but when corners and crossings exist, the best performance will be obtained in the 70 to 150 MHz band. The useful propagation ranges will nevertheless rarely exceed about 250 m when corners are present.

1.3 The monofilar mode

The existence of a cutoff frequency in an empty tunnel and the very adverse effects of obstacles and bends on natural propagation at higher frequencies suggest the removal of the cutoff effect by stringing an axial isolated wire conductor in the tunnel cross-section. The tunnel thereby becomes similar to a two-wire transmission line, the conductors of which are the wire and the tunnel wall. In addition to the waveguide modes the structure can then support a TEM-like transmission line mode which has no cutoff. This was called the coaxial mode in some early papers, because of the resemblance of the transmission line made by the tunnel and the wire to a coaxial cable, but the name monofilar mode is now generally accepted. As the specific attenuation of transmission lines in general increases with frequency, it is expected that this system will allow the use of relatively low frequencies propagating with a rather low specific attenuation.

Fig. 1.5 may give a good though very approximate idea of the main properties of the monofilar mode. As shown, the transmission line current flows along the monofilar wire conductor and returns along the wall. The lines of the electric field are shown as continuous lines. Fixed base stations will be galvanically connected between the wire and the ground, while mobile transceivers will be coupled to the electromagnetic field of the monofilar mode.

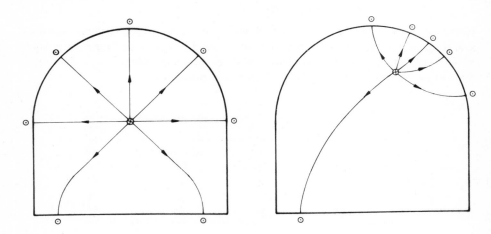

Fig. 1.5   *Current distribution and electric field lines of the monofilar mode*

In a practical installation, gauge considerations will not allow the radio electrician to string the monofilar wire in the middle of the tunnel cross-section, but close to the wall. Such an eccentric location has an unfavourable effect for two reasons which appear clearly on Fig. 1.5. In the first instance the return current is no longer distributed more or less uniformly around the tunnel periphery but it tends to concentrate in those parts of the wall which are close to the wire. Things happen as if the cross-section of the return conductor was reduced. In transmission line terms the resistance of the return conductor is increased and this yields an increase of the specific attenuation of the monofilar mode. This dependence of the specific attenuation on the distance from the wire to the wall and on the frequency is evidenced by the measurements made by Deryck (1973, 1979) in the Lanaye tunnel and shown in Fig. 1.6.

The other effect of an eccentric location of the monofilar wire is that the field lines and the electromagnetic energy tend to be concentrated between the wire and the proximate wall. The result is that portable radio sets located somewhere in the middle of the tunnel cross-section or near the opposite wall will only be weakly coupled to the monofilar mode. This will be expressed later by an eccentricity loss which may in some cases amount to 30 to 40 dB. This is a very serious drawback, for this loss has to be subtracted from the maximum allowable loss: once for base-station-to-mobile and twice for mobile-to-mobile communication links.

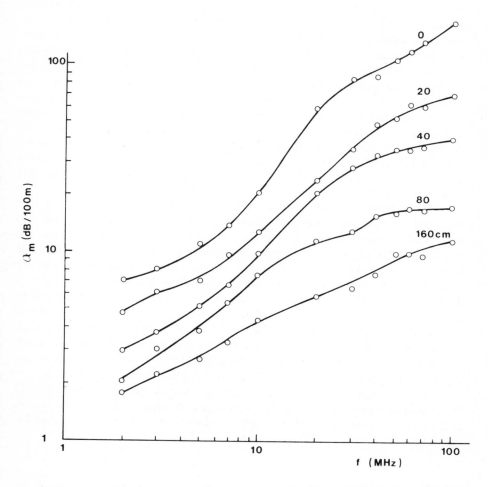

Fig. 1.6   *Measured specific attenuation of the monofilar mode in the Lanaye tunnel as a function of frequency and distance from the cable to the tunnel wall*

The picture of the monofilar mode given by Fig. 1.5 and the subsequent discussion of the properties of this mode are valid for a highly conducting wall, i.e. well below the transition frequency, and provided the skin depth in the wall is much smaller than the tunnel cross-section and than the distance from the wire to the wall. This last condition is in most cases not satisfied and the previous conclusions are thus only qualitative. Formulas similar to eqns. 1.11 to 1.14 exist for transmission lines carrying a TEM-like mode, but for the same reason, they do not give a good agreement with experimental or exact theoretical results.

An example where this qualitative picture fails was observed in salt and potash mines. In spite of the very low conductivity of the wall material the specific attenuation of the monofilar mode was significantly lower than for more conductive walls, which is somewhat surprising. The paradox can easily be explained however. As the skin depth in the wall is very large, the return current of the monofilar mode is distributed in a large area around the tunnel cross-section, whatever the monofilar wire location. The specific resistance of the return path, which is the dominant source of the specific attenuation, is the inverse product of this large area by the conductivity. It is thus independent of the wire location and may be lower than for a very eccentric location in a more conductive tunnel. For the same reason the eccentricity effect will be less marked in such cases. From the discussion of this rather exceptional situation, it appears that a low wall conductivity, but also the use of low frequencies, may have advantages when the monofilar wire conductor must be strung very close to the tunnel wall.

In many practical situations, axial conductors exist in the tunnel, e.g. power lines, trolley wires, water pipes, etc. These conductors are in general not suitable to guide electromagnetic waves with a low specific attenuation, because they may contain longitudinal impedances or connections to the ground at radio frequencies. It will thus be necessary to string a dedicated monofilar wire conductor to support the monofilar mode. However if n axial conductors exist in the tunnel, there will be n TEM-like modes. Any discontinuity along such a structure creates inadvertent mode conversions and results in some loss. This effect as well as the field distribution of the useful mode are dependent on the relative locations of the monofilar wire and of the other conductors and may be reduced by proper design. On the other hand, and in contrast to effects for the waveguide modes, transversal conductors and other obstacles present in the tunnel have little influence on the monofilar mode.

This enumeration of problems shows that, although the general characteristics of the monofilar wire technique are fairly simple, its use requires some experience and skill. A better knowledge of the theoretical aspects of monofilar mode propagation is also needed.

1.4 The long induction loop

A large induction loop installed around a building is sometimes used for paging applications. A similar technique may be used in mines. In the particular case of a tunnel the loop wire is strung near to opposite walls. The length/width ratio of the loop is much larger than unity and the loop perimeter is frequently larger than one wavelength. The structure thus operates as a transmission line rather than as a common induction loop and we will consider it from this point of view. Base stations are connected galvanically to the transmission line, while mobile transceivers are coupled to the magnetic field of the transmission line by means of a loop or ferrite antenna. The transmission line is terminated into its characteristic impedance to avoid standing waves. As will become obvious from the discussion, the use of this technique is restricted to frequencies below a few hundreds of kilohertz.

As was stated in the previous paragraph, the structure can support two TEM-like modes. If the two wires are perfectly symmetric with respect to the tunnel cross-section, these are the classical balanced and unbalanced modes. In the latter the transmission line current flows in equal parts along the two wires and returns along the tunnel wall. In the former the current flows along one wire and returns along the other one, while the tunnel wall carries no *net* longitudinal current.

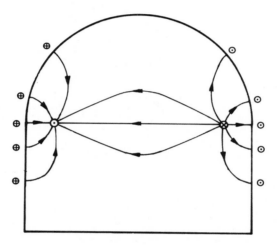

Fig. 1.7   *Current distribution and electric field lines for the long induction loop*

For this reason one should expect the balanced mode to have a much lower specific attenuation than the unbalanced mode. That this is not true can be seen by comparing Fig. 1.7 with Fig. 1.5. Because of the close proximity of the transmission line wires to the tunnel wall, the interwire capacity is much smaller than the wire-to-wall capacity, which means that there is a very weak coupling between the wires. The balanced and unbalanced modes can thus be seen as twice the monofilar mode existing for a single wire, with opposite and equal phases, respectively. Although no *net* current flows in the wall for the balanced mode, there is nevertheless a total current since the opposite-phase monofilar mode current distributions only partly cancel each other. If no cancellation occurs at all, the balanced and unbalanced modes both have the same specific attenuation as the monofilar mode of a single wire, as can be seen easily from Fig. 1.7. Theoretical calculations (Mahmoud, 1974a) confirm this trend. We made measurements of the specific attenuation of the balanced mode in the Lanaye tunnel below 1 MHz. The wires were strung at 50 cm from the wall. The results are shown in Fig. 1.8 and are in fairly good agreement with those of Fig. 1.6 extrapolated below 1 MHz.

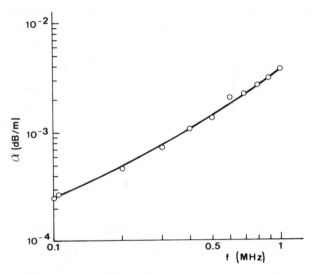

Fig. 1.8   *Specific attenuation of a long induction loop in the Lanaye tunnel*

This tendency is reinforced when exact symmetry of the two wires cannot be maintained. The usefulness of the long induction loop is thus not much better than that of a monofilar wire.

1.5 Leaky feeders

1.5.1 Principle

The term leaky feeder covers a variety of open or semi-open transmission lines. Historically the first leaky feeder was the two-wire ribbon feeder which is very popular as an antenna feeder in television reception. When such a transmission line is strung parallel to the axis of a tunnel, we have a structure analogous to the long induction loop with a balanced and an unbalanced mode. However the situation will be very different because the wire spacing is very small compared with the distance to the tunnel wall. As the electromagnetic field of the balanced mode decreases roughly as the square of the distance from the observation point to the transmission line, this mode induces negligible currents into the wall, and hence its propagation parameters, including the specific attenuation, will not differ much from those of the transmission line in free space. The counterpart is that this mode is not directly useful for radio communications in the tunnel because it provides an extremely weak coupling to the antennas of mobile transceivers. The unbalanced mode has inverse properties and does not differ much from the monofilar mode of a single wire located at the same place as the transmission line. This is why the balanced and unbalanced modes have been called the bifilar and monofilar modes, respectively, in the leaky feeder technique.

To understand why such a transmission line can improve radio communications in the tunnel, we must remember that strictly balanced and unbalanced modes cannot exist independently unless the two wires are perfectly symmetrical with respect to the tunnel wall. Any local asymmetry converts some energy from the one into the other. If such asymmetries are numerous, the monofilar mode will be regenerated by the bifilar mode and will exhibit an apparent specific attenuation equal to that of this mode, with a slight increase due to the mode conversion process. This seems to be an interesting principle for establishing long range radio communications in tunnels, for it promises the coupling properties of the monofilar mode with the low specific attenuation of the bifilar mode.

The ribbon feeder has however a very severe drawback. Its specific attenuation increases rapidly when it is covered with dust or moisture. To obviate this, several coaxial cables with an imperfect shield have been designed. Some of them are shown in Fig. 1.9. The plastic jacket surrounding the cable provides a good protection against the penetration of moisture into the cable and, as a very large part of the power is confined under the shield, the transmission line mode is not seriously affected by a layer of dust and moisture on the external surface of the cable. From a practical viewpoint this is a sufficient reason to devote more attention to the coaxial leaky cables than to the ribbon feeder. Another and perhaps more important characteristic of coaxial leaky feeders is the fact that the leakage field does not decrease as the square of the distance but rather as the distance itself.

A fundamental question in relation to leaky coaxial cables is how to characterise the transfer of electromagnetic energy through the imperfect shield. We will report some theoretical approaches to this very difficult problem. For most structures however, with the exception of the axial-slot cable, and for moderate frequencies, the imperfect shield can be modelled by a transfer inductance that will be denoted m_t. This somewhat questionable parameter relates the axial electric field on the shield surface E_z and the current carried by the latter by the equation

$$E_z = j\omega m_t I$$

The measurement of the transfer inductance requires some care but can easily be made in the laboratory (Krügel, 1956). For this purpose a short section of coaxial cable is fixed inside a metal tube of equal length to make a triaxial structure. The coaxial cable is fed at one end by a generator and terminated into its characteristic impedance. At the generator end the metal tube is closed by a metal plate which is soldered to the cable shield. At the other end, although this is not strictly necessary, the tube is carefully connected to the cable shield through the characteristic impedance of the coaxial cable consisting of the tube and shield. The voltage V

appearing on this load is measured and the transfer impedance is obtained by $Z = V/Id$, where I is the generator current and d is the length of the structure, which must be very short compared with the wavelength. At low frequencies the transfer impedance may comprise a resistive part, which is the specific resistance of the shield, but at higher frequencies the imaginary part $(j\omega m_t)$ should be dominant. Typical values of the transfer inductance of leaky coaxial cables range from 1 to 40 nH m^{-1}.

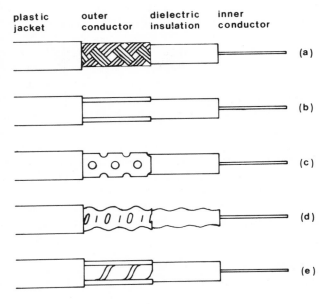

Fig. 1.9   *Some leaky coaxial cables*
(a) Loosely braided
(b) Longitudinal slot
(c) Perforated tube
(d) Helically formed and grinded
(e) Helical tape and longitudinal slot

1.5.2 Leaky feeders below the tunnel cutoff

We have briefly explained in the previous section how the ribbon feeder works as a leaky feeder. Things are somewhat different for leaky coaxial cables. Again the structure can support two TEM-like modes. The first is the perturbed transmission line mode and will be referred to as the coaxial mode. Some authors also use the name bifilar mode. It has the main part of its energy confined under the cable shield but with some leakage outside it. As long as this leakage remains weak, the power dissipated into the tunnel wall remains small and the specific attenuation does not differ much from that of a well-shielded coaxial cable with the same characteristics. The second mode is called the monofilar mode and its properties are just the reverse. Essentially it uses the cable shield and the tunnel wall as conductors but has some leakage under the cable shield. Its specific attenuation is not very different from that of a single wire transmission using a monofilar wire conductor instead of the leaky cable.

As was briefly mentioned in the previous section, leaky coaxial cables exhibit an important difference when compared with a ribbon feeder. For the latter, assuming perfect symmetry with respect to the tunnel wall, the two wires carry exactly equal but opposite currents for the balanced or bifilar mode; no net current flows along the transmission line and this is why the electromagnetic field decreases as the square of the distance from the observation point to the transmission line. This does not remain exactly true if the two wires are not symmetrical with respect to the tunnel wall; however obtaining a noticeable asymmetry requires the ribbon feeder to be located very close to the tunnel wall. In a leaky coaxial cable, on the contrary, the two conductors are intrinsically asymmetrical. Consequently the coaxial mode current which flows along the inner conductor returns mainly along the shield, but also partly along the wall. The transmission line thus carries a net current, with the result that the leakage field decreases more slowly with the distance from the observation point to the cable. The distribution of the coaxial mode leakage field in the tunnel cross-section is in fact that of a monofilar mode travelling at the same velocity as the coaxial mode. In practice, below the tunnel cutoff, this distribution will not differ greatly from that of the main field of the monofilar mode. Conversely the leakage field of the monofilar mode inside the coaxial cable has approximately the same distribution as the main field of the coaxial mode (Delogne and Safak, 1975).

This behaviour of the leakage fields is of fundamental importance in the use of leaky coaxial cables. Indeed any source located on one side of the shield must necessarily excite both modes, though at different power levels. For instance, if a generator is connected at the input of the cable, nearly all the power will be delivered to the coaxial mode but the monofilar mode will also be excited. As we will see in Chapter 3, the initial power levels are such that the leakage field of the coaxial mode into the tunnel space is equal to the field of the monofilar mode. The specific attenuation for the latter is however in most applications several times higher than for the former and, at some distance from the generator, the leakage field of the coaxial mode is the dominant part of the electromagnetic field in the tunnel space. This example shows that the coaxial mode leakage is the mechanism used for base-station-to-mobile communications, while the monofilar mode is useless here.

In mobile-to-mobile communications the transmitter excites the monofilar mode at a higher level than the coaxial mode and the leakage of the latter in the tunnel space will become dominant only at a rather large distance. There is no simple answer to the question of whether the monofilar or the coaxial mode is the mechanism used in a long distance mobile-to-mobile communication: this depends on a number of factors as we will see later.

1.5.3 Leaky feeders above the tunnel cutoff

Well above the cutoff frequency of the tunnel the working principle described above fails. The monofilar mode has a specific attenuation which is much higher than that of the waveguide modes and it is therefore not very useful. On the other hand the leakage field of the coaxial mode no longer has the same distribution as the monofilar mode and plays another role.

To understand this, let us first consider a leaky feeder strung in free space. This is basically a slow wave structure and consequently (Collin and Zucker, 1969) it does not radiate unless discontinuities are introduced along it. But as it comprises two conductors, it can support two slow waves having no cutoff. The first is the coaxial mode, with the main part of the energy confined under the cable shield but with some leakage outside it. Its phase constant β_c is approximately given by

$$\beta_c \simeq k_0 \sqrt{\kappa} \qquad (1.18)$$

where κ is the dielectric constant of the cable insulating material. The second mode has the opposite energy distribution. It is a sort of Goubau wave with leakage under the cable shield; its phase constant lies typically in the range

$$\beta_g = k_0 \ldots 1 \cdot 02 k_0 \qquad (1.19)$$

depending on the cable jacket thickness and dielectric constant.

The field of both modes outside the cable can be expanded in cylindrical harmonics by a formula of the type

$$\sum_{m=0}^{\infty} A_m \begin{array}{c} \sin \\ \cos \end{array} m\phi \, K_n (u\rho) \qquad (1.20)$$

where $n = m$ or $(m + 1)$ according to the field component considered; K_n is the modified Bessel function of the second kind and

$$u = (\beta^2 - k_0^2)^{1/2} \qquad (1.21)$$

The asymptotic formula for large t

$$K_n(t) \simeq [\pi/(2t)]^{1/2} \, e^{-t} \qquad (1.22)$$

shows that the fields outside the cable are in practice confined within an effective radius given by

$$\rho_e = 1/u \qquad (1.23)$$

Practical values of ρ_e for both modes are given in Table 1.2 for a cable with $\kappa = 1 \cdot 6$, $\beta_c = 1 \cdot 26 k_0$ and $\beta_g = 1 \cdot 005 k_0$.

Any discontinuity along this structure will produce some mode conversion, but also radiation in the sense of a continuous spectrum of spherical waves. A generator connected at the cable input has a similar effect, exciting both guided modes but also radiation.

Let us now assume that this cable is strung in a tunnel. In most practical cases,

Table 1.2

f	ρ_{ec}	ρ_{eg}
MHz	m	m
30	2.05	15.90
100	0.62	4.77
150	0.41	3.18
450	0.14	1.06

the effective radius of the Goubau wave is larger than the distance between the cable and the wall and this wave becomes the classical monofilar mode with a somewhat modified field distribution. If a generator is connected to the cable input, the excited monofilar mode and radial radiation (or waveguide modes) are rapidly damped and there remains only the coaxial mode. Now, if the distance between the cable and the wall is larger than ρ_{ec}, this mode is virtually unaffected by the tunnel. Consequently an observer located outside the effective radius will receive an extremely weak signal. Thus the cable does not radiate.

But if the wall comes into the effective radius of the coaxial mode, all wall irregularities and inhomogeneities will cause radiation. This is the only reason why leaky feeders radiate. The process is random in nature. Note that the cable attachments produce a similar effect. It can be shown that if the scattering points are numerous and statistically independent, the field follows a Rayleigh distribution: indeed it is the result of the addition of numerous elementary contributions with random phases.

It has become common practice in leaky feeder techniques at VHF and UHF to characterise the cable radiation by a coupling loss; this quantity is usually defined as the ratio of the power of the coaxial mode to the power received by a dipole antenna located at a specified distance from the cable. As the radiated field is random, the coupling loss should be specified as a value which is not exceeded at a certain percentage of places. In tunnel applications the coupling loss does not depend much on the distance from the receiving antenna to the cable.

This discussion has presented the coupling loss as an experimental parameter; it depends on a number of factors among which are the distance from the cable to the tunnel wall, the roughness of the latter, the mounting hardware, the frequency and, of course, the cable characteristics. It is thus not desirable to deviate significantly from the recommended mounting instructions of a leaky coaxial cable without proceeding to some measurements.

It is obvious that a strong radiation requires an important interaction between the coaxial mode leakage and the wall irregularities and cable attachments. These objects are not very efficient radiators and the process of diffraction and radiation necessarily involves an increase of the coaxial mode specific attenuation. To give an order of magntidue of this effect at 450 MHz, a coupling loss smaller than 95 dB at 99% of places currently yields a doubling of the cable attenuation; this is a very severe drawback. This theory was indeed fully confirmed by a series of measurements taken in the subways of Paris: it was observed that a leaky cable no longer radiated at 450 MHz when it was strung at about 30 cm from the tunnel wall using thin nylon strings; at the same time the specific attenuation, which was doubled when the cable was against the wall, returned to the value of an equivalent non-leaky cable. A further conclusion of this analysis is that no particular leaky cable structure offers definite advantages for use well above the tunnel cutoff; the only difference between the various types of cable, apart from internal parameters like the specific attenuation, is the relative intensity of the leakage; this can however be compensated by a suitable mounting.

1.6 Mode converters

In the leaky feeder technique the transfer of energy from inside the cable to the tunnel space occurs continuously. In contrast with this method, it is possible to use a well-shielded coaxial cable along which discrete mode converters are inserted with a regular or irregular spacing. In such a cable the coaxial mode propagating inside the cable and the monofilar and waveguide modes propagating outside it are obviously completely independent. Creating a local energy exchange between the latter and the former thus requires the opening of the shield in some way.

In the first system of this kind, which has frequently been called the INIEX/ Delogne system because it was proposed by the author as a consultant of the Belgian Institut National des Industries Extractives (INIEX), the opening is an annular slot consisting of a complete interruption of the cable outer conductor. Refined theories of the working of such a slot are available and will be examined later in this book. Below the tunnel cutoff however a simplified quasi-static analysis of the slot can be obtained by considering the cable and the tunnel as two imbricated transmission lines having a common conductor. The latter is interrupted over a short length. The problem thus reduces to a circuit calculation which is suggested by Fig. 1.10a and b, where Z_m and Z_c are the characteristic impedances of the monofilar and coaxial modes, respectively. The only fundamental difficulty in this respect is to define and estimate the value of Z_m. Fortunately it appears that this quantity varies very slowly with the electrical and geometrical parameters of the structure. A comparison with the exact electromagnetic solutions available between 1 and 50 MHz has shown that calculations based on the value of 377 Ω for Z_m never yields an error above 1 dB for the mode conversion factor.

A naked slot, like the one shown in Fig. 1.10, does not provide a good impedance match and a low insertion loss for the coaxial mode. Indeed, the external load impedance $2Z_m$ 'seen' by the slot is rather high and most power flowing in the coaxial mode is reflected back inside the cable. As the objective of the system is to make use of the low specific attenuation of the coaxial mode to extend the range, it is necessary to improve the impedance match and to reduce the insertion loss to a very small value. This can be obtained by adding some lumped circuit elements to the slot. Various circuits may be used and will be described in Chapter 3.

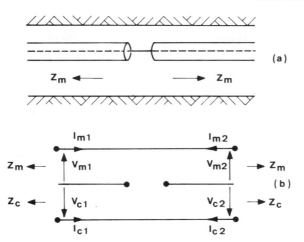

Fig. 1.10   *Principle of the annular-slot mode converter*
(a) Pictorial view
(b) Equivalent circuit

This system has been used intensively in the HF band and at lower frequencies. The mode converter is in general designed to convert about 10% of the power from the coaxial mode into the monofilar mode, which is paid for by an insertion loss of a few tens of decibels for the coaxial mode. The spacing between the mode converters is typically several hundred metres. This technique provides excellent flexibility in the design of a system because the mode converters' parameters and spacing can be varied along the path as a function of the tunnel cross-section, acceptable cable location, distance to the base station and so on. The only delicate point in the design task is to estimate the characteristics of the monofilar mode, namely the specific attenuation and the eccentricity loss; however an optimistic calculation, as well as changes due to the evolution of the underground workings, can still be corrected *in situ* by inserting additional mode converters.

Above the tunnel cutoff the slot will also excite the waveguide modes or, equivalently, natural propagation. It appears that the slot can be considered as an antenna fed by a part of the coaxial mode power. The radiation pattern has of course rotational symmetry about the cable axis. It exhibits a strong maximum at about $10°$ to this axis, which is an excellent characteristic for launching natural propagation in the tunnel. The antenna gain in this optimum direction amounts to about 10 dB.

Another type of mode converter or radiator, according to the frequency range, is obtained by inserting a short length of a leaky cable into a well-shielded coaxial cable. This device has very similar characteristics to the annular slot, with an additional directional property. Mode converters have also been designed for the bifilar line.

1.7 Historical survey

The early attempts to use radio under the earth's surface can be traced back to the beginning of the century and, because of the available technology, were restricted to low frequencies. The first known reports on the subject emanated from the US Bureau of Mines (Colburn *et al.*, 1922; Isley *et al.*, 1928). The conclusion of the second paper was extremely pessimistic and stated: 'these experiments are described, not to encourage further work along this line, but to prevent future investigators from wasting time in what the Bureau's engineers deem an unpromising field'. Eve *et al.*, (1929) reported that remote broadcasting transmitters could be received at 100 m depth and noted that the lower frequencies (20 to 30 kHz) were more favourable. There is little doubt that they observed a through-the-earth propagation. These early investigators mentioned a possible propagation of radio waves along existing conductors but, apart from a suggestion made by Keith-Murray (1933), they did not consider a systematic use of what is presently known as the monofilar mode.

Further reasons for discarding the use of radio underground were the weight, size and power requirements of radio equipment available at that time. Substantial progress was made in this respect during World War II; simultaneously the use of higher frequencies became possible. New tests were undertaken but, surprisingly, very little information was exchanged between experimenters using low frequencies, mainly in mine workings, on the one hand and VHF or UHF, mainly for railway tunnel applications, on the other hand. This mutual ignorance persisted up to about 1970 and no synthesis of many available results was made before this.

In spite of Isley *et al's* 1928 recommendations new tests were carried out by the US Bureau of Mines (Coggleshall *et al.*, 1948), still using low frequencies, but they did not lead to new conclusions. Somewhat later however one of the experimenters (Felegy, 1953) followed by new adepts (Tretiakov, 1959) recommended a systematic use of the monofilar mode, a technique which was well known in surface applications like railway transmission. (Radio pioneers will remember that wire telegraphy with a ground return, which does not differ substantially from the monofilar mode, was widely used during World War I.) At the same time the induction loop technique was already used and specific equipment had been developed for mining applications (Wyke and Gill, 1955). Ten years later the monofilar mode was used in French mines (de Watteville, 1964) and the adverse effect of the wall proximity on its attenuation had been observed (Savkine, 1964). The first theoretical analysis of the monofilar mode (Gabillard, 1969) was however not yet available.

Tests at VHF and UHF were also carried out during this period. Shanklin (1974) observed that the specific attenuation of natural propagation in a 2·4 m diameter test tunnel was about 1.8 dB/100 m at 150 MHz. He improved this value by means of a single wire strung along the tunnel, thus discovering the monofilar mode. He further reduced the attenuation to very low values by stringing a set of reflector wires near to the tunnel wall. Lovell Foot (1950) reported interesting results obtained in railway tunnels; he was the first to note the decrease of the specific attenuation when the frequency was raised above the cutoff, a property which was rather obvious from waveguide theory.

An important step in the evolution of subsurface radio communications was the fortuitous discovery by Monk and Winbigler (1956) that a braided coaxial cable or a twin-wire line seemed to radiate continuously. The leaky feeder phenomenon remained however unexplained for more than ten years. Still in 1965 Farmer and Shepherd tried to explain it for the twin-wire line by a formula which is valid for the radiation of a short length of transmission line, but not in the present case. They reported that the leaky feeder radiation was enhanced by twisting the transmission line, a conclusion which will be contradicted later and is now considered as erroneous. This lack of theoretical support did not prevent the recommendation of the leaky feeder technique by Liégeois (1968) in a very thorough report on the state of radio communications in mines. Various types of leaky feeders were indeed developed and used to provide radio communications with moving trains, for instance in Germany (Hilscher and Plischke, 1968) and in Japan (Mikoshiba and Nurita, 1969).

The rather pragmatic approach used up to about 1968 changed drastically when the European Commission for Steel and Coal and the US Bureau of Mines decided to support important research projects on the subject of radio communications in mines. It was indeed recognised that reliable radio communications and control systems were key factors in increasing productivity and safety in the mining industry. Research groups including theoreticians as well as experimenters were started in various countries and a close cooperation arose between them. Besides numerous papers presented at several congresses or published in the scientific literature, four scientific meetings were exclusively devoted to the subject of subsurface radio communications:

Thru-the-Earth Electromagnetics Workshop, held at the Colorado School of Mines, USA, 15—17 August 1973.
Radio: Roads, Tunnels, Mines, held at Liège, Belgium, 1—5 April 1974.
Symposium on Leaky Feeder Radio Communication Systems, held at the University of Surrey, UK, 9—10 April 1974.
Electromagnetic Guided Waves in Mine Environments, held at the Institute for Telecommunications Sciences, Boulder, USA, 28—30 March 1978.

Many papers presented in these meetings are listed in the bibliography.

The material of this book is essentially based on the work performed by these groups over the last ten years. Here we limit ourselves to a brief summary of the main results which were obtained. We have already mentioned the first theoretical approach to the monofilar mode by Gabillard (1969). To be complete we must also cite the first analysis of the waveguide modes due to Comstock (1971).

Experimental work on leaky feeders supported by a coupled line model analysis was carried out from 1970 onwards by Martin at the National Coal board, UK, in cooperation with the group of Professor Q. V. Davis at the University of Surrey.

This author concluded very early on that frequencies of about 10 to 100 MHz were optimum for leaky feeder applications and oriented his work towards the use of in-line repeatered systems. Professor Gabillard and his coworkers at the University of Lille, France, and at CERCHAR (Centre d'Etudes et de Recherches Charbonnières) recommended later the use of loosely braided coaxial cables below 10 MHz in non-repeatered systems. The divergent conclusions of these groups cannot be justified by propagation considerations but rather by system and operational aspects.

The mode converter techniques were mainly developed by the author and the group of R. Liégeois, L. Deryck and R. De Keyser at INIEX (Institut National des Industries Extractives), Liège, Belgium. The annular slot converter was analysed in 1970. The actual working principle of a twin-wire line used as a leaky feeder was nicely demonstrated in Deryck's Ph.D. thesis (1973). Mode converters for this type of transmission line were proposed in the same document. Deryck's analysis of the twin-wire line, according to which radiation was due to inadvertent mode conversions, was extended by the author (Delogne and Safak, 1975) to the various types of leaky coaxial cable and led him to propose (1976a) the use of leaky sections located at discrete places in a non-leaky cable.

Although these works were supported by some theoretical analyses of idealised models, the main achievements in this respect were obtained by J. R. Wait, D. A. Hill and their coworkers in the USA. Exact numerical solutions were obtained for numerous problems in various geometries: natural propagation in a coal seam, attenuation of the monofilar, bifilar and waveguide modes in rectangular and circular tunnels, the dedicated-wire technique, the quasi-static analysis of leaky feeders, etc. Important contributions of other US workers, namely Emslie, Lagace, etc. should also be mentioned here.

Interesting experimental results, mainly on leaky feeders at UHF, were obtained by Cree and Giles at British Railways, Sniter, Jouan and the author at the Régie Autonome des Transports Parisiens, and also by several Japanese authors. Application of the leaky feeder technique to guided radar was made by Mackay and Beal (1978a, b).

Chapter 2

Electromagnetic theory of subsurface propagation

2.1 Introduction

In this chapter we will state and solve the electromagnetic problem associated with propagation in some idealised subsurface cylindrical structures. The term cylindrical means that the geometrical and electrical parameters of the medium do not depend on a coordinate z which is in most cases the distance along the axis of a tunnel. Although the structures considered here are highly idealised compared with actual subsurface environments, the related electromagnetic problem is among the most sophisticated. The reason for this is that we have to do with at least two media, one of which has a finite conductivity. Consequently simple solutions consisting for instance of transverse electric (TE) or transverse magnetic (TM) waves cannot generally match the boundary conditions. In many cases all of the six scalar components of the electric and magnetic fields are non-zero. The solutions described in this chapter are mainly due to J. R. Wait, D. A. Hill and coworkers. We have added some personal comments and have tried to underline some characteristic properties of the exact solutions in order to support approximate methods which will be used in subsequent chapters.

The main problem which will be investigated here is to solve Maxwell's equations

$$\operatorname{curl} E = -j\omega\,\mu\blacklozenge\,H - J_{\mathrm{ma}} \tag{2.1}$$

$$\operatorname{curl} H = (\sigma + j\omega\,\epsilon_0\,\kappa)\,E + J_{\mathrm{a}} \tag{2.2}$$

where the following notation is used:

E, H: electric and magnetic fields, respectively.
$\mu_0 = 4\pi \times 10^{-7}$, the permeability of the medium, assumed to be equal in all cases to that of free space.
$\epsilon_0 = 8 \cdot 85 \times 10^{-12}$, the free-space permittivity.
σ, κ: conductivity and dielectric constant of the medium, respectively.
$J_{\mathrm{ma}}, J_{\mathrm{a}}$: applied magnetic and electric current densities, respectively.

We will use a complex permittivity ϵ defined by

$$j\omega\,\epsilon \;=\; \sigma + j\omega\,\epsilon_0\,\kappa \tag{2.3}$$

The complex propagation constant γ, complex wave number k and intrinsic impedance η of the medium are defined by

$$\gamma \;=\; jk \;=\; (j\omega\,\mu_0 \times j\omega\,\epsilon)^{1/2} \tag{2.4}$$

$$\eta \;=\; \omega\,\mu_0/k \tag{2.5}$$

The term 'applied' used for the current densities J_{ma} and J_a means that they are considered as given, i.e. as the source of the electromagnetic field. The reader who is not familar with this formulation will point out that magnetic currents do not actually exist. This is undoubtedly true. A reason for using applied magnetic currents is that some electric current distributions may be replaced by an equivalent but simpler magnetic current density. For instance a small loop of area dA and unit normal n carrying a current I is equivalent to a magnetic Hertz dipole of moment (Stratton, 1941)

$$I_m\,ds \;=\; -j\omega\,\mu_0\,I\,dA\,n \tag{2.6}$$

The form of Maxwell's equations 2.1 and 2.2 is valid in a homogeneous medium. The structures considered in this chapter are a juxtaposition of homogeneous media and these equations apply to each of them provided the appropriate values of σ and κ are used. Uniqueness theorems require some boundary conditions to be satisfied at the interface of different media and to infinity. Specifically the tangential components of the electric and magnetic fields must be continuous at an interface.

Boundary conditions to infinity require a detailed discussion which will be carried out in a later section. It is obvious that in any physically realisable solution, sources of electromagnetic fields are located at a finite distance and, as the external medium considered in this chapter has always a finite non-zero conductivity, that the solution of Maxwell's equations must vanish to infinity in all directions. However we will frequently be concerned with propagation modes of a cylindrical structure. Propagation modes are solutions of sourceless $(J_{ma} = 0, J_a = 0)$ Maxwell's equations with a dependence on the longitudinal coordinate z expressed by a factor $\exp(-\Gamma z)$, where

$$\Gamma \;=\; \alpha + j\beta \tag{2.7}$$

is the mode propagation constant, α and β being the specific attenuation and phase constants, respectively. If we assume that α is positive, which generally also yields β positive, the mode fields will be infinite for $z = -\infty$. Such solutions obviously assume that a source with infinite power exists to infinity in the negative direction of the z-axis. A fundamental question occurring here is to know whether or not we must impose these solutions to decrease to infinity in directions other than the negative z-axis. The answer to this question is positive, but it is known that modes violating this condition (improper modes) may in some cases accurately represent the actual fields in a finite region of space.

2.2 Electromagnetic potentials

2.2.1 General formulation

As was mentioned in the previous section, we will frequently be concerned with problems where all six scalar components of the electric and magnetic fields are non-zero. The usefulness of electromagnetic potentials comes from the fact that such fields can frequently be obtained from potentials with only one or two non-zero scalar components. The importance of potentials in the present work is such that a detailed discussion of the derivation will be useful for an adequate choice of the non-zero components.

Maxwell's equations 2.1 and 2.2 have been written with electric and magnetic source currents. As these equations are linear we may treat separately the two types of source. Let us first assume that we apply only electric currents. We will denote this case by primed variables. Hence

$$\text{curl } E' = -j\omega \, \mu_0 \, H' \tag{2.8}$$

$$\text{curl } H' = j\omega \, \epsilon \, E' + J_a \tag{2.9}$$

As the divergence of a curl is zero, eqn. 2.8 implies

$$\text{div } H' = 0 \tag{2.10}$$

This equation is identically satisfied if we obtain H as the curl of a vector potential π'. Putting

$$H' = j\omega \, \epsilon \, \text{curl } \pi' \tag{2.11}$$

and inserting this into eqn. 2.8 yields

$$\text{curl } (E' - k^2 \, \pi') = 0 \tag{2.12}$$

and, consequently, the quantity between brackets is the gradient of some scalar function V':

$$E' = k^2 \, \pi' + \text{grad } V' \tag{2.13}$$

Inserting eqns. 2.11 and 2.13 into 2.9 yields the equation

$$\text{curl curl } \pi' = k^2 \, \pi' + \text{grad } V' + (j\omega \, \epsilon)^{-1} \, J_a \tag{2.14}$$

Using the identity

$$\text{curl curl } \pi' = \text{grad div } \pi' - \nabla^2 \, \pi' \tag{2.15}$$

this equation may be rewritten

$$\nabla^2 \, \pi' + k^2 \, \pi' = \text{grad } (\text{div } \pi' - V') - (j\omega \, \epsilon)^{-1} \, J_a \tag{2.16}$$

So far there is no constraint on π' and V' other than that they satisfy this equation jointly. We may however note that π' is not completely defined by eqn. 2.11. Indeed, as the curl of a gradient vanishes identically, we may still add the gradient of an arbitrary function to π'. As results from eqn. 2.13 the scalar potential V' will then change. Specifically, if π' and V' are suitable potentials, the same electromagnetic fields are obtained from potentials

$$\pi'_0 = \pi' + \text{grad } \psi \tag{2.17}$$

$$V'_0 = V' - k^2 \psi \tag{2.18}$$

where ψ is an arbitrary scalar function. We say that E', H' are invariant under a *gauge transformation*.

When the gauge is chosen such that

$$\text{div } \pi' = V \tag{2.19}$$

eqn. 2.16 simplifies to

$$\nabla^2 \pi' + k^2 \pi' = -(j\omega \epsilon)^{-1} J_a \tag{2.20}$$

and π' is called an *electric-type Hertz potential*, since it is due to the applied electric current. The magnetic field can be obtained from π' by eqn. 2.11 while eqn. 2.9 yields the electric field

$$E' = \text{curl curl } \pi' - (j\omega \epsilon)^{-1} J_a \tag{2.21}$$

The case where only magnetic source currents J_{ma} exist can be treated by duality, using double primed variables and replacing $E', H', J_a, \epsilon, \mu_0, \pi'$ by H'', $E'', -J_{ma}, -\mu_0, -\epsilon, \pi''$, respectively. When both electric and magnetic source currents exist, additivity allows us to solve the problem with an electric-type Hertz potential π' and a *magnetic-type Hertz potential* π'' satisfying equations

$$\nabla^2 \pi' + k^2 \pi' = -(j\omega \epsilon)^{-1} J_a \tag{2.22}$$

$$\nabla^2 \pi'' + k^2 \pi'' = -(j\omega \mu_0)^{-1} J_{ma} \tag{2.23}$$

The fields are then obtained by

$$E = \text{curl curl } \pi' - j\omega \mu_0 \text{ curl } \pi'' - (j\omega \epsilon)^{-1} J_a \tag{2.24}$$

$$H = \text{curl curl} \pi'' + j\omega \epsilon \text{ curl} \pi' - (j\omega \mu_0)^{-1} J_{ma} \tag{2.25}$$

This is the most general representation of the electromagnetic fields using two vector potentials. The reader may feel this formulation to be unnecessarily complicated because we have now to work with six scalar components and intricate formulas. One may indeed observe that E' and H'' satisfy equations

$$\nabla^2 E' + k^2 E' = j\omega \mu_0 J_a \tag{2.26}$$

$$\nabla^2 H'' + k^2 H'' = j\omega \epsilon J_{ma} \tag{2.27}$$

and that it seems simpler to solve these equations directly. This simplification however is only apparent because, as we will see in the next section, many problems can be solved with only one or two adequately chosen scalar components of the potentials.

2.2.2 Representation in terms of two scalars

In many problems we have to write down a general solution of Maxwell's equations in a finite and sourceless ($J_a = 0, J_{ma} = 0$) volume. Equations like 2.22 and 2.23 or 2.26 and 2.27 are then homogeneous and the existence of non-zero scalar components of the fields is due to external sources or to boundary conditions. It has been shown (Jones, 1964) that, in the most general case, the electromagnetic field can be obtained from two Cartesian components of potentials. We will demonstrate this for the slightly more general case where a source exists with only one Cartesian component.

Let us assume that an electric current density parallel to the z-axis

$$J_a = J(x, y, z) u_z \tag{2.28}$$

is applied. We showed in the previous section that the problem could be solved using only an electric-type Hertz potential π' with three components π'_x, π'_y, π'_z and calculating the fields by

$$E' = \text{curl curl } \pi' - (j\omega \epsilon)^{-1} J u_z \tag{2.29}$$

$$H = j\omega \epsilon \text{ curl } \pi' \tag{2.30}$$

The proof is based on the fact that π'_x and π'_y satisfy the Helmholtz equation

$$\left(\frac{\partial^2}{\partial x^2} + \frac{\partial^2}{\partial y^2} + \frac{\partial^2}{\partial z^2} + k^2\right) \pi'_x, \pi'_y = 0 \tag{2.31}$$

and on the observation that the gauge eqn. 2.19 still does not completely define π'. Indeed, if (π'_0, V'_0) satisfy this gauge, so do

$$\pi' = \pi'_0 + \text{grad } \psi \tag{2.32}$$

$$V' = V'_0 - k^2 \psi \tag{2.33}$$

provided ψ is a solution of the Helmholtz equation

$$\nabla^2 \psi + k^2 \psi = 0 \tag{2.34}$$

Let θ be any solution of the equation

$$\frac{\partial^2 \theta}{\partial z^2} + k^2 \theta = \frac{\partial \pi'_{0x}}{\partial x} + \frac{\partial \pi'_{0y}}{\partial y} \tag{2.35}$$

Applying the operator

$$\nabla_t^2 = \frac{\partial^2}{\partial x^2} + \frac{\partial^2}{\partial y^2} \tag{2.36}$$

to both sides of eqn. 2.35 and taking into account that π'_{0x} and π'_{0y} satisfy the Helmholtz equation, we get

$$\left(\frac{\partial^2}{\partial z^2} + k^2\right) \left(\nabla_t^2 \theta + \frac{\partial \pi'_{0x}}{\partial x} + \frac{\partial \pi'_{0y}}{\partial y}\right) = 0 \tag{2.37}$$

Let θ_a be a particular solution of eqn. 2.35. As it necessarily satisfies eqn. 2.37, we may write

$$\nabla_t^2 \, \theta_a = - \frac{\partial \pi'_{0x}}{\partial x} - \frac{\partial \pi'_{0y}}{\partial y} + A_a \, (x, y) \, e^{-jkz} + B_a \, (x, y) \, e^{jkz} \qquad (2.38)$$

where A_a and B_a are any functions.

The general solution of eqn. 2.35 is

$$\theta = \theta_a + A(x, y) \, e^{-jkz} + B(x, y) \, e^{jkz} \qquad (2.39)$$

where A and B are arbitrary functions. One has, using eqn. 2.38,

$$\nabla_t^2 \, \theta = - \frac{\partial \pi'_{0x}}{\partial x} - \frac{\partial \pi'_{0y}}{\partial y} + (A_a + \nabla_t^2 \, A) \, e^{-jkz} + (B_a + \nabla_t^2 \, B) \, e^{jkz} \quad (2.40)$$

One may still choose A and B such that

$$\nabla_t^2 \, A = - A_a \qquad (2.41)$$

$$\nabla_t^2 \, B = - B_a \qquad (2.42)$$

The resulting solution θ will be denoted ψ. This solution is such that, from eqns. 2.35 and 2.40:

$$\frac{\partial^2 \psi}{\partial z^2} + k^2 \, \psi = \frac{\partial \pi'_{0x}}{\partial x} + \frac{\partial \pi'_{0y}}{\partial y} \qquad (2.43)$$

$$\frac{\partial^2 \psi}{\partial x^2} + \frac{\partial^2 \psi}{\partial y^2} = - \frac{\partial \pi'_{0x}}{\partial x} - \frac{\partial \pi'_{0y}}{\partial y} \qquad (2.44)$$

It thus satisfies eqn. 2.34 and may be used to transform $(\boldsymbol{\pi}'_0, V'_0)$ into $(\boldsymbol{\pi}', V')$ according to eqns. 2.32 and 2.33. It now appears from eqns. 2.32 and 2.44 that π'_x and π'_y are such that

$$\frac{\partial \pi'_x}{\partial x} + \frac{\partial \pi'_y}{\partial y} = 0 \qquad (2.45)$$

As a result it is still possible to derive π'_x and π'_y from a single scalar function f by

$$\pi'_x = \frac{\partial f}{\partial y} \qquad (2.46)$$

$$\pi'_y = - \frac{\partial f}{\partial x} \qquad (2.47)$$

This function must obviously satisfy

$$\nabla^2 f + k^2 \, f = 0 \qquad (2.48)$$

Let us now consider the vector

$$f = f \, u_z \qquad (2.49)$$

which satisfies

$$\nabla^2 f + k^2 f = 0 \tag{2.50}$$

The transverse part of π' is given by

$$\pi'_t = \operatorname{curl} f \tag{2.51}$$

The electromagnetic field which will be calculated from π'_t by eqns. 2.24 and 2.25, using eqn. 2.50 and the identity 2.15, is given by

$$E_1 = -j\omega \mu_0 \operatorname{curl} (j\omega \epsilon f) \tag{2.52}$$

$$H_1 = \operatorname{curl} \operatorname{curl} (j\omega \epsilon f) \tag{2.53}$$

Comparing this with eqns. 2.24 and 2.25, it is obvious that the same result can be obtained by using a magnetic-type Hertz potential

$$\pi'' = j\omega\epsilon f \tag{2.54}$$

A very important conclusion follows: *when, in some volume filled with a homogeneous medium, only a z-directed electric current is applied, the electromagnetic field can always be derived from z-directed electric-type and magnetic-type Hertz potentials. As obviously results from the proof, this statement remains valid when only a z-directed magnetic current is applied and, of course, when no electric or magnetic currents at all are applied.*

2.2.3 Adequate choice of potentials

The above statement is extremely interesting for we shall always be concerned with cases where the applied electric or magnetic current densities, if any, are restricted to one Cartesian component. It is then still possible to solve the problem using electric-type and magnetic-type Hertz potentials with one non-zero component along the same Cartesian axis. Frequently there will be no applied current density at all. The choice of this axis is then free. In some cases it is also possible to use two Cartesian components of the same, either electric or magnetic, potential or a component of π' and another component of π''. There thus remains some freedom which can be used to simplify the mathematical analysis.

In order to facilitate the discussion, we have gathered in Appendix A a complete set of formulas for the Cartesian coordinate system. Appendix B contains formulas for the cylindrical coordinate system when the potentials π'_z and π''_z are used. These functions will frequently be denoted

$$U = \pi'_z; \qquad V = \pi''_z \tag{2.55}$$

In order to illustrate how an adequate choice of a pair of potentials can be made, we will discuss a particularly useful example. Suppose that we are trying to find guided modes characterised by a propagation factor $\exp(-\Gamma z)$ in a structure containing a plane interface between two media. We will assume that the interface lies in the (x, z) plane and that the media for $y < 0$ and $y > 0$ have parameters (ϵ_1, k_1)

and (ϵ_2, k_2). The boundary conditions at the interface require the components E_x, E_z, H_x and H_z to be continuous, which entails the continuity of ϵE_y and of H_y. These conditions are obviously equalities between functions of x. We use formulas A.13 to A.30.

Let us first see if the boundary conditions can be satisfied by a single potential component:

(a) π_x'

This requires the continuity of

$$\frac{\partial^2 \pi_x'}{\partial x^2} + k^2 \pi_x' \quad , \quad \frac{\partial \pi_x'}{\partial x} \quad , \quad \epsilon \frac{\partial \pi_x'}{\partial y}$$

As the second condition implies

$$\pi_x'(x, -0) = \pi_x'(x, +0) + C \tag{2.56}$$

where C is a constant, the first condition requires

$$k_1^2 \, \pi_x'(x, -0) = k_2^2 \, \pi_x'(x, +0) \tag{2.57}$$

These equations show that π_x' will be independent of x. Consequently the use of π_x' is restricted to the case where all the variables are independent of x, the non-zero field components being (E_x, H_y, H_z).

(b) π_y'

This requires the continuity of

$$\frac{\partial^2 \pi_y'}{\partial x \partial y} \quad , \quad \frac{\partial \pi_y'}{\partial y} \quad , \quad \epsilon \, \pi_y' \quad , \quad \epsilon \frac{\partial \pi_y'}{\partial x}$$

The second and third conditions entail the first and fourth ones, respectively. Two boundary conditions are thus still required. This yields solutions with $H_y = 0$.

(c) π_z'

This requires the continuity of

$$\frac{\partial \pi_z'}{\partial x} \quad , \quad (k^2 + \Gamma^2) \, \pi_z' \quad , \quad \epsilon \frac{\partial \pi_z'}{\partial y}$$

The second condition is obviously incompatible with the first, unless all variables are independent of x. This type of solution has non-zero field components (E_y, E_z, H_x).

Solutions derived from a single component of $\boldsymbol{\pi}''$ can be considered as dual from the previous ones. In particular π_y'' yields solutions with $E_y = 0$. Solutions derived from π_x'' and π_z'' are suitable when all the variables are independent of x. They yield non-zero field components which are (E_y, E_z, H_x) and (E_x, H_y, H_z), respectively. However it can be seen that solutions obtained from π_y' include those derived from π_z' and π_x'. A similar remark applies to π_y'' as compared with π_z'' and π_x''. Consequently, only π_y' and π_y'' are interesting as single potential components.

To show the limitation of solutions derived from a single component, let us consider a problem with a stratification parallel to the (x, z) plane and suppose that we have n media and thus $(n-1)$ interfaces. Using either π_y' or π_y'' will require $2(n-1)$ interface boundary conditions. As each of the two external media imposes a condition to infinity, we finally have $2n$ conditions. Now it appears that the general solution of the equation

$$\left(\frac{\partial^2}{\partial x^2} + \frac{\partial^2}{\partial y^2} + k^2 + \Gamma^2\right) \pi = 0 \tag{2.58}$$

by separation of variables in a homogeneous medium contains two constants to define π as a function of y. We have thus $2n$ unknowns which are completely defined by our $2n$ conditions. Consequently we cannot add any supplementary condition and the use of either π_y' or π_y'' alone would not allow the solution of problems such as that of the juxtaposition of two different stratifications with an interface in the (y, z) plane. It would also not allow the solution of the problem of the addition of a longitudinal wire to the structure. In such cases it will be necessary to use two scalar potential components.

We will not dwell on a discussion of the 15 possible choices of a pair of scalar components, but formulate some useful observations. We justified the choice of parallel components of $\boldsymbol{\pi}'$ and $\boldsymbol{\pi}''$ in Section 2.2.2. For interfaces parallel to the z-axis, if the pair (π_z', π_z'') is used, the boundary conditions involve only these functions and their first derivatives. Moreover the continuity of E_z and H_z at these interfaces are uncoupled because π_z' and π_z'' yield $H_z = 0$ and $E_z = 0$, respectively. This pair is thus of interest. Furthermore it is well adapted to the case of cylindrical coordinates. In some cases however other pairs may be of interest too, but it must be stressed that this depends on the approach which is chosen to solve the problem.

2.3 Thin-cable approximation

As we have seen in the first chapter, many systems developed to guide e.m. waves along tunnels involve cables or wires which are drawn parallel to the cable axis. An exact solution of the relevant e.m. problem would require the solution of Maxwell's equations inside and outside the cable and the expression of boundary conditions at the cable surface. This is a titanic work, even in the simplest cases, and some simplifying assumptions are needed.

In this section, we will devote some attention to the fact that the diameter of wires and leaky feeders used in subsurface radio communications is generally much smaller than the distance from the cable to the tunnel wall and than the wavelength. Specifically we will show that, provided some conditions are met, the cable can be modelled by a very simple surface impedance condition. This provides an extreme simplification, as the need to solve Maxwell's equations in the interior of the cable is circumvented.

For this purpose let us consider a cable with radius ρ_1, surrounded within a radius ρ_2 by a homogeneous and sourceless medium with permittivity ϵ. Although the analysis will be carried out for a mode with propagation factor $\exp(-\Gamma z)$, it can be extended to other field configurations by decomposing the latter into a summation of terms containing factors $\exp(-jhz)$, i.e. by a Fourier transform. The solution of Maxwell's equations for $\rho_1 < \rho < \rho_2$ may be written in the form of eqns. B.19 to B.24 of Appendix B, where A_m, B_m, P_m and Q_m are as yet arbitrary constants. For better consistency with the notations used in the following sections, we will write

$$v = \sqrt{-k^2 - \Gamma^2} \tag{2.59}$$

where k is the complex wave number of the medium for $\rho_1 < \rho < \rho_2$.

We will investigate the influence of tangential electric field components at the cable's external surface

$$\Phi_m = E_{\phi m}(\rho_1) \tag{2.60}$$

$$\Psi_m = E_{zm}(\rho_1) \tag{2.61}$$

on the electromagnetic field for $\rho \gg \rho_1$.

Writing eqns. B.21 and B.22 at $\rho = \rho_1$, solving these equations for A_m and B_m, inserting the result into eqns. B.19 and B.24, and using the notation

$$f_m(v\rho) = I_m(v\rho) - \frac{I_m(v\rho_1)}{K_m(v\rho_1)} K_m(v\rho) \tag{2.62}$$

$$g_m(v\rho) = I_m(v\rho) - \frac{I'_m(v\rho_1)}{K'_m(v\rho_1)} K_m(v\rho) \tag{2.63}$$

we obtain after tedious calculations

$$E_{\phi m}(\rho) = j\omega\mu_0 P_m g'_m(v\rho) - \frac{j\Gamma m}{\rho} Q_m f_m(v\rho) + A_{\phi\phi m}\Phi_m + A_{\phi zm}\Psi_m \tag{2.64}$$

$$E_{zm}(\rho) = -v^2 Q_m f_m(v\rho) + A_{zzm}\Psi_m \tag{2.65}$$

$$H_{\phi m}(\rho) = \frac{-j\Gamma m}{\rho} P_m g_m(v\rho) - j\omega\epsilon\, v\, Q_m f'_m(v\rho) + B_{\phi\phi m}\Phi_m + B_{\phi zm}\Psi_m \tag{2.66}$$

$$H_{zm}(\rho) = -v^2 P_m g_m(v\rho) + B_{z\phi m}\Phi_m + B_{zzm}\Psi_m \tag{2.67}$$

with

$$A_{\phi\phi m} = \frac{K'_m(v\rho)}{K'_m(v\rho_1)} \tag{2.68}$$

$$A_{\phi zm} = -\frac{j\Gamma m}{v}\left[\frac{K'_m(v\rho)}{v\rho_1 K'_m(v\rho_1)} - \frac{K_m(v\rho)}{v\rho K_m(v\rho_1)}\right] \tag{2.69}$$

$$A_{zzm} = \frac{K_m(v\rho)}{K_m(v\rho_1)} \tag{2.70}$$

$$B_{\phi\phi m} = -\frac{j\Gamma m}{j\omega\mu_0} \frac{K_m(v\rho)}{v\rho\, K'_m(v\rho_1)} \tag{2.71}$$

$$B_{\phi zm} = -\frac{1}{j\omega\mu_0\, v} \left[\frac{(\Gamma m)^2}{v\rho\, v\rho_1} \frac{K_m(v\rho)}{K'_m(v\rho_1)} + k^2 \frac{K'_m(v\rho)}{K_m(v\rho_1)} \right] \tag{2.72}$$

$$B_{z\phi m} = -\frac{v}{j\omega\mu_0} \frac{K_m(v\rho)}{K'_m(v\rho_1)} \tag{2.73}$$

$$B_{zzm} = \frac{\Gamma m}{\omega\mu_0} \frac{K_m(v\rho)}{v\rho_1\, K'_m(v\rho_1)} \tag{2.74}$$

Here, we are interested in the field values at some distance from the cable, either to obtain the field distribution in the tunnel space or to express boundary conditions at $\rho = \rho_2$. As the cable is assumed to be thin, we will examine the values of eqns. 2.68 to 2.74 for $\rho \gg \rho_1$. We may further assume that $|v\rho_1| \ll 1$, i.e. that the cable radius is small compared with the transverse wave number.

Let us first examine the modes without rotational symmetry, $m > 0$. We may use the approximations of modified Bessel functions for small arguments (see Appendix C):

$$I_m(x) \simeq \frac{x^m}{2^m\, m!}$$

$$K_m(x) \simeq \frac{2^{m-1}\,(m-1)!}{x^m} \qquad \text{for } |x| \ll 1$$

to show that the coefficients 2.68 to 2.74 are very small, which is not true for the coefficients of P_m and Q_m in eqns. 2.64 to 2.67. Specifically it can be shown that, if $|v\rho| \ll 1$, the coefficients 2.68 to 2.74 are given by quantities like $(\rho_1/\rho)^m$, $(\rho_1/\rho)^{m+1}$ and $m(\rho_1/\rho)^m$, which are very small if $\rho \gg \rho_1$. This statement is *a fortiori* valid if $|v\rho|$ is not small. Consequently we may expect that only the P_m and Q_m terms in eqns. 2.64 to 2.67 are significant provided $|v\rho_1| \ll 1$ and $\rho \gg \rho_1$.

This is however not true for the rotational symmetric term, $m = 0$. Indeed, using the approximation

$$K_0(x) \simeq -\ln\frac{Cx}{2}; \qquad |x| \ll 1$$

where $C = 1\cdot7810 \ldots$, one can show that A_{zz0} and $B_{\phi z0}$ behave as $1/\ln(v\rho_1)$ while the other coefficients 2.68 to 1.74 vanish or behave as $(v\rho_1)$.

From this discussion we may conclude that in the limit of a vanishing cable radius, i.e. for $|v\rho_1| \ll 1$, and as long as we are interested in the calculation of the field in regions $\rho \gg \rho_1$, only the rotational symmetric component $\Psi_0 = E_{z0}(\rho_1)$

of the electric field at the cable surface has a significant contribution. The rotational symmetric part $\phi_0 = E_{\phi 0}(\rho_1)$ and the asymmetric components $\Phi_m = E_{\phi m}(\rho_1)$ and $\Psi_m = E_{zm}(\rho_1)$ have a negligible influence. This assumption will be referred to as the *thin-cable approximation*. It obviously does not mean that the rotational asymmetric part of the field does not exist, even at the cable outer surface. In particular the medium or the sources located at $\rho > \rho_2$ may force these components to exist and in many cases to have comparable amplitudes for some values of ρ. We must however observe that it is possible to imagine cable structures which would imply a large value of the neglected term $E_{\phi 0}(\rho_1)$, but this is generally not the case in leaky feeders.

The thin-cable approximation has the advantage that it reduces the study of the cable interior to searching for a relation between the rotational symmetric parts E_{z0}, $H_{\phi 0}$ and H_{z0} at the cable outer surface. As can be expected for a cable carrying some longitudinal currents, it turns out in most cases that H_{z0} is zero or negligible and that the cable can be modelled by a surface impedance relation of the type

$$E_{z0}(\rho_1) = Z_0 H_{\phi 0}(\rho_1) \tag{2.75}$$

The quantity $I = 2\pi\rho_1 H_{\phi 0}(\rho_1)$ will be referred to as the total current carried by the cable. Defining the *specific external impedance*

$$z_0 = \frac{Z_0}{2\pi\rho} \tag{2.76}$$

we may rewrite eqn. 2.75 as

$$E_{z0} = z_0 I \tag{2.77}$$

The surface impedance Z_0 is not an intrinsic characteristic of the cable, for it depends on the external medium through the mode propagation constant Γ.

2.4 Surface impedance of a thin conducting wire

We will now consider the simple case of a cable consisting of a conducting wire of radius c. In order to obtain the surface impedance, we need to concentrate on the electromagnetic field inside the wire, for which we use the following notation:

$$\gamma_i = j\omega\mu\hat{\sigma} \tag{2.78}$$

$$\hat{\sigma} = \sigma + j\omega\epsilon \tag{2.79}$$

$$w = \sqrt{\gamma_i^2 - \Gamma^2}, \qquad \text{Re } w \geqslant 0 \tag{2.80}$$

Note that we will not assume here that $\mu = \mu_0$ for the wire material.

The solution of Maxwell's equations inside the wire is given by eqns. B.18 to B.24 in which we have to use w instead of u. The constants A_m and B_m must be zero to keep the fields finite on the wire axis. On eliminating the constants P_m and

Q_m between eqns. B.19 to B.24, it can easily be shown that, anywhere inside the conducting wire, we have

$$E_{zm} = Z_{zzm} H_{zm} + Z_{z\phi m} H_{\phi m} \tag{2.81}$$

$$E_{\phi m} = Z_{\phi zm} H_{zm} + Z_{\phi\phi m} H_{\phi m} \tag{2.82}$$

where

$$Z_{zzm} = \frac{-jm \, \Gamma\rho \, I_m(w\rho)}{\hat{\sigma}\rho \, w\rho \, I'_m(w\rho)} \tag{2.83}$$

$$Z_{z\phi m} = \frac{w\rho \, I_m(w\rho)}{\hat{\sigma}\rho \, I'_m(w\rho)} \tag{2.84}$$

$$Z_{\phi\phi m} = \frac{jm \, \Gamma\rho \, I_m(w\rho)}{\hat{\sigma}\rho \, w\rho \, I'_m(w\rho)} \tag{2.85}$$

$$Z_{\phi zm} = \frac{-(\gamma_i\rho)^2 \, I'_m(w\rho)}{\hat{\sigma}\rho \, w\rho \, I_m(w\rho)} + \frac{m^2 \, (\Gamma\rho)^2 \, I_m(w\rho)}{\hat{\sigma}\rho \, (w\rho)^3 \, I'_m(w\rho)} \tag{2.86}$$

As the boundary conditions at an interface require the continuity of the tangential components of the fields, these relations, wherein ρ is replaced by c, hold on the external surface of the wires. Hence we have obtained a surface impedance matrix for each cylindrical harmonic. This matrix depends on the mode propagation constant Γ through the transverse wave number w. The result is valid whatever the external medium and the sources located therein. We can now introduce some approximations.

As was explained in the previous section, in the thin-wire approximation we are interested solely in the rotational symmetric element $Z_{z\phi 0}$, denoted Z_0. This surface impedance is given by eqn. 2.84 with $m = 0$ and $\rho = c$. We may further assume that $|\Gamma| \ll \gamma_i$. This approximation is justified in all cases where the wire is a good conductor. A good conductor is indeed defined by the property

$$\sigma \gg \omega\epsilon \tag{2.87}$$

which implies

$$\gamma_i \simeq \frac{1+j}{\delta} \tag{2.88}$$

where

$$\delta \simeq \sqrt{\frac{2}{\omega\sigma\mu}} \tag{2.89}$$

is the skin depth in the conductor. Our assumption is thus that the mode wavelength is much larger than the skin depth in the conductor material. In this case, $w \simeq \gamma_i$ and eqn. 2.84 yields

$$Z_0 \simeq \frac{\gamma_i c \, I_0(\gamma_i c)}{\sigma c \, I_0'(\gamma_i c)} \tag{2.90}$$

This value has now become independent of the mode propagation constant. Two extreme cases may be considered. First, if the wire radius is small compared with the skin depth, $|\gamma_i c|$ is a small quantity. Using the approximations

$$I_0(x) \simeq 1$$
$$I_0'(x) \simeq \frac{x}{2} \tag{2.91}$$

for $|x| \ll 1$, we get the trivial static limit

$$Z_0 \simeq 2/(\sigma c) \tag{2.92}$$

If, on the other hand, $|\gamma_i c|$ is a large quantity, the use of the large-argument approximation

$$I_0(x) \simeq \frac{e^x}{\sqrt{2\pi x}}, \qquad |x| \gg 1 \tag{2.93}$$

yields

$$Z_0 \simeq \eta_i \tag{2.94}$$

where

$$\eta_i = \sqrt{\frac{j\omega\mu}{\hat{\sigma}}} \simeq \sqrt{\frac{j\omega\mu}{\sigma}} \tag{2.95}$$

is the intrinsic impedance of the conductor material. This specific external impedance is thus given by

$$z_0 \simeq \frac{\eta_i}{2\pi c} \tag{2.96}$$

We now may comment somewhat further on the thin-wire approximation. As we have explained in the previous section, this approach is used in some applications where the quantities of interest depend mainly on $Z_0 = Z_{z\phi 0}$. This does not mean that the other neglected surface impedances are much smaller than this quantity. As a proof of this statement, we may consider a conducting wire for which we have $|\Gamma| \ll \gamma_i$ and $|\gamma_i c| \gg 1$. Then for all cylindrical harmonics for which $|\gamma_i c| \gg m$, it is easy to show from eqns. 2.83 to 2.86 that

$$|Z_{zzm}| \quad , \qquad |Z_{\phi\phi m}| \ll |\eta_i|$$
$$Z_{z\phi m} \simeq -Z_{\phi z m} \simeq \eta_i$$

For these cylindrical harmonics, the surface impedance is thus isotropic and equal to Z_0.

2.5 Transfer properties of leaky feeders

★ *2.5.1 The transfer impedance concept*

Coaxial cables have been used for the transmission of electromagnetic energy for many years. The screening effect of the outer conductor provides obvious advantages. Perfect shielding can be obtained by using a plain outer conductor provided the shield thickness is large compared with the skin depth in the conductor material. This condition cannot easily be fulfilled at low frequencies: skin depth in copper at $100\,\text{kHz}$ is about $0 \cdot 2\,\text{mm}$. Under such conditions, signals propagating inside the coaxial cable give rise to an axial voltage on the outer surface of the shield and are thereby coupled to the external space. The inverse process also occurs by reciprocity. In order to get a simple picture of the phenomenon, we may imagine that the cable is drawn parallel to a return path which may be referred to as the structural ground. The cable shield can then be seen as the common return path of two transmission lines. In the framework of the present book these transmission lines convey what we have called the coaxial and monofilar modes.

As a current flowing inside the coaxial cable gives rise to an axial voltage in the outer transmission line, and vice versa, we expect to get coupled transmission line equations of the type

$$-\frac{dV_c}{dz} = z_c I_c + z_t I_m \tag{2.97}$$

$$-\frac{dI_c}{dz} = y_c V_c \tag{2.98}$$

$$-\frac{dV_m}{dz} = z_m I_m + z_t I_c \tag{2.99}$$

$$-\frac{dI_m}{dz} = y_m V_m \tag{2.100}$$

where the subscripts c and m refer to coaxial and monofilar, respectively.

Alternatively we may write eqns. 2.97 and 2.99 as

$$-\frac{dV_c}{dz} = (z_c - z_t) I_c - z_t I_s \tag{2.101}$$

$$-\frac{dV_m}{dz} = (z_m - z_t) I_m - z_t I_s \tag{2.102}$$

where

$$I_s = -(I_c + I_m) \tag{2.103}$$

is the total current carried by the cable shield in the positive direction of the z-axis. Equivalent circuits of a line element corresponding to the use of either eqns. 2.97–2.99 or 2.101–2.102 are shown in Figs. 2.1a and b, respectively. For completeness we have also shown a transverse coupling admittance y_t which is sometimes used, with a corresponding modification of eqns. 2.98–2.100.

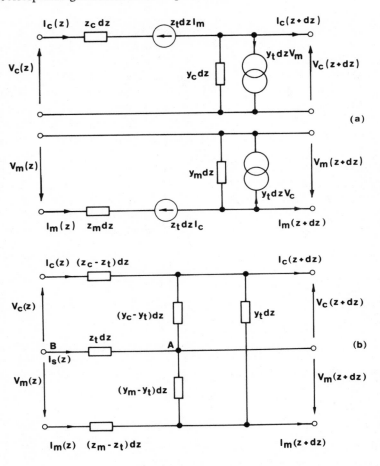

Fig. 2.1   *Two equivalent circuits of an elementary length dz of two coupled transmission lines*

A cable shield is said to be *electrically thin* if the tangential electric field is continuous through the shield. For a plain shield this obviously requires the shield thickness to be small compared with the skin depth. For such shields, the circuit diagram of Fig. 2.1b provides a simple physical interpretation of the impedance elements. Indeed, according to this representation the axial voltage drop is the voltage which appears between points A and B. The axial electric field along the shield is given by

$$E_z = \frac{-(V_A - V_B)}{dz} = z_t I_s \qquad (2.104)$$

and the *specific transfer impedance* z_t appears to be the impedance of a unit length of the imperfect shield. Similarly $(z_c - z_t)$ and $(z_m - z_t)$ appear to be the specific series impedances of the inner and outer circuits in the hypothetic situation where the actual shield is removed and replaced by a perfect conductor. For electrically thick shields, Fig. 2.1 may remain valid but giving a physical interpretation to Fig. 2.1b is not allowed. Indeed, points located on the inner and outer surfaces of the shield are not equipotential and we may not say that a point like B is actually common to the inner and outer circuits.

The model of coupled lines shown in Fig. 2.1 was first developed by Schelkunoff (1934) for coaxial cables with a plain but imperfectly conducting shield. In this case the model is not necessarily restricted to electrically thin shields. For practical reasons, namely mechanical flexibility, the shield of coaxial cables is frequently braided. It then contains numerous small apertures through which an additional leakage of electromagnetic energy can occur. In leaky feeders like those shown in Fig. 1.9, this mechanism is deliberately enhanced.

As a first and rough approach to this problem, we will show that the apertures give rise to an additional inductive transfer impedance. For this purpose let us consider a periodic distribution of apertures in a infinitely thin conducting plane shield, as shown in Fig. 2.2 We will assume that the aperture shape and location are symmetric with respect to the z-axis. We further suppose that the shield carries a mean current density K_{z0} in the direction of the z-axis and that this quantity does not depend on x. The lines of current are obviously distorted by the apertures, as shown in Fig. 2.2a. Fig. 2.2b shows a magnified view of the reference aperture located at the origin of the axes.

The current distribution induces a magnetic field which, by symmetry, has only a y-component inside the apertures. Let us assume that the current distribution $K(x, z)$ is known and that $H_y(x, 0, z)$ inside the reference aperture may be calculated from it by magnetostatic methods. This seems to be justified provided the period of the structure is small compared with the wavelength. Obviously $H_y(x, 0, z)$ is an odd function of x and it is negative for $x > 0$. Maxwell's equation

$$\text{curl } E = -j\omega\mu_0 H \qquad (2.105)$$

shows that E_x and E_z are odd and even functions of x, respectively. As a result, the average value of the tangential electric field in the apertures is parallel to the z-axis. No tangential electric field can of course exist on the screen outside the apertures.

Let us now consider average values on one period of the structure. These quantities are defined by expressions like

$$f_0 = \frac{1}{A_p} \int_{\text{period}} f \, dS \qquad (2.106)$$

where A_p is the area of one period. They are the zero-order terms of a two-dimensional Fourier series. We obtain in particular

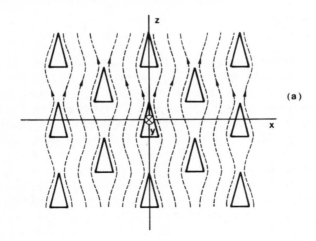

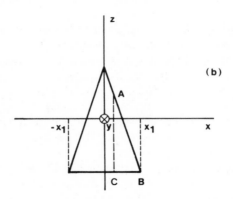

Fig. 2.2 *Plane shield with a periodic distribution of apertures*

$$K_{z0}u_z = \frac{1}{A_p} \int K(x, z) \, dS \tag{2.107}$$

$$E_{z0} = \frac{1}{A_p} \int_{aperture} E_z \, dS \tag{2.108}$$

Part of the last integral can be calculated by applying the integral form of eqn. 2.105 to the contour ABC of Fig. 2.2b. As the electric field is normal to AB and BC, which run along a metallic surface, this yields

$$\int_C^A E_z(x, 0, z) \, dz = -j\omega\mu_0 \int_{ABC} H_y(x', 0, z') \, dx' \, dz' \tag{2.109}$$

This quantity has still to be integrated from $(-x_1)$ to x_1 to obtain the integral in the second part of eqn. 2.108. As the right-hand side of eqn. 2.109 is proportional to the average current density K_{z0}, this provides a relation of the type

$$E_{z0} = j\omega M_s K_{z0} \tag{2.110}$$

On the other hand the average values of the magnetic field above and below the screen are tangential and x-oriented. They undergo a discontinuity given by

$$K_{z0} = H_{x0} (y - 0) - H_{x0} (y + 0) \tag{2.111}$$

Eqn. 2.110 thus becomes a transfer condition for the electromagnetic fields:

$$E_{z0} = j\omega M_s [H_x (y - 0) - H_x (y + 0)] \tag{2.112}$$

It can easily be seen that M_s is a positive quantity. It is called the *surface transfer inductance*, while

$$Z_s = j\omega M_s \tag{2.113}$$

is called the *surface transfer impedance*, with units of henries and ohms, respectively. If we abstract the coordinate system, eqn. 2.112 may be written

$$E_t = Z_s n_2 \times (H_{t2} - H_{t1}) \tag{2.114}$$

for a plane screen separating two media referred to by subscripts 1 and 2; here n_2 is the unit normal to the screen, oriented toward medium 2.

We now can extend the concept of surface transfer inductance to perforated coaxial-cable shields. As long as the size and period of the apertures are small compared to the shield radius b, we may expect that the fields near to the shield are not significantly modified by the curvature and that we have the same surface transfer inductance as for a plane shield. Eqn. 2.114 then becomes

$$E_{z0} = Z_s [H_{\phi 0} (b + 0) - H_{\phi 0} (b - 0)] \tag{2.115}$$

The magnetic field discontinuity on the right-hand side of this equation is obviously related to the screen current I_s by

$$2\pi b [H_{\phi 0} (b + 0) - H_{\phi 0} (b - 0)] = I_s \tag{2.116}$$

Combining these equations and comparing the result with eqn. 2.104 yields the *specific transfer impedance*

$$z_t = \frac{Z_s}{2\pi b} \tag{2.117}$$

which consists of a *specific transfer inductance*

$$m_t = \frac{M_s}{2\pi b} \tag{2.118}$$

These quantities have units of $\Omega \, m^{-1}$ and $H \, m^{-1}$, respectively.

If the size and lateral period of the apertures are not small compared with the shield radius, we may expect that relations of type 2.115 to 2.118 will still exist, but the transfer parameters can obviously no longer be obtained from those of a plane shield. It is also clear that, particularly in this case, the fields depart significantly from rotational symmetry. In some problems we may nevertheless restrict our interest to the rotational-symmetric part, as we have seen in Section 2.3. Finally, if the axial size of the apertures is not small compared with the wavelength, it is to be expected that a surface transfer impedance defined by eqn. 2.115 will depend on the mode wave number.

2.5.2 A critical review of previous work

The discussion of the transfer impedance concept carried out in the previous paragraph was deliberately restricted to a quasi-static analysis since we intended to provide a simple introduction to a difficult subject. As we will see later, this simplified approach has the merit of providing results which are not worse than those of some more sophisticated methods.

Several questions nevertheless require clarification. To be a really significant and useful concept, the surface transfer impedance should be an intrinsic parameter of the shield. This requires that its value should remain unchanged when other parameters like the dielectric constant of the cable insulation, the diameter of the inner conductor or the electrical and geometrical parameters of the external medium are varied. Krügel (1956) measured the transfer impedance of a large number of braided cables. He found that the transfer impedance of a given shield does not vary when the internal parameters (dielectric constant, inner conductor radius) are changed. The measurements were made on cables with rather efficient shields, having a large number of very small holes. Krügel's conclusions thus agrees with the qualitative picture given in the previous section, but there is no justification for extrapolating it to shields with large holes.

In the context of leaky feeder applications a related question is: is the transfer inductance dispersive, i.e. does it depend on the mode propagation constant and, in particular, does it take the same value for the monofilar and coaxial modes? Classical measurements, like those reported by Krügel (1956) and many other authors, were made on samples much shorter than the wavelength and cannot provide an answer to this question. Test on longer samples would not be useful since that kind of measurement necessarily involves two modes of propagation. Furthermore interpretation of the results would depend on the use of some transmission line model.

It is felt that the possible dispersive character of the transfer inductance can only be proved by the exact solution of Maxwell's equations for some configurations. Unfortunately this is an extremely intricate task. To the author's knowledge, if we except Schelkunoff's (1934) pioneering work on plain shields, three attempts have been made in this direction. Wait (1976b) developed the mathematical model of a loosely braided cable, the shield consisting of multiwire counterwound helices. The numerical results (Hill and Wait, 1980b) show that, when the number of wires in each helix is larger than about four, the transfer inductance is independent of the frequency and of the mode propagation constant. Similar results were obtained for the simpler case of unidirectional helices (Hill and Wait, 1980a). In both cases, the calculations were made for a cable drawn into free space. The monofilar mode then becomes a sort of Goubau wave but it is not expected that the general conclusion would be modified if the cable were drawn into a tunnel. The conclusion is that the transfer inductance is an intrinsic property of a shield with apertures (i.e. frequency and propagation constant independent) when the spatial periods of the aperture distribution become small compared with the cable perimeter and with the mode wavelength. Delogne and Laloux (1980) have studied the problem of the axially slotted cable. Although the convergence of their numerical results was not quite satisfactory, they seem to indicate that the transfer impedance is mode dependent, at least for slot angles of 90° and more. This is not too surprising for this kind of structure.

Several authors have tried to answer these questions by means of transmission line models. A good review of these approaches was given by Casey and Vance (1978). The reader interested in a thorough bibliographic analysis is referred to this paper, as we will limit ourselves to a discussion of the methods which have been used. Analyses of the penetration through small apertures in cable shields are often based on Bethe's (1944) famous paper on the scattering of a plane wave by a small hole in a perfectly conducting plane. The use of this method will be discussed in Section 2.5.4. Bethe showed that the scattered fields are the same as those radiated by two dipoles in free space: an electric dipole perpendicular to the aperture and a magnetic dipole parallel to the aperture. The dipole strengths are related to the incident electric and magnetic fields (defined as the fields that would exist on a screen without hole) by electric and magnetic polarisabilities, respectively.

Using this approach, Kaden (1957) obtained a coupled transmission line model involving a specific transfer inductance, but also a specific mutual capacitance between the centre conductor and the structure outside the shield. These transmission line parameters result from the magnetic and electric dipoles, respectively. This method was refined and applied to coupled transmission line theory by various investigators, namely Latham (1972), Vance (1975), Lee and Baum (1975) and Casey (1976).

The use of an electric dipole and thus of capacitive coupling in the transmission line equations is highly questionable. The electric dipole is related to the average normal electric field inside the aperture, while the magnetic dipole is related to the average tangential electric field. It is known that specifying the tangential electric field on the boundary of a volume ensures uniqueness of the Maxwell's equations' solution in that volume. Consequently the use of an electric dipole, i.e. of the normal electric field, is not required to describe the radiation of a hole. On the contrary the electric dipole is completely defined once the magnetic dipole is given.

The situation bears many similarities with the well-known problem of radiation by apertures. Kottler's formulas (Stratton, 1941) allow us to calculate the fields inside a volume from a knowledge of $n \times E$, $n \times H$, $n \cdot E$ and $n \cdot H$ on the boundary and the equivalence theorem tells us that these surface fields radiate as superficial electric and magnetic current and charge densities given by

$$K = -n \times H$$
$$K_m = n \times E$$
$$\sigma = -\epsilon n \cdot E \tag{2.119}$$
$$\sigma_m = -\mu n \cdot H$$

where n is the external normal to the boundary. However these quantities may not be given independently for this would violate the uniqueness theorem. Indeed, they are uniquely defined either by $n \times E$ or by $n \times H$. The advantage of Kottler's formulas is that they are valid for arbitrary surfaces, while the formulas for the calculation from either $n \times E$ or $n \times H$ alone must be found for each particular geometry; this is in general a difficult task. The counterpart is that the use of Kottler's formulas implicitly supposes that the complete problem has been solved beforehand!

This question has generally been overlooked in the literature on the shielding efficiency of coaxial cables. For instance, Lee and Baum's (1975) modal analysis of braided-shield coaxial cables is based on the equivalence theorem and is consequently a decomposition of Kottler's formulas into a summation over the eigenmodes of a coaxial cable. This is undoubtedly correct,[*] but the simultaneous use of magnetic and electric dipoles derived from the plane wave scattering by a hole in an infinite plane is not allowed, even for a single small hole in a coaxial cable.

For the study of shields containing a large number of holes, many authors start with the excitation of the TEM mode by an isolated hole and merely add the contributions of the individual holes to this mode. For this method to be valid, it is required that no coupling can occur through the higher-order modes of the coaxial cable. These modes are obviously below cutoff, but some of them have a specific attenuation which is of the order of $(2b)^{-1}$, where b is the shield radius. Neglecting the higher-order modes is only allowed when the spatial periods of the aperture distribution (in the ϕ- and z-directions) are large compared with the cable radius. This condition is practically never met.

A related question is the following one. Most authors, even if they did not explicitly add the contribution of individual holes to the TEM mode, have attempted to derive transmission line models in which the line voltage and current are those of this mode. This approach leads to the use of a transverse coupling admittance. It must be stressed that, when the distance between holes is not large, the fields at all the points inside the cable may significantly differ from the TEM mode. The latter does not exist as such. Even if we restrict our attention to the rotational-symmetric part of the fields, the existence of an E_{z0} component requires a departure of $E_{\rho 0}$ and $H_{\phi 0}$ from the transverse field distribution of the TEM mode. The use of transmission line equations in which the voltage and current are those of the TEM mode

---

[*] In part III of the paper, Lee and Baum however note that only the magnetic current density is needed to describe the effect of an aperture. Consequently they drop the electric current terms. This is not allowed for it is equivalent to postulating that the average values of $n \times H$ and $n \cdot E$ in the aperture are zero. Subsequent results are erroneous.

The use of transmission line equations in which the voltage and current are those of the TEM mode consequently does not appear to be adequate. An alternative formulation seems to be required. It will be explained in the next section.

2.5.3 Exact transmission line model for leaky coaxial cables

In this section, we will establish an exact transmission line model for the fields inside a leaky coaxial cable. We denote by a and b the radii of the inner conductor and of the internal surface of the outer conductor, respectively, and by ϵ and κ the permittivity and dielectric constant of the cable insulation. As we are here not directly interested in the calculation of a transfer impedance, we assume that the tangential components of the electric field on the inner face of the shield

$$E_\phi(b, \phi, z) = F(\phi, z) \tag{2.120}$$

$$E_z(b, \phi, z) = G(\phi, z) \tag{2.121}$$

are given and first proceed to the calculation of the fields inside the cable.

The solution of this problem may be derived from z-oriented electric and magnetic Hertz potentials satisfying a homogeneous Helmholtz equation for $a < \rho < b$. We use a Fourier series and transform of the various quantities as for instance

$$U(\rho, \phi, z) = \sum_{m=-\infty}^{\infty} e^{jm\phi} \frac{1}{2\pi} \int_{-\infty}^{\infty} \mathscr{U}_m(u\rho) e^{jhz} \, dh \tag{2.122}$$

where

$$u = \sqrt{-k_0^2 \kappa + h^2} \tag{2.123}$$

In order to satisfy the boundary conditions $E_\phi = E_z = 0$ at $\rho = a$, the pertinent functions are

$$P_m \mathscr{V}_m(u\rho) = P_m \left[I_m(u\rho) - \frac{I'_m(ua)}{K'_m(ua)} K_m(u\rho) \right] \tag{2.124}$$

$$Q_m \mathscr{U}_m(u\rho) = Q_m \left[I_m(u\rho) - \frac{I_m(ua)}{K_m(ua)} K_m(u\rho) \right] \tag{2.125}$$

The electromagnetic fields may be written as

$$\mathscr{E}_{\rho m} = \frac{\omega\mu_0 \, m}{\rho} P_m \mathscr{V}_m(u\rho) + jh \, u \, Q_m \mathscr{U}'_m(u\rho) \tag{2.126}$$

$$\mathscr{E}_{\phi m} = j\omega\mu_0 \, u \, P_m \mathscr{V}'_m(u\rho) + \frac{j^2 h \, m}{\rho} Q_m \mathscr{U}_m(u\rho) \tag{2.127}$$

$$\mathscr{E}_{zm} = -u^2 \, Q_m \mathscr{U}_m(u\rho) \tag{2.128}$$

$$\mathscr{H}_{\rho m} = jh \, u \, P_m \mathscr{V}'_m(u\rho) - \frac{\omega\epsilon \, m}{\rho} Q_m \mathscr{U}_m(u\rho) \tag{2.129}$$

$$\mathscr{H}_{\phi m} = \frac{j^2 h m}{\rho} P_m \mathscr{V}_m(u\rho) - j\omega\epsilon u Q_m \mathscr{U}'_m(u\rho) \tag{2.130}$$

$$\mathscr{H}_{zm} = -u^2 P_m \mathscr{V}_m(u\rho) \tag{2.131}$$

The remaining unknowns are the coefficients P_m and Q_m. They may be obtained by expressing the boundary conditions 2.120 and 2.121. This yields

$$P_m = \frac{bu^2 \mathscr{F}_m(h) - hm \mathscr{G}_m(h)}{j\omega\mu_0 u^3 b \mathscr{V}'_m(ub)} \tag{2.132}$$

$$Q_m = \frac{-\mathscr{G}_m(h)}{u^2 \mathscr{U}_m(ub)} \tag{2.133}$$

where $\mathscr{F}_m(h)$ and $\mathscr{G}_m(h)$ are the coefficients of the expansion of $F(\phi, z)$ and $G(\phi, z)$, respectively, according to eqn. 2.122.

The electromagnetic fields inside the coaxial cable can be obtained by calculating the inverse Fourier transforms of eqns. 2.126 to 2.131. This task can be completed if we specify the functions $F_m(h)$ and $G_m(h)$. In an actual leaky feeder problem, this would require the calculation of the fields outside the cable and the expression of the continuity of the tangential components inside the shield apertures. We will not undertake such an insurmountable task, but restrict our attention to the derivation of transmission line equations.

The first step is to consider only the rotational-symmetric components, which are given by

$$\mathscr{E}_{\rho 0} = -jh \frac{\mathscr{U}'_0(u\rho)\mathscr{G}_0(h)}{u \mathscr{U}_0(ub)} \tag{2.134}$$

$$\mathscr{E}_{\phi 0} = \frac{\mathscr{V}_0(u\rho) \mathscr{F}_0(h)}{\mathscr{V}'_0(ub)} \tag{2.135}$$

$$\mathscr{E}_{z 0} = \frac{\mathscr{U}_0(u\rho)\mathscr{G}_0(h)}{\mathscr{U}_0(ub)} \tag{2.136}$$

$$\mathscr{H}_{\rho 0} = \frac{jh}{j\omega\mu_0} \frac{\mathscr{V}'_0(u\rho) \mathscr{F}_0(h)}{\mathscr{V}'_0(ub)} \tag{2.137}$$

$$\mathscr{H}_{\phi 0} = j\omega\epsilon \frac{\mathscr{U}'_0(u\rho)\mathscr{G}_0(h)}{u \mathscr{U}_0(ub)} \tag{2.138}$$

$$\mathscr{H}_{z 0} = \frac{u \mathscr{V}_0(u\rho) \mathscr{F}_0(h)}{j\omega\mu_0 \mathscr{V}'_0(ub)} \tag{2.139}$$

The next step is the inverse Fourier transformation and is less obvious. Most authors, like Lee and Baum (1975), extracted the contribution of a pole at $h = k_0 \sqrt{\kappa}$ yielding the TEM mode. It can indeed be shown that this is a pole of the quantity $\mathscr{U}'_0(u\rho)/[u \mathscr{U}_0(ub)]$, i.e. of $\mathscr{E}_{\rho 0}$ and $\mathscr{H}_{\phi 0}$. The other field components do not have a pole at this value, excepted of course if it is deliberately introduced in the excitation functions $\mathscr{F}_0$ and $\mathscr{H}_0$. The residue of $\mathscr{E}_{\rho 0}$ and $\mathscr{H}_{\phi 0}$ at this pole has the ρ^{-1} dependence of the TEM mode and the transmission line voltage and current which can be obtained this way are obviously those of this mode. We have already explained in the previous section why this approach is highly questionable. We can now reformulate the discussion in mathematical terms.

Extraction of a pole in the inverse Fourier transformation is frequently possible but it yields only a part of the solution. The usefulness of this method is restricted to the cases where this part is dominant. In some cases the remaining part of the inverse Fourier transform can even cancel the pole contribution. This is undoubtedly the case when holes in the cable shield provide a coupling with a lossy external medium, for the pole contribution gives rise to a lossless propagation and must necessarily be cancelled. We thus look for another approach.

Writing transmission line equations implicitly assumes that line voltage and current are defined. In the present case, the definition of the line voltage is not unique since the electric field is not irrotational, excepted for the possible contributions of the poles at $h = \pm k_0 \sqrt{\kappa}$. The definition of the coaxial mode voltage by

$$V_c(z) = \frac{1}{2\pi} \int_0^{2\pi} d\phi \int_a^b E_\phi(\rho, \phi, z)\, d\rho$$

$$= \int_a^b E_{\rho 0}(\rho, z)\, d\rho \tag{2.140}$$

is of course the simplest one: it involves only the field derived from the rotational-symmetric part of the electric potential U, but is not restricted to the TEM part of the field. The idea of using the current carried by the internal conductor

$$I_i(z) = \int_0^{2\pi} H_\phi(a, \phi, z)\, a\, d\phi \tag{2.141}$$

as the line current does not appear to be adequate. It can be seen, indeed, that we cannot derive transmission line equations between $V(z)$ and $I_i(z)$ from eqns. 2.134, 2.136 and 2.138.

Instead we propose to use a modified definition of the transmission line current. We define it as a quantity proportional to the flux of magnetic induction in the coaxial line. The proportionality constant is chosen in such a way as to obtain the current of the inner conductor in the case of a perfectly shielded cable. We consequently define the coaxial mode current by

$$I_c(z) = \frac{2\pi}{\ln(b/a)} \int_a^b H_{\phi 0}(\rho, z)\, d\rho \tag{2.142}$$

The Fourier transforms of the line voltage and current defined in this way can be obtained from eqns. 2.134 and 2.138. Taking into account $\mathscr{V}_0(ub) = 0$ yields

$$\mathscr{V}_c(h) = \frac{-jh\,\mathscr{G}_0(h)}{u^2} \tag{2.143}$$

$$\mathscr{I}_c(h) = \frac{2\pi j\omega\epsilon}{\ln(b/a)} \frac{\mathscr{G}_0(h)}{u^2} \tag{2.144}$$

From these relations we get

$$-jh \ \mathscr{V}_c(h) = j\omega l_0 \ \mathscr{I}_c(h) - \mathscr{G}_0(h) \tag{2.145}$$

$$-jh \ \mathscr{I}_c(h) = j\omega c_0 \ \mathscr{V}_c(h) \tag{2.146}$$

where

$$l_0 = \frac{\mu_0}{2\pi} \ln(b/a) \tag{2.147}$$

$$c_0 = \frac{2\pi\epsilon}{\ln(b/a)} \tag{2.148}$$

are the specific inductance and capacitance of a perfectly shielded coaxial cable, respectively. As multiplying by (jh) in the Fourier transform corresponds to a derivation, the inverse transformation of eqns. 2.145 and 2.146 yields the transmission line equations:

$$-\frac{d\,V_c}{dz} = j\omega l_0 \ I_c - E_{z0}\,(b,z) \tag{2.149}$$

$$-\frac{d\,I_c}{dz} = j\omega c_0 \ V_c \tag{2.150}$$

This derivation based on the use of a modified definition of the transmission line current has the merits of being simple and rigorous. Simplicity results from the fact that there is no need to use a capacitive coupling term. This conclusion puts an end to a long-standing controversy. Also note that the specific inductance l_0 and capacitance c_0 are those of a perfectly shielded cable; this justifies the physical interpretation of Fig. 2.1b, given in Section 2.5.1.

The use of definition 2.142 instead of 2.141 for the line current will not entail any additional difficulty in problems where transmission line models are used. Indeed, if we consider, for instance, the junction of a leaky cable with a non-leaky one, the continuity of H_ϕ at the junction requires the continuity of $I_c(z)$ as well as that of $I_i(z)$.

The reader may observe that eqns. 2.149 and 2.150 are immediate consequences of Maxwell's equations for transverse magnetic fields with rotational symmetry:

$$\frac{\partial E_{\rho 0}}{\partial z} = -j\omega\mu_0 \ H_{\phi 0} + \frac{\partial E_{z0}}{\partial \rho} \tag{2.151}$$

$$\frac{\partial H_{\phi 0}}{\partial z} = -j\omega\epsilon \ E_{\rho 0} \tag{2.152}$$

Relating $E_{z0}(b,z)$ to the shield current I_s however remains necessary to justify the transmission line eqn. 2.101.

2.5.4 Theoretical prediction of transfer impedance

As we explained in Section 2.5.2, exact calculation of scattering through the apertures of an imperfect shield is an extremely difficult task. Resorting to approximate methods thus appears as a need in order to easily obtain numerical values.

The method currently used is based on Bethe's (1944) theory of scattering through small apertures. Bethe was interested in calculating the coupling of cavities through small holes, a problem which bears some resemblance to the present one. We will devote some attention to this approach since it appears to provide one of the most powerful methods of calculating the transfer impedance of braided and perforated shields. We will also discuss some possible improvements.

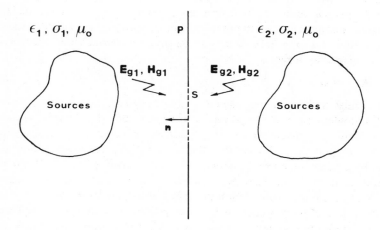

Fig. 2.3   *Scattering by hole in an infinite plane separating two homogeneous media*

Let us consider an infinitely thin and perfectly conducting screen P separating two media of different characteristics (Fig. 2.3). The screen is provided with an aperture S. Sources may exist on either side of the screen. They would, if there were no aperture, produce fields (E_{g1}, H_{g1}) in medium 1 and (E_{g2}, H_{g2}) in medium 2. These fields are assumed to be known and are called the 'generator' fields. It is clear that E_{g1}, E_{g2} are normal to P and that H_{g1}, H_{g2} are tangential to P. We will denote by n the unit normal to the screen, oriented toward medium 1.

The total fields in the presence of the aperture will be written

$$\left.\begin{array}{l} E = E_{g1} + E_{s1} \\ H = H_{g1} + H_{s1} \end{array}\right\} \quad \text{in medium 1} \tag{2.153}$$

$$\left.\begin{array}{l} E = E_{g2} + E_{s2} \\ H = H_{g2} + E_{s2} \end{array}\right\} \quad \text{in medium 2} \tag{2.154}$$

where (E_{s1}, H_{s1}) and (E_{s2}, H_{s2}) are the scattered fields. They are solution of Maxwell's equations in their respective half-spaces and must satisfy the following boundary conditions on P:

$$E_{s1t} = E_{s2t} = 0 \qquad\qquad \text{outside the aperture} \qquad (2.155)$$

$$E_{s1t} = E_{s2t} \qquad\qquad \text{in the aperture} \qquad (2.156)$$

$$H_{s1t} - H_{s2t} = H_{g2t} - H_{g1t} \qquad \text{in the aperture} \qquad (2.157)$$

where the subscript t indicates the tangential components.

We will indicate how to solve this problem. Let us start from the still unknown value E_{at} of E_{s1t} and E_{s2t} inside the aperture and calculate E_{s1} and H_{s1} at all points of medium 1. We are faced with the problem of calculating the fields in a homogeneous half-space knowing the tangential component E_{at} on the boundary. We know that a discontinuity of the tangential components of the electric field occurs at the crossing of a superficial density of magnetic current. Consequently a surface density of magnetic current

$$K_m = n \times E_{at} \qquad (2.158)$$

applied against the conducting plane P without aperture would radiate fields E_{s1}, H_{s1}. This problem can easily be solved by using the method of images. As the image of a magnetic current in front of a perfectly conducting plane has opposite tangential components, it is seen that the fields E_{s1}, H_{s1} can be calculated as those radiated by a magnetic current density $2\,K_m$ in an infinite space with the characteristics of medium 1. Similarly E_{s2}, H_{s2} can be calculated as the fields radiated by $(-2\,K_m)$ in an infinite medium with the characteristics of medium 2.

Having solved these problems, it becomes possible to set up an integral equation for K_m by expressing boundary condition 2.157. This method can be used to solve any problem of scattering through a hole in an infinite screen. If the hole size is small compared with the wavelength we may consider that the generator fields are constant in the aperture. More important is the fact that the integral equation for K_m relates this quantity to magnetic fields on the hole surface. Consequently near-field approximations may be used; this means that the integral equation may be approximated by its magnetostatic equivalent. Furthermore the electric parameters (conductivity, permittivity) of the media on either side of the screen do not appear in the equation. As a result K_m and thus the tangential electric field in the aperture are not influenced by these parameters. This is however not true for the normal electric field (Van Bladel, 1970).

In our search for the transfer impedance of perforated shields, we are interested in averaged quantities. Of particular importance in view of eqn. 2.108 is the average tangential electric field

$$S E_0 = \int_S E_{at}\, dS = -\int_S n \times K_m\, dS \qquad (2.159)$$

The problem of solving the integral equation for K_m for small holes is a classical one in electromagnetic theory. The problem is the calculation of the magnetic moment m which is related to $(2\,K_m)$ by

$$2 \int_S K_m \, dS = j\omega\mu_0 \, m \qquad (2.160)$$

The magnetic moment is related to the generator fields by a relation of the type

$$m = \alpha_m \, (H_{g2} - H_{g1}) \qquad (2.161)$$

The dyadic α_m is the magnetic polarisability of the aperture. Combining the last three equations we obtain

$$S E_0 = \frac{j\omega\mu_0}{2} \, n \times [\alpha_m \cdot (H_{g1} - H_{g2})] \qquad (2.162)$$

Numerical values of the magnetic polarisability have been published in the literature for various aperture shapes. It is clear that α_m is a symmetric dyadic when the aperture has two symmetry axes. For circular holes with radius a, one has

$$\alpha_m = \frac{8a^3}{3} \, I \qquad (2.163)$$

where I is the identity dyadic (Bethe, 1944). For an ellipse with major axis along the x-axis (Collin, 1960):

$$\alpha_m = \frac{2\pi a^3}{3} \left[\frac{e^2}{K(e) - E(e)} \, u_x u_x + \frac{e^2 (1 - e^2)}{E(e) - (1 - e^2) K(e)} \, u_y u_y \right] \qquad (2.164)$$

where $e = [1 - (b/a)^2]^{1/2}$ is the eccentricity, a is the major axis, b is the minor axis and

$$K(e) = \int_0^{\pi/2} \frac{d\phi}{(1 - e^2 \sin^2 \phi)^{1/2}} \qquad (2.165)$$

$$E(e) = \int_0^{\pi/2} (1 - e^2 \sin^2 \phi)^{1/2} \, d\phi \qquad (2.166)$$

are Legendre's complete normal elliptic integrals.

Cohn (1951), using electrolytic tank measurements, obtained polarisabilities of the rectangle, the rectangle with half-circular ends, the rosette, the dumb-bell, the H-shaped aperture, and the cross with half-circular ends. De Smedt and Van Bladel (1980) have circulated the polarisabilities of apertures with the four different shapes shown in Fig. 2.4. They write the magnetic polarisability dyadic in the form

$$\alpha_m = S^{3/2} \, (\nu_{mx} \, u_x u_x + \nu_{my} \, u_y u_y) \qquad (2.167)$$

where S is the aperture area. The dimensionless parameters ν_{mx} and ν_{my} are shown in Figs. 2.5 and 2.6 where, by convention, it is assumed that the major axis is parallel to the x-axis.

Next we proceed to the application of these results to the approximate calculation of the transfer impedance of cable shields with holes. It is clear that this requires a cautious approach. Let us first consider an isolated hole. As long as the hole size is small compared with the shield radius and with the distance of the hole from the inner conductor, we may hope that the magnetic field resulting from a

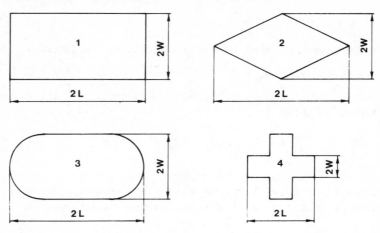

Fig. 2.4   *Apertures, the magnetic polarisabilities of which are shown on Figs. 2.5 and 2.6: rectangle, diamond, rectangle with circular ends, cross*

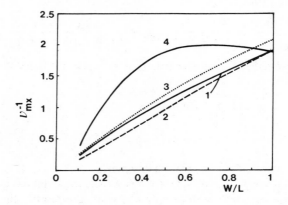

Fig. 2.5   *Inverse of dimensionless polarisability ν_{mx} as a function of the aspect ratio: refer to Fig. 2.4 for notation*

prescribed tangential electric field in the aperture will not differ too much from the value that it would have for a hole in a plane screen, at least in the immediate vicinity of the hole. We must also adequately define the generator fields. At the cable exterior, for reasons explained in Section 2.3, the generator field may be taken as the rotational-symmetric component which would exist on a cable without a hole. Inside the cable, provided we are not close to the end of the cable, the generator field is that of the TEM mode.

In leaky cables however we have to deal with a multitude of closely spaced

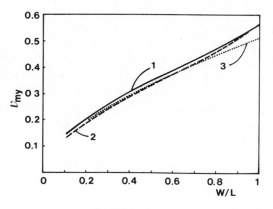

Fig. 2.6 *Dimensionless polarisability v_{my} as a function of the aspect ratio: refer to Fig. 2.4 for notation*

holes. The way to manage this difficulty is to consider that the generator fields of a given hole, taken as the reference, are those that would exist in the presence of all other holes. Remembering that the spacing between holes is in general too small to allow neglect of higher-order modes in the coupling process, we see that we run into extreme complexity. The argument however can be used as a justification of a brute force attack. As we have closely spaced holes, we may hope that removing one of them will not substantially modify the average field. Consequently we use the rotational-symmetric component of the actual magnetic field as the generator field. Obviously we cannot expect very good accuracy for this method.

Eqn. 2.162 may now be written for this specific case as

$$S E_0 = \frac{j\omega\mu_0}{2} \alpha_{\phi\phi} [H_\phi (b + 0) - H_\phi (b - 0)] u_z \qquad (2.168)$$

where b is the shield radius. If n is the number of holes per unit length of cable, the average axial electric field along the screen is given by $E_{z0} = n S E_0/(2\pi b)$. Consequently the surface transfer impedance defined by eqn. 2.115 is given by

$$Z_s = \frac{j\omega\mu_0 n \alpha_{\phi\phi}}{4\pi b} \qquad (2.169)$$

The specific transfer inductance results from eqns. 2.117 and 2.118 and is given by

$$m_t = \frac{\mu_0 n \alpha_{\phi\phi}}{8\pi^2 b^2} \qquad (2.170)$$

This result is reported by Kaden (1957), Vance (1975), Lee and Baum (1975) and others. The reader will note that these authors use an 'effective' magnetic polarisability defined by $\alpha_{\phi\phi e} = \alpha_{\phi\phi}/2$.

A good test of the validity of the present method is obtained by comparing the theoretical predictions with available experimental data. The latter have been taken mainly from Krügel (1956). Fig. 2.7 defines the geometrical parameters of a braided shield: the angle α of the carriers with the cable axis; the width d and the spacing g of the carriers; and the distance a between two parallel carriers. To these parameters we must add the mean radius b of the shield. Table 2.1 gives a list of cable shields with the experimental value m_{exp} of the specific transfer inductance. The next column gives the theoretical value m_{th} resulting from eqns. 2.167 and 2.170, using the data of Figs. 2.4 to 2.6.

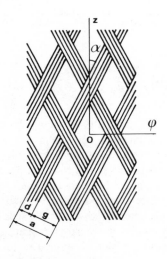

Fig. 2.7   *Geometrical parameters of a braided shield*

Another method of calculating m_t was given in a previous paper (Delogne and Safak, 1975). Unfortunately, this paper contains erroneous results.[*] The corrected results are shown as m_{unif} and m_{sing} in Table 2.1.

It is seen that the agreement between experimental and theoretical predictions is not quite satisfactory. There are many reasons for this, some of which have already been mentioned. There is no doubt that the interaction between neighbouring holes is not properly taken into account. Furthermore the thickness of the braid wires is not always negligible compared with the aperture size. Consequently, some shields shown in the table may not be considered as electrically thin. Indeed, each aperture

---

[*] In the third line of Section 2.3, one should read that there are $(2 \cos\alpha/a)$ strips per unit length in the x-direction. The left-hand side of eqn. 13 should be $(2 I_1 \cos^2\alpha/a)$ and final formula 15 should be replaced by

$$Z_{ST} = \frac{-2j\omega\mu_0 \, \text{tg} \, \alpha}{\pi a d} \, F$$

Table 2.1 *Comparison of theoretical and experimental values of specific transfer inductance for braided shields*

Cable number (Krügel)	Shield wire diameter	b	g	a	α	m_{exp}	m_{th}	m_{sing}	m_{unif}
	mm	mm	mm	mm	deg		nH m^{-1}		
1	0·15	5·15	0·642	2·142	37·4	0·7	0·51	0·53	0·24
2	0·20	5·20	0·754	2·154	37·7	0·48	0·81	0·82	0·27
3	0·25	5·25	0·669	2·169	37·9	0·2	0·69	0·58	0·27
4	0·20	3·50	0·527	1·327	43·6	1·1	1·28	1·41	0·75
5	0·25	4·30	0·410	1·410	51·2	1·75	0·56	0·79	0·35
6	0·25	3·75	0·429	1·429	43·3	0·45	0·56	0·64	0·29
7	0·25	4·30	0·704	1·704	40·8	0·7	1·40	1·44	0·78
8	0·25	3·75	0·688	1·688	30·7	0·2	1·24	1·08	0·58
9	0·25	3·75	0·831	1·831	21·2	0·08	1·60	1·25	0·72
10	0·15	2·05	0·110	0·710	48·6	0·5	0·082	0·133	0·041
11	0·15	2·30	0·210	0·810	47·7	0·95	0·375	0·50	0·21
13	0·15	3·65	0·671	1·271	48·3	3·6	3·14	3·88	2·43
15	0·20	2·95	0·529	1·529	48·7	1·5	1·38	1·72	0·84
16	0·20	2·45	0·454	1·054	56·8	5·3	3·05	4·67	2·61
17	0·20	2·10	0·586	1·186	44·0	2·3	3·92	4·06	2·46
18	0·20	1·60	0·468	1·068	31·8	0·4	2·51	2·27	2·31
19	0·15	2·05	0·215	0·815	59·6	4·4	0·73	1·39	0·58
21	0·15	1·55	0·230	0·830	47·0	1·2	0·71	0·87	0·38
23	0·15	1·225	0·228	0·828	30·6	0·75	0·61	0·54	0·23
24	0·15	2·90	0·323	1·223	57·5	3·3	0·68	1·24	0·52
25	0·15	2·05	0·177	1·077	48·0	0·7	0·15	0·23	0·07
26	0·15	3·65	0·516	1·416	60·4	5·7	1·86	3·56	1·80
27	0·15	2·30	0·316	1·216	47·7	1·2	0·57	0·75	0·31
28	0·15	3·65	1·010	1·910	48·2	4·9	4·73	5·79	3·66
29	0·15	1·75	0·311	1·211	28·2	0·8	0·463	0·43	0·18
	0·20	4·40	2·141	2·941	30·0	58	26·0		
	0·20	4·40	3·464	4·264	45·0	45	27·8		
	0·20	4·40	4·283	5·083	60·0	41	24·1		

behaves as a waveguide below cutoff and the attenuation of waves propagating through them may be significant. This effect is probably less important for the cables described in the last three lines of Table 2.1. These cables have been specially designed to be leaky (Degauque *et al.*, 1978). From this comparison we may conclude that, at the present state of knowledge, theoretical calculations of specific transfer impedance can only provide a rough estimation.

2.5.5 Experimental determination of transfer impedance

Because of the inaccuracy of theoretical predictions of the transfer impedance, it is interesting to determine this quantity by experiment. The principle of the measurement is shown in Fig. 2.8. A short length of the cable under test is coaxially fixed inside a metal tube. At one end, the metal tube is short-circuited with the cable shield. This end of the cable under test is connected to a generator provided with a means of measuring the current delivered to the cable. At the other end, the cable under test is terminated into a matched load. Impedance matching is not critical but in general this makes the measurement of the generator current easier. At this end the tube is left open and the voltage which appears between the metal tube and the cable shield is measured with a high-impedance voltmeter.

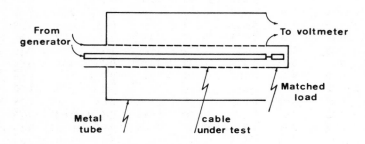

Fig. 2.8   *Test set-up for the measurement of* $|Z_T|$

Provided the length of the test section is small compared with the wavelength, no current flows along the metal tube and opposite currents equal to the generator current I flow along the shield and the internal conductor. The magnetic field between the cable shield and the metal tube is thus zero. Consequently the voltage V measured by the generator is equal to the voltage drop along the shield and we have

$$V = |z_t| L I \tag{2.171}$$

where L is the length of the test section. This allows an easy determination of the absolute value of the specific transfer impedance.

It is useful to measure $|z_t|$ as a function of frequency. Typical shapes of the curve $|z_t(f)|$ are shown in Fig. 2.9 and may be explained as follows. At sufficiently high frequency the transfer of electromagnetic energy through the shield is due only to the apertures and the transfer impedance consists only of the transfer inductance ($z_t = j\omega m_t$). At low frequency however this impedance is very small and the transfer impedance is mainly due to the finite conductivity of the shield material. It is interesting to compare the low-frequency part of the curves with the result obtained in Schelkunoff's (1934) pioneering work on plain metallic shields. When the shield thickness t is small compared with its radius b, the specific transfer impedance may be approximated by

$$z_t \simeq \frac{1}{2\pi b t\, \sigma}\; \frac{\gamma T}{\mathrm{sh}\,(\gamma T)} \tag{2.172}$$

where $\gamma = j\omega\mu\sigma$ and σ and μ are the conductivity and permeability, respectively, of the shield material. The first factor in this expression is the d.c. resistance of the

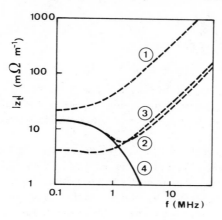

Fig. 2.9   *Typical transfer impedance curves for the d.c. resistance and the transfer inductance for (1) both high, (2) moderate and low, (3) both low. Solid curve (4) for copper tube 0·2 mm thick*

shield. The decrease of $|z_t|$ with frequency starts when the skin depth becomes of the order of t; in this part of the curve the plain shield may obviously not be considered as electrically thin. For actual shields with apertures a dip may occur in the curve of $|z_t (f)|$ between the low- and high-frequency approximations, particularly for small transfer inductances. The curves closely follow an expression of the type

$$z_t = R_{dc} \frac{\gamma T}{sh (\gamma T)} + j\omega m_t$$

One final remark may be made. Krugel (1956) has measured the transfer impedance of various coaxial cables with the same shield, but with different internal parameters (inner conductor diameter and dielectric constant of insulation). He found that the transfer impedance does not depend on the internal parameters. This confirms our previous observation that z_t is an intrinsic property of the shield when the latter contains a large number of small apertures. In particular this provides a proof that z_t will not depend on the propagation constant for such shields.

2.5.6 External surface impedance of a leaky coaxial cable

When we are dealing with leaky feeders used in subsurface communication systems, we may use the thin-cable approximation which has been justified in Section 2.3. Let us recall that the quantity of interest is the surface impedance defined by eqn. 2.75. This means that we could have restricted our attention in the previous paragraphs to the following problem: given the rotational-symmetric component $H_{\phi 0}$ $(b + 0)$ on the external surface of a leaky coaxial cable, calculate the rotational-symmetric component E_{z0} $(b + 0)$ of the electric field on the same surface and thus the external surface impedance Z_0 defined by

$$E_{z0} (b + 0) = Z_0 H_{\phi 0} (b + 0) \tag{2.173}$$

or the specific external impedance

$$z_0 = E_{z0} (b + 0) / [2\pi b H_{\phi 0} (b + 0)] \tag{2.174}$$

This problem, which should be solved for a mode with propagation factor exp $(- \Gamma z)$, may seem simpler than trying to prove the validity of the transfer impedance concept in a quite general situation. It nevertheless remains quite sophisticated.

For this reason we shall calculate Z_0 assuming that the shield can be described by a transfer impedance. We follow the analysis of Wait and Hill (1975b). Before defining the structure, let us consider in cylindrical coordinates a homogeneous medium located between radii ρ_1 and ρ_2. With the notation of Appendix B, we may write for the rotational symmetric fields in this region

$$E_{z0} (\rho) = -u^2 V_0 (u\rho) \tag{2.175}$$

$$H_{\phi 0} (\rho) = -j\omega\epsilon u V_0' (u\rho) \tag{2.176}$$

with

$$V_0 (u) = Q I_0 (u\rho) + B K_0 (u\rho) \tag{2.177}$$

If this region is located inside a thin cable, i.e. if $|u\rho| \ll 1$, we may approximate this function and its derivative by

$$V_0 (u\rho) \simeq Q + B \ln \frac{Cu\rho}{2} \tag{2.178}$$

$$V_0' (u\rho) \simeq B/(u\rho) \tag{2.179}$$

where $C = 1\cdot 7810 \ldots$ Let us consider the quantity

$$z_0 (\rho) = \frac{E_{z0} (\rho)}{2\pi\rho H_{\phi 0} (\rho)} \tag{2.180}$$

We may write

$$\frac{2\pi j\omega\epsilon z_0 (\rho)}{u^2} \simeq \frac{Q}{B} + \ln \frac{Cu\rho}{2} \tag{2.181}$$

Considering two radii ρ_1 and ρ_2 and eliminating the ratio Q/B, we obtain the general relation

$$z_0 (\rho_2) \simeq z_0 (\rho_1) + \frac{u^2}{2\pi j\omega\epsilon} \ln \frac{\rho_2}{\rho_1} \tag{2.182}$$

This relation is very useful for obtaining the external specific impedance of a leaky cable. The inner conductor, with conductivity σ and permeability μ, has a radius a. Then we have, according to eqn. 2.94,

$$z_0\left(a\right) \simeq \frac{1}{2\pi a}\sqrt{\frac{j\omega\mu}{\sigma}} \tag{2.183}$$

If the cable insulation has a wave number k and a permittivity ϵ, we obtain

$$z_0\left(b-0\right) \simeq z_0\left(a\right) - \frac{k^2 + \Gamma^2}{2\pi j\omega\epsilon}\ln\frac{b}{a} \tag{2.184}$$

If the specific transfer impedance of the shield is z_t, we may write eqns. 2.115 and 2.117 as

$$\frac{1}{z_0\left(b+0\right)} = \frac{1}{z_0\left(b-0\right)} + \frac{1}{z_t} \tag{2.185}$$

Finally the shield may be coated by a dielectric layer with wave number k_c, permittivity ϵ_c and external radius c. We then have, using eqn. 2.182

$$z_0\left(c\right) \simeq z_0\left(b+0\right) - \frac{k_c^2 + \Gamma^2}{2\pi j\omega\epsilon_c}\ln\frac{c}{b} \tag{2.186}$$

Eqns. 2.183 to 2.186 may be represented by the ladder network of Fig. 2.10, with

$$z_{ba} \simeq -\frac{k^2 + \Gamma^2}{2\pi j\omega\epsilon}\ln\frac{b}{a} \tag{2.187}$$

$$z_{cb} \simeq -\frac{k_c^2 + \Gamma^2}{2\pi j\omega\epsilon_c}\ln\frac{c}{b} \tag{2.188}$$

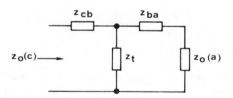

Fig. 2.10   *The equivalent ladder network for the external specific impedance of a leaky cable with external coating*

It is instructive to consider the value of the external specific impedance for a particular case. Let us assume that the internal conductor is perfectly conducting, that the specific transfer impedance consists of an inductance m_t and that there is no external coating. Remembering the expression 2.147 for the specific inductance of a perfectly shielded cable, it is easy to show that

$$z_0\left(b+0\right) = j\omega\,\frac{m_t\,l_0\left(1 + \Gamma^2/k^2\right)}{m_t + l_0\left(1 + \Gamma^2/k^2\right)} \tag{2.189}$$

Neglecting the specific attenuation, $\Gamma \simeq j\beta$, it is seen that the specific external impedance consists of a reactance. This reactance has a zero for $\beta = k$ and a pole for $\beta = k \, (l_0 + m_t)/l_0$ and is a decreasing function of β. It is positive if the wave is faster than the TEM mode of the perfectly shielded cable $(\beta < k)$ or if $\beta > k \, (l_0 + m_t)/l_0$ and negative between these two limits.

2.6 Natural propagation in a low-conductivity layer

Having prepared the required mathematical tools in the previous sections, we are now ready to consider the propagation of electromagnetic waves in some subsurface structures. By natural propagation, we mean that there is no wire or leaky feeder specially installed to guide e.m. waves. The simplest of these structures is a plane layer of relatively low conductivity embedded between two semi-infinite media of higher conductivity. This situation is characteristic of coal seams as we have seen in Section 1.1. We will start with the analysis of some waveguiding properties of the layer and then consider the excitation of guided modes by elementary sources.

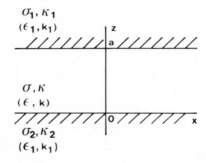

Fig. 2.11   *Low-conductivity layer embedded between two semi-infinite half-spaces of higher conductivity*

2.6.1 Guided plane waves

The geometric and electrical parameters of the layer are defined in Fig. 2.11. In this section, we are interested in plane waves, with no dependence of the fields on the x-coordinate but with dependence on the y-coordinate by a factor $\exp(-\Gamma y)$. These modes can be obtained from a z-oriented electric Hertz potential $\pi'_z = U(z) \exp(-\Gamma y)$, where U satisfies the equation

$$\frac{d^2 U}{dz^2} - u^2 \, U = 0 \tag{2.190}$$

with

$$u = \sqrt{-k^2 - \Gamma^2} \tag{2.191}$$

By convention, we use for u the square root with the determination

$$-\pi/2 < \arg u \leqslant \pi/2 \tag{2.192}$$

The fields may be obtained by

$$E_y = -\Gamma \frac{dU}{dz} \tag{2.193}$$

$$E_z = -\Gamma^2 U(z) \tag{2.194}$$

$$H_x = -j\omega \epsilon \Gamma U(z) \tag{2.195}$$

where

$$j\omega\epsilon = j\omega\epsilon_0 \kappa + \sigma \tag{2.196}$$

All these quantities need to be used with subscripts 1 or 2 for the external half-spaces.

The solution of eqn. 2.190 yielding decreasing fields in the lower and upper half-spaces is given by

$$U = \begin{cases} A\,e^{-u_1 z} & \text{for} \quad z > a & (2.197a) \\ B\,e^{uz} + C\,e^{-uz} & \text{for} \quad 0 < z < a & (2.197b) \\ D\,e^{u_2 z} & \text{for} \quad z < 0 & (2.197c) \end{cases}$$

where A, B, C, D are arbitrary constants. These may be determined by requiring the continuity of E_y and H_x, i.e. of $\Gamma\,dU/dz$ and $k^2 U$ at the interfaces. This yields a linear system of four homogeneous equations with four unknowns. The mode equation is obtained by setting the determinant of this system to zero.

With the notation

$$R_1 = \frac{k_1^2 u - k^2 u_1}{k_1^2 u + k^2 u_1} \tag{2.198}$$

$$R_2 = \frac{k_2^2 u - k^2 u_2}{k_2^2 u + k^2 u_2} \tag{2.199}$$

the mode equation is found to be

$$R_1 R_2 = e^{2ua} \tag{2.200}$$

The quantity R_1 is recognised as the reflection factor of plane waves with a propagation factor $\exp(-\Gamma y - uz)$ on the interface $z = a$, while R_2 is relative to the interface $z = 0$. As u, u_1 and u_2 are even functions of the unknown Γ, the roots of this equation occur in opposite pairs $\pm\Gamma^{(0)}, \pm\Gamma^{(1)}, \pm\Gamma^{(2)} \ldots$, where by convention $-\pi/2 < \arg \Gamma^{(n)} \leqslant \pi/2$.

Before giving some numerical results, we would like to discuss some limiting cases. For perfectly conducting boundaries, k_1 and k_2 become infinite, the reflection factors R_1, R_2 tend to unity and the solutions of eqn. 2.200 are $ua = jn\pi$. The zero-order mode is the classical TEM mode of the parallel-plate waveguide. The specific attenuation of this mode is the same as for plane waves in an infinite space of conductivity σ and dielectric constant κ. The modes with $n \neq 0$ are the TM waveguide modes. They have critical frequencies which correspond to a seam thickness equal to n times the half-wavelength in this medium. Below this frequency they suffer very high attenuation and above it the attenuation remains larger than that of the TEM mode. The latter has no critical frequency. It is clear that, when $|k_1| \gg k$ and $|k_2| \gg |k|$, the modes of the actual waveguide are similar to those of this limiting case. This occurs below the transition frequencies of the external media, i.e. for $\omega\epsilon_0 \kappa_1 \ll \sigma_1$ and $\omega\epsilon_0 \kappa_2 \ll \sigma_2$, provided $\sigma_1 \gg \sigma$ and $\sigma_2 \gg \sigma$. In the limiting case, the fields of the even (odd) modes are even (odd) functions of z in the seam and the TEM mode is independent of z.

As expected, the specific attenuation of the dominant zero-order mode increases with frequency. In actual subsurface propagation problems it takes acceptable values only below a few megahertz. In this frequency range, the higher-order modes are below cutoff and are consequently useless. This is the main reason why we will not devote attention at present to the TE modes which all have critical frequencies.

The main field components of the TM modes are E_z and H_x with a slight longitudinal component. The transverse components are even (odd) functions of z for the even (odd) modes, at least for the limiting case or when the external media have the same characteristics.

Another limiting case occurs when the waveguide thickness tends to zero and when the external media are identical. It can be seen from eqn. 2.200 that we have then $u_1 \simeq 0$. Consequently the specific attenuation of the dominant mode is equal to that of plane waves in the external medium. Actually when the waveguide thickness is small, the specific attenuation lies between that of the internal medium and those of the external half-spaces.

A further comment on the mode equation concerns the choice of determinations 2.192 for the square roots u_1 and u_2. We will discuss this question assuming that $k_1 = k_2$. This choice is of course arbitrary but, once it was made, we had to discard the solutions $\exp(u_1 z)$ for $z > a$ and $\exp(-u_1 z)$ for $z < 0$ in order to avoid solutions which grow exponentially in the upper and lower half-spaces. If we keep the form 2.197 of the solution, the modes with Re $u_1 \leqslant 0$ have exponentially increasing fields in the upper and lower half-spaces. Such modes are called improper. Solutions of the mode equation with Re $u_1 \leqslant 0$ undoubtedly exist. For instance, when a solution (u, u_1) with $|ua| \ll 1$ exists, it can be seen that $(ju, -u_1)$ is also an approximate solution of the mode equation.

Improper modes obviously cannot have a physical existance. It is however known that they may play an important role in the solution of some excitation problems, particularly when they are located close to the branch cuts in the complex plane, but of course on the wrong Riemann sheath, or near to the saddle-point in another complex plane (Collin and Zucker, 1969; Baños, 1966). The possible influence of improper modes has not yet been investigated in the context of subsurface propagation.

Some values of the attenuation of the zero-order mode have been published (Wait, 1976a) for a coal seam embedded between two semi-infinite rock half-spaces having identical electrical parameters. We performed a systematic numerical analysis of this mode. Eqn. 2.200 has been solved by Newton—Raphson's iteration method for a wide choice of electrical parameters and for frequencies between 100 kHz and 100 MHz. From a theoretical point of view, it is interesting to note that, for some combinations of electrical parameters and frequency, it happens that no root is found, i.e. no guided mode exists. Specifically, when the conductivities σ and σ_1 are both very low, the zero-order mode disappears above a few megahertz. This is not surprising since the coal and rock media are then lossy dielectrics ($\omega\epsilon \gg \sigma$) rather than conductors and the coal seam height is much smaller than the wavelength.

For normal values of the rock conductivity however the zero-order mode always exists. A first conclusion of this systematic analysis is that the rock dielectric constant κ_1 has a very limited influence on the mode propagation constant. Consequently the value $\kappa_1 = 2\kappa$ has been used for all results presented below. These results are shown in Fig. 2.12a to c for three seam heights $a = 2$, 4 and 8 m, two values of the coal dielectric constant $\kappa = 4$ and 9, and two values of the rock conductivity $\sigma_1 = 0\cdot1$ and $0\cdot01$ S m^{-1}. Six curves are shown on each diagram: the five lower curves, from bottom to top, give the mode attenuation for conductivity ratios σ_1/σ of 5000, 1000, 500, 100 and 50, and the upper curve gives the attenuation of plane waves in the rock medium.

Comments on the results are as follows. Below a frequency corresponding roughly to the cutoff of the first-order mode, i.e. when the seam height is smaller than half the wavelength in coal, the attenuation of the zero-order mode increases with frequency and, to some extent, with the resistivity of the coal medium. It may be verified that the attenuation is most frequently much higher than that of plane waves in an infinite coal medium. This is not surprising, particularly at the low-frequency end of the curves: since the skin depth in the rock medium is large, attenuation mainly occurs in this medium. Actually, obtaining a mode attenuation close to that of plane waves in coal requires a high rock conductivity rather than a high value of the ratio σ_1/σ. The dependence of the mode attenuation on the coal conductivity is reduced when frequency increases.

As expected the attenuation varies inversely with seam thickness. The curves suggest that relatively low attenuations can be obtained at high frequencies for thick seams. The reader should however keep in mind the evolution of the electrical parameters of natural media with frequency as shown in Fig. 1.1.

2.6.2 Excitation by a vertical electric dipole

Waves with vertical polarisation can be excited by a vertical electric dipole. We thus consider such a dipole with moment Ids located at a height z_0 on the z-axis within the low-conductivity seam of Fig. 2.11. The problem is mathematically similar to

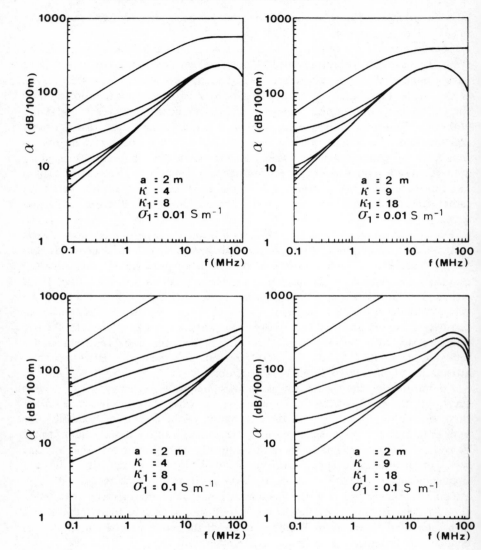

Fig. 2.12a   *Specific attenuation of the dominant mode in a coal seam of thickness 2 m*

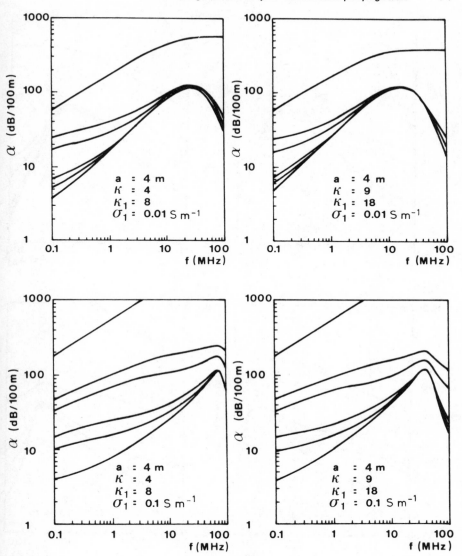

Fig. 2.12b   *Specific attenuation of the dominant mode in a coal seam of thickness 4 m*

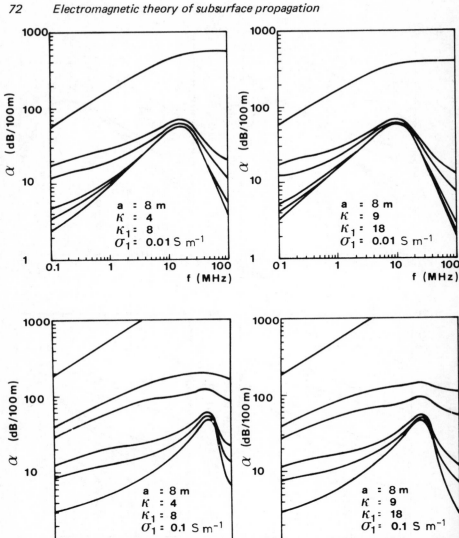

Fig. 2.12c   *Specific attenuation of the dominant mode in a coal seam of thickness 8 m*

the radiation of a vertical electric dipole located between a flat earth and the iono-
sphere. The solution is available (Wait, 1972), but we will derive it in detail in order
to correct some inaccuracies and also because the mathematical methods used are
powerful tools for the solution of several subsurface propagation problems. This
approach will allow us to be more concise in the following sections.

The method used here consists of first calculating the radiation of the dipole in
an infinite homogeneous space and then of modifying this result to account for
reflection and refraction at the interfaces. The solution of the first problem yields
the primary fields. Using a vertical electric Hertz potential, we have

$$\pi_z^{(p)} = M \exp\left(-jkR/R\right) \tag{2.201}$$

where R is the distance to the dipole and

$$M = \frac{Ids}{4\pi j\omega\epsilon} \tag{2.202}$$

This is a very classical result. However it is not in an adequate form since spherical
coordinates are ill-suited to the geometry of our problem. Cartesian or cylindrical
coordinates are undoubtedly preferable. We thus recalculate the radiation of a
vertical electric dipole in a homogeneous space, using Cartesian coordinates. The
vertical electric Hertz potential satisfies the equation

$$\left(\frac{\partial^2}{\partial x^2} + \frac{\partial^2}{\partial y^2} + \frac{\partial^2}{\partial z^2} + k^2\right) \pi_z^{(p)} = \frac{Ids}{j\omega\epsilon} \delta(x)\,\delta(y)\,\delta(z-z_0) \tag{2.203}$$

In order to solve this equation we consider a double Fourier transform representation

$$\pi_z^{(p)}(x, y, z) = \frac{1}{4\pi^2} \int_{-\infty}^{\infty}\int_{-\infty}^{\infty} V(k_x, k_y, z) \exp\left[j(xk_x + yk_y)\right] dk_x\, dk_y \tag{2.204}$$

The Fourier transform satisfies the differential equation

$$\left(\frac{\partial^2}{\partial z^2} + h^2\right) V = \frac{Ids}{j\omega\epsilon}\delta(z-z_0) \tag{2.205}$$

where h is defined by

$$h = (k^2 - k_x^2 - k_y^2)^{1/2} \tag{2.206}$$

with the arbitrary choice of the determination

$$-\pi < \arg h \leqslant 0 \tag{2.207}$$

The technique for solving eqn. 2.205 is classical. Since this equation is homogeneous for $z \neq z_0$, its general solution is composed of two different linear combinations of the exponentials $\exp(\pm jhz)$ for $z > z_0$ and $z < z_0$. We thus have four
arbitrary constants. Two of them are determined by the boundary conditions at

infinity. The two remaining constants are obtained by expressing that V is continuous at $z = z_0$ and that $\partial V/\partial z$ undergoes a discontinuity $Ids/(j\omega\epsilon)$, as required by the right-hand side of eqn. 2.205. Thus we obtain

$$V(k_x, k_y, z) = \frac{-Ids}{2j\omega\epsilon\, jh} \exp\left(-jh\,|z - z_0|\right) \tag{2.208}$$

The next task is to calculate the double inverse Fourier transform eqn. 2.204. This is an integral over the whole (k_x, k_y) plane. Advantage can be gained from the observation that V is a function of $(k_x^2 + k_y^2)$ through the variable h by carrying out this integration in polar coordinates (λ, ψ) defined by

$$\begin{aligned} k_x &= \lambda \cos\psi \\ k_y &= \lambda \sin\psi \end{aligned} \tag{2.209}$$

This yields

$$\pi_z^{(p)} = \frac{-Ids}{8\pi^2 j\omega\epsilon} \int_0^\infty \frac{\exp\left(-jh\,|z - z_0|\right)}{jh} \lambda\,d\lambda \int_0^{2\pi} \exp\left[j\lambda\rho\cos(\psi - \phi)\right] d\psi \tag{2.210}$$

where (ρ, ϕ) are the polar coordinates in the (x, y) plane. The ψ-integral can easily be evaluated using the Anger–Jacobi formula

$$e^{j\alpha\,\cos\xi} = \sum_{n=-\infty}^{\infty} j^n \exp(jn\xi) J_n(\alpha) \tag{2.211}$$

Hence

$$\pi_z^{(p)} = \frac{-Ids}{4\pi j\omega\epsilon} \int_0^\infty J_0(\lambda\rho)\, \frac{\exp(-jh\,|z - z_0|)}{jh} \lambda\,d\lambda \tag{2.212}$$

This result can be seen as an integral along the positive real axis of the complex λ-plane. Note that the definition 2.206 now defines h as a function of λ by

$$h = \sqrt{k^2 - \lambda^2} \tag{2.213}$$

still with the determination 2.207. The integral can be transformed into one along the path C as shown on Fig. 2.13 by observing that

$$J_0(\lambda\rho) = \tfrac{1}{2}\left[H_0^{(2)}(\lambda\rho) - H_0^{(2)}(\lambda\rho\, e^{-j\pi})\right] \tag{2.214}$$

thereby obtaining

$$\pi_z^{(p)} = \frac{-Ids}{2\pi j\omega\epsilon} \int_C H_0^{(2)}(\lambda\rho)\, \frac{\exp(-jh\,|z - z_0|)}{jh} \lambda\,d\lambda \tag{2.215}$$

The results 2.212 and 2.215 are classical. They can be traced back to Sommerfeld. The primary potential is now in an adequate form with which to start the analysis of the dipole radiation in the low-conductivity seam. The secondary potential $\pi_z^{(s)}$

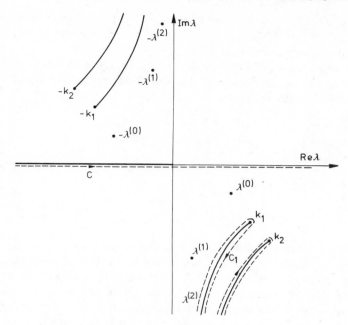

Fig. 2.13   *Singularities of the integrand of eqn. 2.228 in the complex λ-plane*

is due to multiple reflection and refraction of the primary potential at the interfaces $z = 0$ and $z = a$. In each of the media $z < 0$, $0 < z < a$ and $z > a$, it satisfies an equation of type 2.203, but without a source term and with an adequate value of k. This homogeneous equation can be solved by a Fourier transform technique similar to that used for eqn. 2.203. Consequently, after the same manipulations, we may write the secondary potential in the form

$$\pi_z^{(s)} = \frac{-Ids}{2\pi j\omega\epsilon} \begin{cases} \int_C H_0^{(2)}(\lambda\rho) \quad A(\lambda)\, e^{-jh_1 z}\, \dfrac{\lambda d\lambda}{jh_1} & \text{for } z > a \\[3mm] \int_C H_0^{(2)}(\lambda\rho) \quad [B(\lambda)\, e^{jhz} + C(\lambda)\, e^{-jhz}]\, \dfrac{\lambda d\lambda}{jh} & \\[2mm] & \text{for } 0 < z < a \\[3mm] \int_C H_0^{(2)}(\lambda\rho) \quad D(\lambda)\, e^{jh_2 z}\, \dfrac{\lambda d\lambda}{jh_2} & \text{for } z < 0 \end{cases}$$

$$(2.216)$$

where h_1 and h_2 are defined as functions of λ by

$$h_1 = \sqrt{k_1^2 - \lambda^2}\,; \qquad\qquad -\pi < \arg h_1 \leqslant 0 \qquad\qquad (2.217)$$

$$h_2 = \sqrt{k_2^2 - \lambda^2}\,; \qquad\qquad -\pi < \arg h_2 \leqslant 0 \qquad\qquad (2.218)$$

and A, B, C, D are as yet unknown functions of λ. The arbitrary choice of determinations 2.217 and 2.218 has been duly taken into account in eqn. 2.216 to satisfy the boundary conditions at infinity.

The unknown functions are to be determined from the boundary conditions at the interfaces. The continuity of the tangential field components requires that the total potential satisfies the conditions

$$k^2 \, \pi_z \, (a - 0) \; = \; k_1^2 \, \pi_z \, (a + 0)$$

$$\frac{\partial \pi_z \, (a - 0)}{\partial z} \; = \; \frac{\partial \pi_z \, (a + 0)}{\partial z}$$

(2.219)

$$k^2 \pi_z \, (+ 0) \; = \; k_2^2 \, \pi(- 0)$$

$$\frac{\partial \pi_z \, (+ 0)}{\partial z} \; = \; \frac{\partial \pi_z \, (- 0)}{\partial z}$$

(2.220)

Since these conditions are to be met for all values of (x, y), they are also required for the double Fourier transforms. Consequently they may be expressed directly for the integrands of eqns. 2.215 and 2.216, since these integrals are the particular form of the double Fourier transforms for the rotational-symmetric case. From a methodological viewpoint, this remark is essential since it provides the mathematical justification for the procedure. It also helps us in pointing out that some heuristic approaches to the problem may be erroneous. For instance it is possible to rewrite eqn. 2.215 as an integral along the real axis of the h-plane

$$\pi_z^{(\mathrm{p})} \; = \; \frac{Ids}{4\pi^2 \, \omega\epsilon} \; \int_{\mathrm{real \; axis}} H_0^{(2)} \, (\lambda\rho) \exp(- jh \, | z - z_0 |) \, dh$$

which resembles a Fourier transform on the z-variable, with λ now defined as a function of h. One can then attempt to write the secondary potential as similar integral forms in the h-plane and to satisfy the boundary condition by equating the integrands. This approach has no mathematical justification. Actually it leads to a result which is the integral of an odd function and is consequently null.[*]

Let us now return to the correct formulation of the problem. The algebraic calculations to obtain the unknown functions are lengthy, but straightforward. We define the reflection factors of the interfaces by

$$R_1 \, (\lambda) \; = \; \frac{k_1^2 h - k^2 h_1}{k_1^2 h + k^2 h_1}$$

(2.221)

$$R_2 \, (\lambda) \; = \; \frac{k_2^2 h - k^2 h_2}{k_2^2 h + k^2 h_2}$$

(2.222)

---

[*]Wait (1962) states the problem correctly by integrating in his C-plane, which is our λ-plane. However he is erroneous in stating that the integral may be evaluated along the real axis of his S-plane, which is the h-plane.

and obtain

$$B(\lambda) = -\exp(-jhz_0) + \frac{\exp(-jhz_0) + R_1 \exp[-jh(2a-z_0)]}{\Delta} \exp(jha)$$

$$(2.223)$$

$$C(\lambda) = -\exp(jhz_0) + \frac{\exp(jhz_0) + R_2 \exp(-jhz_0)}{\Delta} \exp(jha) \quad (2.224)$$

$$A(\lambda) = \frac{k^2(1+R_1)}{k_1^2 R_1} C(\lambda) \quad (2.225)$$

$$D(\lambda) = \frac{k^2(1+R_2)}{k_2^2 R_2} B(\lambda) \quad (2.226)$$

$$\Delta(\lambda) = \exp(jha) - R_1 R_2 \exp(-jha) \quad (2.227)$$

Note that the expressions of the reflection factors are identical to eqns. 2.198 and 2.199 with the substitutions $u = -jh$, $u_1 = -jh_1$, $u_2 = -jh_2$, $\Gamma = j\lambda$.

Hence we concentrate on the region $0 < z < a$ only. The total potential may now be written

$$\pi_z(x, y, z) = \frac{-M}{2j} \int_C H_0^{(2)}(\lambda\rho) \frac{F(\lambda)}{\Delta(\lambda)} \frac{\lambda d\lambda}{h} \quad (2.228)$$

where

$$F(\lambda) = [e^{jhz_<} + R_2 e^{-jhz_<}][e^{jh(a-z_>)} + R_1 e^{-jh(a-z_>)}] \quad (2.229)$$

and $z_<$ and $z_>$ denote the smallest and the largest of the heights z and z_0, respectively.

The integral 2.228 can be evaluated by function theoretic methods. In order to do this, we need to identify the singularities of the integrand in the λ-plane (Fig. 2.13). In the first instance we note that the Hankel function is two-valued; the reason for this is the existance of a logarithm function in the definition of the Neumann function $N_0(\lambda\rho)$. Consequently a branch cut must be drawn along the negative real axis. According to eqn. 2.214, the integration contour C must run below this branch cut. The functions $h_1(\lambda)$ and $h_2(\lambda)$ defined by eqns. 2.217 and 2.218 are also two-valued. According to the chosen determinations, we must draw branch cuts corresponding to the forbidden boundaries arg $h_1 = \pi$ and arg $h_2 = \pi$. It can be seen that these branch cuts are parts of the hyperbolas passing through the points of affixes $\pm k_1$ and $\pm k_2$ and having the real and imaginary axes as asymptotes. Actually the branch cuts are those parts of the hyperbolas that run along the imaginary axis. The definition 2.213 of h also requires a branch cut. However it can be seen that the integrand of eqn. 2.228 remains unchanged if h is replaced by $(-h)$: this results from quantities 2.221, 2.222, 2.227 and 2.229 considered as functions of h. Consequently the two-valued character of h is not maintained for the integrand of eqn. 2.228 and the branch cut is immaterial.

Furthermore the integrand of eqn. 2.228 has poles at the zeros of $\Delta(\lambda)$. Actually the equation $\Delta(\lambda) = 0$ is the mode equation 2.200 provided we replace u, u_1, u_2 and λ by $(-jh)$, $(-jh_1)$, $(-jh_2)$ and $(-j\Gamma)$, respectively. It is easy to see that the roots occur in opposite pairs $\pm\lambda^{(0)}$, $\pm\lambda^{(1)}$, ..., where, by convention, $-\pi < \arg \lambda^{(i)} \leqslant 0$.

So far the integral 2.228 has been calculated along the C-contour. The behaviour of $H_0^{(2)}(\lambda\rho)$ and $F(\lambda)$ is such that this contour may be completed by portions of circles of infinite radius located in the lower half-plane. The contour may further be transformed into C_1 shown on Fig. 2.13, provided we extract the contributions of the residues at the poles $\lambda^{(0)} = -j\Gamma^{(0)}$, $\lambda^{(1)} = -j\Gamma^{(1)}$ .... The integral along the remaining contour C_1 gives the unguided radiation, while the pole contribution yields the guided modes. Keeping only the latter, we may write

$$\pi_z = -\pi M \sum_{n=0}^{\infty} \frac{H_0^{(2)}(\lambda\rho) F(\lambda)\lambda}{h \dfrac{d\Delta}{d\lambda}} \tag{2.230}$$

After some manipulation and replacing $\lambda^{(n)}$ by $-j\Gamma^{(n)}$, we obtain

$$\pi_z = -\frac{M}{a} \sum_{n=0}^{\infty} \frac{\Lambda}{R_2} K_0(\Gamma\rho) f(z) f(z_0) \tag{2.231}$$

where

$$f(z) = e^{jhz} + R_2 e^{-jhz} \tag{2.232}$$

$$\Lambda = \left[1 - j \frac{h}{2\lambda a R_1 R_2} \frac{\partial(R_1 R_2)}{\partial\lambda}\right]^{-1} \tag{2.233}$$

The various terms of the expansion, including the factor Λ, need of course to be evaluated at $\lambda = \lambda^{(n)}$. The main components of the fields follow from the formulas of Appendix B and are given by

$$E_z = \frac{+M}{a} \sum_{n=0}^{\infty} \frac{\Lambda \Gamma^2}{R_2} K_0(\Gamma\rho) f(z) f(z_0) \tag{2.234}$$

$$H_\phi = \frac{-j\omega\epsilon M}{a} \sum_{n=0}^{\infty} \frac{\Lambda\Gamma}{R_2} K_1(\Gamma\rho) f(z) f(z_0) \tag{2.235}$$

In the limit $|\Gamma\rho| \to \infty$, we may use the approximation

$$K_n(\Gamma\rho) \simeq \sqrt{\frac{\pi}{2\Gamma\rho}} \, e^{-\Gamma\rho} \tag{2.236}$$

and retrieve cylindrical waves which are closely related to the plane waves discussed in Section 2.6.1.

These formulas have important applications in several domains of electromagnetic wave propagation, namely for the excitation of waves in the earth-ionosphere wave-guide, along dielectric slabs, etc. Here we are particularly interested in the propagation along a low-conductivity layer like a coal seam. In this application, it may in general be assumed that the upper and lower half-spaces have the same electrical parameters. The mode equation then takes the form

$$R_1 = R_2 = \pm e^{jha} \tag{2.237}$$

where the upper and lower sign apply to the even and odd modes, respectively. This yields

$$\frac{f(z)\,f(z_0)}{R_2} = \begin{cases} 4\cos\,[h(z-a/2)]\,\cos\,[h(z_0-a/2)] \\ 4\sin\,[h(z-a/2)]\,\sin\,[h(z_0-a/2)] \end{cases} \tag{2.238}$$

respectively. Note that h is complex. Furthermore, the excitation factor takes the form (Wait, 1976a)[*]

$$\Lambda = \left[1 + \frac{k_1^2 - k^2}{k_1^2 - k^2 + h^2}\,\frac{\sin(ha)}{ha} \right]^{-1} \tag{2.239}$$

where Λ has to be evaluated for the mode under consideration. An interesting limiting case already discussed in Section 2.6.1 is that of perfectly conducting walls: we then have $\Lambda^{(0)} = 1/2$, $\Lambda^{(n)} = 1$ for $n \neq 0$.

2.6.3 Excitation by a horizontal magnetic dipole

Alternatively, waves with vertical polarisation can be excited by a horizontal magnetic dipole. This type of source is realised as a loop or ferrite antenna with horizontal axis. It is obviously not omnidirectional like the electric dipole: as expected, radiation is maximum at the side of the dipole and zero on the axis. The radiation of the horizontal dipole can be calculated using methods similar to those of the preceding section and the calculation is available (Wait, 1962). However the problem is more complicated since two potential components are required and all six components of the electromagnetic field are non-zero. One of the potential components is necessarily a magnetic vector parallel to the dipole. The other one may be a vertical magnetic potential or an electric potential parallel to the dipole.

A simpler approach yielding a partial answer is to use the results of the fore-going section with the reciprocity theorem. The form of the theorem which will be used here is as follows. Consider two experiments carried out in a medium, not necessarily homogeneous, but isotropic. In the first experiment, electric and magnetic current densities J_a', J_{ma}' are applied simultaneously at finite distance and radiate fields E', H'. In the second experiment, applied current densities J_a'', J_{ma}'' radiate fields E'', H''. Then we have

$$\int (J_a' \cdot E'' - J_{ma}' \cdot H'')\,dV = \int (J_a'' \cdot E' - J_{ma}'' \cdot H')\,dV \tag{2.240}$$

[*] In eqn. 3 of this paper the factor $(kh)^{-1}$ should be replaced by $(2kh)^{-1}$.

where the volume integrals are over all space.

In the present application, we consider for the first experiment a vertical electric dipole located at the point of vector position $r' = (\rho', \phi', z')$:

$$J'_a = I\,ds\,\delta(r - r')\,u_z \tag{2.241}$$

and no magnetic current. The fields (E', H') of this dipole were calculated in Section 2.6.2. In particular, as results from eqns. 2.202 and 2.235, we have

$$H'(0, \phi, z_0) = \frac{I\,ds}{4\pi a} \sum_{n=0}^{\infty} \frac{\Lambda\Gamma}{R_2} K_1\,(\Gamma\rho')\,f(z')\,f(z_0)\,u_{\phi'} \tag{2.242}$$

when we keep only the guided modes.

In the second experiment, the source is a magnetic dipole located at a height z_0 on the z-axis and parallel to the x-axis:

$$J''_{ma} = I_m\,ds\,\delta(r)\,u_x \tag{2.243}$$

The fields (E'', H'') of this source are the unknowns of the problem. Using the reciprocity theorem, we obtain the vertical electric field radiated by this magnetic dipole at the point (ρ', ϕ', z')

$$E''_z(\rho', \phi', z') = \frac{I_m\,ds\cos\phi'}{4\pi a} \sum_{n=0}^{\infty} \frac{\Lambda\Gamma}{R_2} K_1\,(\Gamma\rho')\,f(z')\,f(z_0) \tag{2.244}$$

The other field components cannot all be directly obtained from E''_z. However at some distance from the magnetic dipole, the main field components are E''_z and H''_ϕ, with the relation

$$H''_\phi \simeq \frac{-j\omega\epsilon}{\Gamma}\,E''_z \tag{2.245}$$

In using eqn. 2.244 we have to remember that an infinitesimal loop is equivalent to a magnetic dipole as given by eqn. 2.6.

2.7 Natural propagation in empty tunnels

2.7.1 Introduction: planar air waveguide

Natural propagation of electromagnetic waves in empty tunnels was explained qualitatively in Section 1.2.1, where reference was made to attenuation formulas valid for waveguides with highly conducting walls. This approach is undoubtedly invalid in the VHF and UHF bands for tunnel walls made of rock, concrete, bricks and similar materials. Indeed, the frequency is then well above the transition frequency of these materials (i.e. $\omega\epsilon \gg \sigma$), which consequently behave as lossy dielectrics rather than conductors. It is felt that a more refined analysis is required.

To start this problem, we consider the simplest geometry consisting of a planar air slab located between two conducting half-spaces having the same electrical characteristics. Mathematically this problem is identical with the one solved in Section 2.6. However we are interested here in the case where the slab thickness is several times the wavelength in air, and at frequencies such that the complex dielectric constant of the surrounding medium, denoted by

$$\kappa'_1 = \kappa_1 + \frac{\sigma_1}{j\omega\epsilon_0} \tag{2.246}$$

is close to κ_1. We will assume that the latter takes on a value which is neither close to unity nor large.

The question we want to answer is: can we have modes with propagation constant close to that of plane waves in air, i.e. such that $\Gamma \simeq jk_0$? For this purpose, we rewrite the mode equation 2.200 as

$$e^{2ua} = R^2 \tag{2.247}$$

with the reflection coefficient R given by

$$R = \frac{\kappa'_1 \, ua - u_1 a}{\kappa'_1 \, ua + u_1 a} \tag{2.248}$$

where

$$ua = \sqrt{-(k_0 a)^2 - (\Gamma a)^2} \tag{2.249}$$

$$u_1 a = \sqrt{(k_0 a)^2 (\kappa'_1 - 1) + (ua)^2} \tag{2.250}$$

The assumption $\Gamma \simeq jk_0$ is equivalent to $|ua| \ll k_0 a$. As κ'_1 takes on a moderate value, this entails $|ua| \ll |u_1 a|$ and $|\kappa'_1 \, ua| \ll |u_1 a|$. We may write

$$R = -\frac{1-\xi}{1+\xi} \tag{2.251}$$

where

$$\xi = \frac{\kappa'_1 \, ua}{u_1 a} \tag{2.252}$$

and $|\xi| \ll 1$. It is seen that the reflection factor is close to (-1). Consequently, the mode equation 2.247 can only be satisfied if

$$ua = p\,\pi j + \nu \tag{2.253}$$

where $|\nu| \ll 1$. It will be explained below that $(p-1)$, not p, is the mode order. Using the first two terms of a Taylor series for both sides of the mode equation, we obtain

$$ua \simeq \frac{p\pi j}{1 + 2\kappa'_1/(\mu_1 a)} \tag{2.254}$$

This may further be approximated by

$$ua \simeq \frac{p\pi j}{1 + 2\kappa_1'/(jk_0 a\sqrt{\kappa_1' - 1})} \tag{2.255}$$

where the determination $-\pi/2 < \arg u_1 \leqslant \pi/2$ for proper modes was taken into account.

It can be seen from this equation that for $p \neq 0$, the condition $|ua| \ll k_0 a$ can only be met if $k_0 a \gg 1$ *and only for those modes for which* $p\pi \ll k_0 a$. The propagation constant of these modes can now be obtained from eqns. 2.249 and 2.255. Taking into account that $\kappa_1' \simeq \kappa_1$ and separating into real and imaginary parts, we obtain

$$\beta_n^{(v)} \simeq \sqrt{k_0^2 - (n+1)^2 \, (\pi/a)^2} \tag{2.256}$$

$$\alpha_n^{(v)} \simeq \frac{2(n+1)^2 \, \pi^2}{k_0^2 \, a^3} \, \mathrm{Re} \frac{\kappa_1'}{\sqrt{\kappa_1' - 1}} \tag{2.257}$$

where $n \geqslant 0$ is the mode order. The superscript (v) has been added to α and β to recall that we are studying modes with the main polarisation vertical. Eqn. 2.256 is identical with that of a planar waveguide with perfectly conducting walls. Eqn. 2.257 for the specific attenuation is quite remarkable. It shows that the attenuation is proportional to the square of $(n+1)$ and inversely proportional to the square of frequency and to the third power of the slab thickness. Furthermore it is determined by the dielectric constant κ_1 of the external medium rather than by its electric conductivity.

These results have a simple physical meaning. The modes consist of a plane wave which bounces against the upper and lower half-spaces. The complex incidence angle is given by $\sin \theta = \Gamma/(jk)$ and is close to $\pi/2$. It is known that $R \simeq -1$ at grazing incidence. The attenuation is due to the slight departure from total reflection. The small amount of power which is refracted into the wall is the cause of attenuation; under the condition $\omega\epsilon_0 \kappa_1 \gg \sigma_1$, it is governed by κ_1 rather than by σ_1. The main effect of the wall conductivity is to cause a rapid attenuation of the refracted wave into the external media. This role is nevertheless essential in the propagation mechanism. Indeed, if the external media were lossless dielectrics, the refracted wave would not be attenuated and guided modes would not exist since an infinite power would be required.

The reader probably observed that we discarded the value $p = 0$. This merits some discussion. We have seen that the condition $\Gamma \simeq jk$ entails $|\xi| \ll 1$ and thus $R \simeq -1$. Consequently, we might write eqn. 2.253 where $|v| \ll 1$. For the value $p = 0$, the mode equation 2.247 may be approximated by

$$1 + 2ua \simeq 1 - 4 \frac{\kappa_1' \, ua}{u_1 a}$$

and we have $u_1 a \simeq -2\kappa_1'$. Consequently, this mode is improper since it has $\mathrm{Re}\,(u_1 a) < 0$. The question is one of mode nomenclature. Usually the mode numbering is made in a one-to-one correspondence with those of a waveguide with perfectly conducting walls (Waldron, 1967). If we assume that the conductivity σ_1 is progressively reduced starting from a very high value, then the approximate value of ua evolves from $p\pi$ to $(p + 1)\pi$ for the pth order modes. Consequently, in eqns. 2.253 to 2.257, we must consider that the mode order is $(p - 1)$.

Up to now, attention has been focused on the modes with the main polarisation vertical but the frequency range of interest in this section is such that the modes with horizontal polarisation are also above cutoff. It is easy to show that the relevant mode equation is now

$$e^{2ua} = S^2 \qquad (2.258)$$

where

$$S = \frac{u - u_1}{u + u_1} \qquad (2.259)$$

is the reflection factor for waves with the electric field perpendicular to the plane of incidence. It is easy to repeat the foregoing discussion for these modes. This yields

$$\beta_n^{(h)} \simeq \sqrt{k_0^2 - (n\pi/a)^2} \qquad (2.260)$$

$$\alpha_n^{(h)} \simeq \frac{2 (n\pi)^2}{k^2 a^3} \ \text{Re} \ \frac{1}{\sqrt{\kappa_1' - 1}} \qquad (2.261)$$

Note that $n > 0$ is the mode order.

★ *2.7.2 Geometric optical approach for the planar air waveguide*

Radiation of a dipole located inside the planar air waveguide may be calculated by the method outlined in Section 2.6.2. The foregoing interpretation of the guided modes in terms of reflections on the guide wall suggests that the problem could also be solved by summing the contributions of rays undergoing reflections on the guide walls. This geometrical optical approach has been developed by Mahmoud and Wait (1974c).

We will calculate the radiation of a point source located at a height z_0 along the z-axis of Fig. 2.11. For simplicity, we assume that the electrical characteristics of the two external media are identical. The geometrical description of the multiple reflections undergone by a ray issued from the point source is shown in Fig. 2.14. A doubly infinite set of images of the point source is set up. It can be seen that the rays corresponding to images located at heights $2na \pm z_0$ undergo $2|n|$ and $|n| + |n - 1|$ reflections, respectively, with $-\infty < n < \infty$. They have path lengths given by

$$r_n^\pm = \sqrt{\rho^2 + (2na \pm z_0 - z)^2} \qquad (2.262)$$

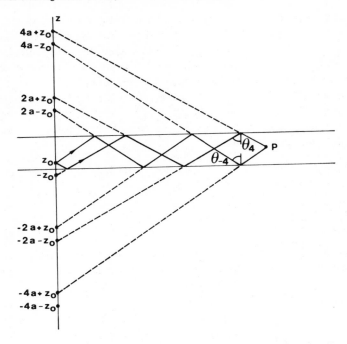

Fig. 2.14   *Geometrical optical approach. The drawing shows the double infinity of images and the two rays undergoing four reflections*

and are reflected under incidence angles $\theta_n^{\pm}$ given by

$$\sin \theta_n^{\pm} = \frac{\rho}{r_n^{\pm}} \qquad (2.263)$$

respectively. The corresponding reflection factors are given by

$$R_n^{\pm} = \frac{\kappa_1' \cos \theta_n^{\pm} - \sqrt{\kappa_1' - \sin^2 \theta_n^{\pm}}}{\kappa_1' \cos \theta_n^{\pm} + \sqrt{\kappa_1' - \sin^2 \theta_n^{\pm}}} \qquad (2.264)$$

for vertical polarisation and

$$S_n^{\pm} = \frac{\cos \theta_n^{\pm} - \sqrt{\kappa_1' - \sin^2 \theta_n^{\pm}}}{\cos \theta_n^{\pm} + \sqrt{\kappa_1' + \sin^2 \theta_n^{\pm}}} \qquad (2.265)$$

for horizontal polarisation. Outside the very near field zone of the source, any component of the primary field may be expressed in the form

$$F_p(\rho, \phi) = f(\phi) \frac{e^{-jk_0 r_0}}{r_0} \qquad (2.266)$$

where

$$r_0 = \sqrt{\rho^2 + (z_0 - z)^2} \tag{2.267}$$

We have neglected the vertical directivity of the source since we are interested in the radiation at large distances from the z-axis. Summing the contributions of the direct and reflected rays, we obtain the total field

$$F(\rho, \phi, z) = f(\phi) \sum_{n=-\infty}^{\infty} \left[(G_n^+)^{2|n|} \frac{\exp(-jk_0 r_n^+)}{r_n^+} + (G_n^-)^{|n|+|n-1|} \right.$$

$$\left. \frac{\exp(-jk_0 r_n^-)}{r_n^-} \right] \tag{2.268}$$

where $G_n^\pm$ stands for $R_n^\pm$ or $S_n^\pm$ according to the case.

It is possible to rearrange eqn. 2.268 into a summation wherein the index is the number p of reflections undergone by the ray. This yields

$$F(\rho, \theta, \phi) = f(\phi) \frac{\exp(-jk_0 r_0^+)}{r_0^+} + \sum_{p=1}^{\infty} \left[(G_s^q)^p \frac{\exp(-jk_0 r_s^q)}{r_s^q} \right.$$

$$\left. + (G_t^q)^p \frac{\exp(-jk_0 r_t^q)}{r_t^q} \right] \tag{2.269}$$

where

$$q = -, s = \frac{-p+1}{2}, \quad t = \frac{p+1}{2} \qquad \text{if } p \text{ is odd}$$

$$q = +, s = -p/2, \quad t = p/2 \qquad \text{if } p \text{ is even}$$

It is also possible to introduce some approximations in order to improve the convergence. The interested reader should refer to Mahmoud and Wait (1974c).

This geometrical optical approach has been developed above on a heuristic basis. It implicitly assumes that the field of a point source located above a flat ground can be obtained as the sum of a direct and a reflected ray. It is known that this requires that the ground wave has a negligible contribution to the field. Therefore it is felt that a firmer justification of the method is required.

For the case of a vertical electric dipole, the exact solution of the present problem is given by the integral 2.228. It is possible to use the Taylor series expansion for the denominator of the integrand:

$$\Delta^{-1} = e^{-jha} \sum_{n=0}^{\infty} (-1)^n R^{2n} e^{-2jhna} \tag{2.270}$$

Using the approximation 2.236 for the modified Bessel function, the integral of each term can then be approximately evaluated by the method of stationary phase. This yields the result 2.268. The accuracy of the method of stationary phase (Morse and Feshbach, 1953) is conditioned by the requirement that R^{2n} is a slowly varying function of λ in the vicinity of the saddle point $\lambda_n^{\pm} = k_0 \sin \theta_n^{\pm}$.

The geometrical optical method has been used to calculate the attenuation of a radio link between two electric dipoles with the following parameter values: $a = 4$ m, $z = z_0 = 1$ m, $f = 450$ MHz, $\kappa_1 = 6$, $\sigma_1 = 0\cdot1$ S m^{-1}. The complex dielectric constant κ_1' of the walls then assumes the value $6 - j3\cdot996$. The results are shown in Fig. 2.15. Fig. 2.15a shows the global behaviour of the attenuation for horizontal dipoles in the distance range $10 < \rho < 500$ m. The diagram represents 5000 calculated points. Standing waves are due to the existence of modes with different velocities. In the case of two modes with longitudinal phase constants β_1, β_2, two successive nodes are separated by a distance d such that $(\beta_1 - \beta_2)d = 2\pi$. On Fig. 2.15a several modes are beating for $\rho < 100$ m but, above this distance, the beating of only two modes is clearly visible. In the present case, it may be verified, using eqn. 2.260, that we observe the beating of modes $n = 1$ and $n = 2$. The global attenuation above 100 m corresponds very well to the propagation factor $\rho^{1/2}$ exp $(-\alpha\rho)$ where α is given by eqn. 2.261 with $n = 1$, since the first-order mode becomes dominant. The horizontal scale was expanded to show the detailed structure (Fig. 2.15b): this phenomenon will be discussed below.

Similar diagrams are shown in Fig. 2.15c and d for vertical electric dipoles. As eqn. 2.257 predicts $\alpha_0 = 8\cdot3$, $\alpha_1 = 33\cdot2$ and $\alpha_2 = 74\cdot7$ dB/100 m, only the zero-order mode should exist after a few tens of metres. The fine structure which is visible in Fig. 2.15 cannot be due to a beating of modes since this phenomenon should decrease with the distance and cannot explain that the spacing between minimums is close to half the free-space wavelength. This indeed supposes the existence of two waves propagating with opposite phase velocities equal to the velocity of light. The fine structure of the diagrams results from the inaccuracy of the method of stationary phase applied to each of the terms of eqn. 2.270. It may indeed be observed that the condition that R^{2n} should be a slowly varying function of λ in the vicinity of the saddle-point $\lambda_n^{\pm} = k_0 \sin \theta_n^{\pm}$ is not satisfied for large n. This is particularly true for vertical polarisation because of the existence of a Brewster angle. Including R^{2n} into the calculation of the saddle-point would probably increase the accuracy, but this has not yet been done.

In spite of its inaccuracy, the geometrical optical approach gives a useful picture of natural propagation phenomena in tunnels well above the cutoff frequency. A further advantage of this method is that it is possible to take any wall roughness into account. According to Beckmann and Spezzichino (1964), for a rough surface with standard deviation σ, the specular reflection coefficients given by Fresnel's equations 2.264 and 2.265 are reduced by a factor exp $(-\phi^2/2)$, where $\phi = 4\pi\sigma \sin\theta/\lambda_0$. This can easily be included in calculations based on eqns. 2.268 or 2.269.

2.7.3 Rectangular tunnel

The geometrical parameters of a rectangular tunnel are shown in Fig. 2.16. An attempt to solve this problem exactly leads to extremely high complexity. Indeed,

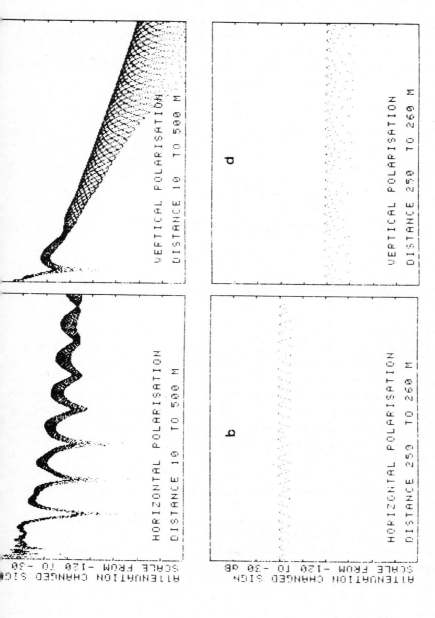

Fig. 2.15 *Attenuation of a radio link between two electric dipoles in a planar air waveguide, calculated by the geometrical optical approach: $a = 4\,m$, $z = z_0 = 2\,m$, $f = 450\,MHz$, $\kappa_1 = 6$ and $\sigma_1 = 0.1\,S\,m^{-1}$*

if we use Cartesian coordinates, it is required to write different analytical expressions for two potentials inside the waveguide and in each of the eight external regions. Boundary conditions must then be expressed at each of the twelve interfaces. Alternatively, if Fourier transforms are used in the analysis, for instance on the x-coordinate, one has to match Fourier transforms on the domains $x < 0$, $0 < x < a$ and $x > a$. This leads to a complex three-part Wiener–Hopf problem with two coupled functions corresponding to the potentials used. Simplifying assumptions or approximate methods are obviously required.

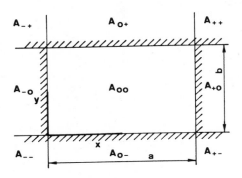

Fig. 2.16   *Geometrical parameters for the rectangular tunnel*

(a) Perfectly conducting side walls
A first simplifying assumption is to consider perfectly conducting side walls. In this case, we assume that the regions $x < 0$ and $x > a$, $-\infty < y < \infty$, are perfectly conducting. We are now left with the only three regions in which different analytical expressions must be used. This model may be a good one for a tunnel bored in a coal seam embedded in highly conducting rock. We will assume that the media for $0 < x < a$ and either $y < 0$ or $y > b$ have a dielectric constant κ_1, a conductivity σ_1 and a complex wave number $k_1 = k_0 \sqrt{\kappa_1'}$ where $\kappa_1' = \kappa_1 + \sigma_1/(j\omega\epsilon_0)$ is the complex dielectric constant.

A complete analysis of the radiation of a vertical or horizontal electric dipole located at a point of coordinates $(x_0, y_0, 0)$ inside the tunnel is available in the literature (Mahmoud and Wait, 1974c). The reader is warned that, for better consistency with other sections of the present book, we have changed the notation used in this article. We will not go into the details of the calculations, since they closely follow the method outlined in Section 2.6.2. We will rather draw attention to some peculiarities and comment on the final results.

Let us consider the case of the horizontal electric dipole of moment Ids parallel to the x-axis. As was shown in Section 2.2, the problem can be solved by means of two potential components π_x' and π_x'' satisfying Helmholtz equations. The equations are homogeneous for π_x' in the regions $y < 0$ and $y > b$, and for π_x'' in all three regions $y < 0$, $0 < y < b$ and $y > b$. This choice of potentials however is not the most suitable one. It can indeed be seen from the formulas in Appendix A that the continuity of E_x, E_z, H_x and H_z at the interfaces $y = 0$, $y = b$ involves the potentials and their first and second derivatives with respect to y. Furthermore some of the conditions involve both potential components, which are thereby coupled.

Alternatively we may consider the four sourceless regions $y < 0$, $0 < y < y_0$, $y_0 < y < b$, $y > b$ and use any pair of parallel potentials.[*] Considerable simplification is obtained from using the pair π_y', π_y''. Indeed, as the derived fields are given by

$$E_x = \frac{\partial^2 \pi_y'}{\partial x \, \partial y} + j\omega\mu_0 \frac{\partial \pi_y''}{\partial z} \tag{2.271}$$

$$E_y = -\left(\frac{\partial^2}{\partial x^2} + \frac{\partial^2}{\partial z^2}\right) \pi_y' \tag{2.272}$$

$$E_z = \frac{\partial^2 \pi_y'}{\partial y \, \partial z} - j\omega\mu_0 \frac{\partial \pi_y''}{\partial x} \tag{2.273}$$

$$H_x = -j\omega\epsilon \frac{\partial \pi_y'}{\partial z} + \frac{\partial^2 \pi_y''}{\partial x \, \partial y} \tag{2.274}$$

$$H_y = -\left(\frac{\partial^2}{\partial x^2} + \frac{\partial^2}{\partial y^2}\right) \pi_y'' \tag{2.275}$$

$$H_z = j\omega\epsilon \frac{\partial \pi_y'}{\partial x} + \frac{\partial^2 \pi_y''}{\partial y \, \partial z} \tag{2.276}$$

the continuity of E_x, E_z, H_x and H_z at the interfaces $y = 0$, $y = b$ simply requires that of $\epsilon\pi_y'$, $\partial\pi_y'/\partial y$, π_y'', $\partial\pi_y''/\partial y$. These conditions do not couple π_y' and π_y'' at the interfaces. This is not surprising since the discussion in Section 2.2.3 had shown that this pair is the simplest choice for a stratification perpendicular to the y-axis.

The counterpart of this simplicity is that π_y' and π_y'' are coupled by the source. Indeed, considering the volume current density of the dipole

$$\mathbf{J} = Ids \, \delta(x - x_0) \, \delta(y - y_0) \, \delta(z) \, \mathbf{u}_x \tag{2.277}$$

as a surface current density in the plane $y = y_0$, we will have to express the conditions

$$H_x(x, y_0 + 0, z) - H_x(x, y_0 - 0, z) = 0 \tag{2.278}$$

$$H_z(x, y_0 + 0, z) - H_z(x, y_0 - 0, z) = Ids \, \delta(x - x_0) \, \delta(z) \tag{2.279}$$

---

[*] Some pairs of non-parallel potentials may also be used. The pair π_y', π_z' was used in a similar problem (Mahmoud and Wait, 1974a; Mahmoud 1974a, b) but these are also coupled at the interfaces.

The derivation further involves a cosine series on the x-coordinate and a Fourier transform on the z-coordinate. The Fourier transform of the final result is obtained as

$$\pi_y' = \frac{4\,Ids}{j\omega\epsilon_0 a}\sum_{m=0}^{\infty}\frac{\delta_m\,(m\pi/a)}{(k_0^2+u_m^2)\Delta_m^{(v)}}\sin\frac{m\pi x}{a}\cos\frac{m\pi x_0}{a}\begin{cases}f_{m-}^{(v)}(y_0)f_{m+}^{(v)}(b-y)\\-f_{m-}^{(v)}(b-y_0)f_{m+}^{(v)}(y)\end{cases}$$

(2.280)

$$\pi_y'' = \frac{4\,Ids}{ja}\sum_{m=0}^{\infty}\frac{h\delta_m}{(k_0^2+u_m^2)\Delta_m^{(h)}}\cos\frac{m\pi x}{a}\cos\frac{m\pi x_0}{a}\begin{cases}f_{m+}^{(h)}(y_0)f_{m+}^{(h)}(b-y)\\f_{m+}^{(h)}(b-y_0)f_{m+}^{(h)}(y)\end{cases}$$

(2.281)

for

$$\begin{cases}y_0<y<b\\0<y<b\end{cases}$$

where h is the variable of the Fourier transform and

$$\delta_m = \begin{cases}\frac{1}{2} & \text{if } m=0\\1 & \text{if } m\neq 0\end{cases}$$

(2.282)

$$u_m = [-k_0^2+(m\pi/a)^2+h^2]^{1/2}$$

(2.283)

$$u_{m1} = [-k_1^2+(m\pi/a)^2+h^2]^{1/2}$$

(2.284)

$$R_m = \frac{k_1^2 u_m-k_0^2 u_{m1}}{k_1^2 u_m+k_0^2 u_{m1}}$$

(2.285)

$$S_m = \frac{u_m-u_{m1}}{u_m+u_{m1}}$$

(2.286)

$$\left.\begin{array}{c}\Delta_m^{(h)}\\\Delta_m^{(v)}\end{array}\right\} = e^{u_m b}-\begin{Bmatrix}S_m^2\\R_m^2\end{Bmatrix}e^{-u_m b}$$

(2.287)

$$f_{m\pm}^{(h)} = (1/2)[e^{u_m y}\pm S_m\,e^{-u_m y}]$$

(2.288)

$$f_{m\pm}^{(v)} = (1/2)[e^{u_m y}\pm R_m\,e^{-u_m y}]$$

(2.289)

Before discussing this result, we will consider the vertical electric dipole. This case is somewhat simpler since it can be solved with a single electric potential π_y'. With the same notation, we find for the Fourier transform

$$\pi_y' = \frac{-4\,Ids}{j\omega\epsilon_0 a}\sum_{m=1}^{\infty}\frac{1}{u_m\,\Delta_m^{(v)}}\sin\frac{m\pi x}{a}\sin\frac{m\pi x_0}{a}\begin{cases}f_{m+}^{(v)}(y)f_{m+}^{(v)}(b-y_0)\\f_{m+}^{(v)}(b-y_0)f_{m+}^{(v)}(y)\end{cases}$$

(2.290)

for

$$\begin{cases} y_0 < y < b \\ 0 < y < y_0 \end{cases}$$

It remains to take the inverse Fourier transforms of the results 2.280, 2.281 and 2.290. This may be done by standard methods of function theory, to obtain a mode expansion of the potentials and subsequently of the fields. The similarity with Section 2.6.2 is complete. Two sets of modes with equations $\Delta_m^{(v)} = 0$ and $\Delta_m^{(h)} = 0$ are obtained. These equations are identical to those obtained for the parallel plate waveguide, except that u and u_1 are now replaced by u_m and u_{m1}.

Modes with equation $\Delta_m^{(h)} = 0$ result from poles of π_y''. As results from eqn. 2.272 the transverse electric field of these modes is horizontal. For this reason they are called $E_{mn}^{(h)}$ modes. Note that all three components of the magnetic field are non-zero. Repeating the discussion made in Section 2.7.1, it can be seen that provided $(k_0 a)^2 \gg [m^2 + n^2] \pi^2$, the (m,n)th order mode ($m \geq 0, n > 0$) has phase and attenuation constants approximately given by

$$\beta_{mn}^{(h)} \simeq \sqrt{k_0^2 - \left(\frac{m\pi}{a}\right)^2 - \left(\frac{n\pi}{b}\right)^2} \tag{2.291}$$

$$\alpha_{mn}^{(h)} \simeq \frac{2(n\pi)^2}{k_0^2 b^3} \, \text{Re} \, \frac{1}{\sqrt{\kappa_1' - 1}} \tag{2.292}$$

with $\alpha_{mn}^{(h)} \ll \beta_{mn}^{(h)}$.

Similarly modes with equations $\Delta_m^{(v)} = 0$ result from poles of π_y'. They have $H_y = 0$ and could be called $H_{mn}^{(h)}$ modes. All three components of the electric field are non-zero. However when the condition $(ka)^2 \gg [m^2 + (n+1)^2] \pi^2$ is fulfilled, with $m > 0, n \geq 0$, they have phase and attenuation constants approximately given by

$$\beta_{mn}^{(v)} \simeq \sqrt{k_0^2 - \left(\frac{m\pi}{a}\right)^2 - \frac{(n+1)^2 \pi^2}{b^2}} \tag{2.293}$$

$$\alpha_{mn}^{(v)} \simeq \frac{2(n+1)^2 \pi^2}{k_0^2 b^3} \, \text{Re} \left[\frac{\kappa_1'}{\sqrt{\kappa_1' - 1}} \right] \tag{2.294}$$

and it can be seen that the dominant component of the transverse electric field is vertical. For this reason we call these modes $E_{mn}^{(v)}$ modes.*

* In the limit of four perfectly conducting side walls, the equations $\Delta_m^{(v)} = 0$ and $\Delta_m^{(h)} = 0$ have the same roots. The eigenvalues are actually degenerate, excepted for either $m = 0$ or $n = 0$. In this degenerated case our E_{mn}^v and E_{mn}^h modes do however not tend toward any of the TE_{mn} or TM_{mn} modes of the ideal waveguide, but rather toward a combination of them. The reason for this is that it is impossible to obtain either $E_z = 0$ or $H_z = 0$ with a single y-oriented potential. The notation $E_{mn}^{(v)}$, $E_{mn}^{(h)}$ used here is intended to recall that the main transverse electric field is vertical or horizontal, respectively. It should not be confused with the notation E_{mn} used for the TM_{mn} modes of the ideal waveguide. The indices (m, n) however have been kept in a one-to-one correspondence with those for the ideal waveguide.

The result of the present analysis is that a vertical electric dipole excites only $E_{mn}^{(v)}$ modes. A horizontal electric dipole excites mainly $E_{mn}^{(h)}$ modes but also, with a smaller amplitude, $E_{mn}^{(v)}$ modes. Common to both types of mode is the fact that the specific attenuation is independent of m well above the cutoff frequency. The specific attenuation of $E_{mn}^{(h)}$ modes may be significantly smaller than that of $E_{mn}^{(v)}$ modes, particularly when the dielectric constant κ_1 is high. Finally, it should be observed that the condition $(k_0 a)^2 \gg (m^2 + n^2) \pi^2$ or $(k_0 a)^2 \gg [m^2 + (n+1)^2]$ π^2 implies that the fields of these modes are approximately TEM.

(b) Approximations for a tunnel with four imperfectly conducting walls

It has been explained that an exact analytical approach to this problem would lead to extreme complexity. Several approximate methods for calculating the mode propagation constants of this structure have been published and will be discussed.

The analysis of Emslie *et al.* (1973) is restricted to the lowest-order $E^{(v)}$ and $E^{(h)}$ modes, but it can easily be extended to higher-order modes. As the structure is symmetrical we will restrict our attention to the $E^{(h)}$ mode. These authors assume that the fields inside the waveguide may be written in the form

$$E_x = E_0 \cos [k_1 (x - a/2)] \cos [k_2 (y - b/2)] \exp(-jk_3 z)$$

$$E_y = 0$$

$$E_z = (jk_1/k_3) E_0 \sin [k_1 (x - a/2)] \cos [k_2 (y - b/2)] \exp(-jk_3 z)$$

$$H_x = k_1 k_2 /(\omega \mu_0 k_3) E_0 \sin [k_1 (x - a/2)] \sin [k_2 (y - b/2)] \exp(-jk_3 z)$$

$$H_y = (k_1^2 + k_3^2)/(\omega \mu_0 k_3) E_0 \cos [k_1 (x - a/2)] \cos [k_2 (y - b/2)] \exp(-jk_3 z)$$

$$H_z = jk_2 /(\omega \mu_0) E_0 \cos [k_1 (x - a/2)] \sin [k_2 (y - b/2)] \exp(-jk_3 z)$$

$$(2.295)$$

where the coordinate wave numbers k_1, k_2, k_3 may be complex and must satisfy the relation $k_1^2 + k_2^2 + k_3^2 = k_0^2$. Actually this is the type of expression which can be obtained by separation of variables with a single scalar magnetic potential π_y''. The reader will remember that this type of solution was obtained for the $E^{(h)}$ modes of the tunnel with perfectly conducting side walls. Similar expressions are written, with adequate notation, for the regions A_{-0}, A_{+0}, A_{0+} and A_{0-} of Fig. 2.16. It then appears that the boundary conditions cannot be exactly satisfied. When the tunnel cross-section is large, approximations similar to those used in previous sections can be made and the boundary conditions can be exactly satisfied for all field components on the horizontal walls and for E_z, H_y on the vertical walls, but not for the axial component H_z. It is however observed that this component is much smaller than H_y and consequently the boundary conditions are approximately satisfied.

A very similar approach was used by Laakman and Steier (1976). These authors however start with more general field expressions which are derived from two scalar potential components. All six components of the fields are now non-zero. Here also it appears that the boundary conditions can be satisfied only in an approximate way. Actually, this analysis is equivalent to that of Emslie *et al.* since the part of the solution which can be obtained from a π_y' potential (assumed to be the zero by Emslie *et al.*) is of second-order magnitude. It is thus not surprising that the approximate results obtained in the two papers are identical.

Common to these analyses is the fact that the fields in the corner regions A_{--}, A_{-+}, A_{+-} and A_{++} of Fig. 2.16 do not enter the derivation of the solution, which therefore neglects the effect of the corners. Actually, the solution does not take the corner regions into account and the latter may even contain a non-homogeneous medium. The authors were consequently allowed to assume that the electrical parameters of the top and bottom walls differed from those of the vertical walls.

Assuming that the horizontal and vertical walls have complex dielectric constants κ_h' and κ_v', respectively, the propagation constant of the $E_{mn}^{(v)}$ mode is given approximately by

$$\Gamma_{mn}^{(v)} = \alpha_{mn}^{(v)} + j\,\beta_{mn}^{(v)} \tag{2.296}$$

$$\alpha_{mn}^{(v)} \simeq \frac{2(m+1)^2\,\pi^2}{k_0^2\,b^3}\,\mathrm{Re}\,\frac{\kappa_h'}{\sqrt{\kappa_h'-1}} + \frac{2(n+1)^2\,\pi^2}{k_0^2\,a^3}\,\mathrm{Re}\,\frac{1}{\sqrt{\kappa_v'-1}}$$

$$\tag{2.297}$$

$$\beta_{mn}^{(v)} \simeq \sqrt{k_0^2 - \left[\frac{(m+1)\pi}{a}\right]^2 - \left[\frac{(n+1)\pi}{b}\right]^2} \tag{2.298}$$

where the mode indices m, n run from zero to infinity. The main components of the transverse field for this mode are given by

$$E_y \simeq \sin\frac{(m+1)\pi x}{a}\,\sin\frac{(n+1)\pi y}{b}\,\exp(-\Gamma_{mn}^{(v)}\,z) \tag{2.299}$$

$$H_x \simeq \sqrt{\frac{\mu_0}{\epsilon_0}}\,E_y \tag{2.300}$$

The results for the $E_{mn}^{(h)}$ mode are obtained by interchanging the x- and y-axes. These approximations are valid under the assumptions

$$k_0 a, k_0 b \gg 1$$

$$|\kappa_v'-1| \gg \frac{(m+1)\pi}{k_0 a} \tag{2.301}$$

$$\left|\frac{\kappa_h'}{\kappa_h'-1}\right| \gg \frac{(n+1)\pi}{k_0 b}$$

These results are remarkable in several respects. The specific attenuation of the $E_{mn}^{(v)}$ mode is obtained by adding the values of α for waveguides with perfectly conducting horizontal walls and with perfectly conducting vertical walls, keeping the other walls unchanged. The main transverse field components are approximately zero on all four walls.

Comparison of this behaviour with that of the fields in a waveguide with perfectly conducting side walls was made by Laakman and Steier (1976) but we will diverge somewhat from the interpretation given by these authors. Here again the question is one of nomenclature of the waveguide modes. Our mode indices (m, n) run from zero to infinity, while those used by all authors referred to here run from unity to infinity. Our choice is justified by a one-to-one correspondence with the usual nomenclature for the waveguide with perfectly conducting side walls. If k_x and k_y are the transverse wave numbers, the quantities $(k_x a, k_y b)$ tend to $(m\pi, n\pi)$ for infinite wall conductivities and to $[(m + 1)\pi, (n + 1)\pi]$ for a large cross-section and dielectric walls. We do not follow Laakman and Steier when they state that the modes with either m or n equal to zero do not exist for a waveguide with large cross-section and dielectric walls. We recall rather that the TM_{00}, TM_{m0} and TM_{0n} modes cannot exist if the walls are perfectly conducting.

Another but still very similar approach to the problem of the rectangular waveguide with four imperfectly conducting walls was given by Andersen *et al.* (1975). These authors assume that the field inside the waveguide is composed of four plane waves with vectorial wave numbers (k_x, k_y, k_z), $(-k_x, k_y, k_z)$, $(k_x, -k_y, k_z)$ and $(-k_x, -k_y, k_z)$ and express that these waves are related by plane wave reflection coefficients on the waveguide walls: for instance, the first wave is reflected into the second one at the interface $x = a$. It can be shown that this formulation does not differ in any way from Laakman and Steier's (1976). These methods implicitly contain the proof that the guided modes of the rectangular structure cannot be found by separation of the variables x and y. Some superiority of Andersen *et al.*'s analysis may be claimed from the fact that they use a r.m.s. minimisation algorithm to find an approximate solution of the overdetermined set of boundary conditions, rather than ignoring some of these.

It is also possible, at least when the complex dielectric constant of the side walls is relatively large, to obtain a good approximation by replacing the boundary conditions on all walls by a surface impedance condition (Wait, 1980).

Finally it must be mentioned that the geometrical optical approach developed by Mahmoud and Wait (1974c) and analysed in Section 2.7.2 for the planar air waveguide has been extended by these authors to the waveguide with four imperfectly conducting walls. This method suffers from the weaknesses mentioned earlier but no alternative way of solving excitation problems has yet been proposed.

2.7.4 Circular tunnel

The geometry of the circular tunnel (Fig. 2.17) lends itself much more easily to exact analytical solution of Maxwell's equations than the rectangular tunnel. The problem involved is formally identical to that of the dielectric waveguide (Stratton, 1941) which has fundamental importance in fibre optics, but we are here interested in the guided modes of this structure for a quite different range of parameters. The inside of the waveguide, of radius a, is filled with air. The external medium has a complex dielectric constant κ_1', a permeability μ_1 and a complex wave number k_1. We will use solutions of Maxwell's equations in the form given in Appendix B (eqns. B.25–B.33) and denote

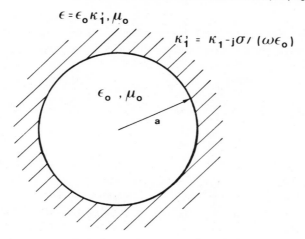

Fig. 2.17   *Geometrical and electrical parameters of a circular tunnel*

$$\Lambda = \sqrt{k_0^2 + \Gamma^2}; \qquad -\pi < \arg \Lambda \leqslant 0 \tag{2.302}$$

$$\Lambda_1 = \sqrt{k_1^2 + \Gamma^2}; \qquad -\pi < \arg \Lambda_1 \leqslant 0 \tag{2.303}$$

The solutions, which remain finite on the tunnel axis and tend toward zero at infinity, are of the form

$$E_\rho = \sum_{m=0}^{\infty} \left[-\frac{j\omega\mu_0 m}{\rho} P_m J_m(\Lambda\rho) - \Lambda\Gamma Q_m J_m'(\Lambda\rho) \right] \cos\left[m(\phi - \phi_m) \right] \tag{2.304}$$

$$E_\phi = \sum_{m=0}^{\infty} \left[j\omega\mu_0 \Lambda P_m J_m'(\Lambda\rho) + \frac{\Gamma m}{\rho} Q_m J_m(\Lambda\rho) \right] \sin\left[m(\phi - \phi_m) \right] \tag{2.305}$$

$$E_z = \sum_{m=0}^{\infty} \Lambda^2 Q_m J_m(\Lambda\rho) \cos\left[m(\phi - \phi_m) \right] \tag{2.306}$$

$$H_\rho = \sum_{m=0}^{\infty} \left[-\Lambda\Gamma P_m J_m'(\Lambda\rho) - \frac{j\omega\epsilon_0 m}{\rho} Q_m J_m(\Lambda\rho) \right] \sin\left[m(\phi - \phi_m) \right] \tag{2.307}$$

$$H_\phi = \sum_{m=0}^{\infty} \left[-\frac{\Gamma m}{\rho} P_m J_m(\Lambda\rho) - j\omega\epsilon_0 \Lambda Q_m J_m'(\Lambda\rho) \right] \cos\left[m(\phi - \phi_m) \right] \tag{2.308}$$

$$H_z = \sum_{m=0}^{\infty} \Lambda^2 P_m J_m(\Lambda\rho) \sin\left[m(\phi - \phi_m) \right] \tag{2.309}$$

where the angles ϕ_m are arbitrary, for the inside of the tunnel. In the external medium the arbitrary constants P_m, Q_m should be replaced by $A_{m1}, B_{m1}; \epsilon_0, \mu_0,$

Λ by ϵ_1, μ_1, Λ_1; and $J_m(\Lambda\rho)$ by $H_m^{(2)}(\Lambda_1\rho)$. The mode equation is obtained by expressing the continuity of the tangential field at $\rho = a$ and writing that the determinant of the system of equations obtained this way is zero. This yields the classical equation

$$\left[\frac{J_m'(\Lambda a)}{\Lambda a\, J_m(\Lambda a)} - \frac{\mu_1}{\mu_0}\frac{H_m^{(2)'}(\Lambda_1 a)}{\Lambda_1 a\, H_m^{(2)}(\Lambda_1 a)}\right]\left[\frac{J_m'(\Lambda a)}{\Lambda a\, J_m(\Lambda a)} - \kappa_1'\frac{H_m^{(2)'}(\Lambda_1 a)}{\Lambda_1 a\, H_m^{(2)}(\Lambda_1 a)}\right]$$

$$= -\left(\frac{m\Gamma}{k_0}\right)^2\left[\frac{1}{(\Lambda a)^2} - \frac{1}{(\Lambda_1 a)^2}\right]^2 \qquad (2.310)$$

For the remainder of this section we will assume that $\mu_1 = \mu_0$.

This equation can of course be solved numerically. Here we are interested in developing approximations valid when the tunnel cross-section is large compared with the wavelength. This is the kind of work which was performed in Section 2.7.1 for the planar waveguide. Actually we follow the analysis made by Marcatili and Schmeltzer (1964) in the context of optical transmission, but we will differ in the mode nomenclature. We examine the following question: assuming that $k_0 a \gg 1$, can we have modes with a propagation constant Γ close to (jk_0), i.e. such that $|\Lambda a| \ll k_0 a$? Again, provided $|\kappa_1'|$ is neither close to nor large compared with unity, this entails $|\Lambda a| \ll |\Lambda_1 a|$, $|\Lambda_1 a| \gg 1$. Now, for those modes for which $|\Lambda_1 a| \gg m$, we may use the large-argument approximations of the Hankel function and of its derivative. This yields

$$\frac{H_m^{(2)'}(\Lambda_1 a)}{H_m^{(2)}(\Lambda_1 a)} \simeq -j \qquad \text{if} \qquad |\Lambda_1 a| \gg m \qquad (2.311)$$

The right-hand side of eqn. 2.310 may be approximated by $(m/\Lambda^2 a^2)^2$. Consequently this equation reduces to the approximate form

$$\left[\frac{J_m'(\Lambda a)}{J_m(\Lambda a)} + j\frac{\Lambda a}{\Lambda_1 a}\right]\left[\frac{J_m'(\Lambda a)}{J_m(\Lambda a)} + j\kappa_1'\frac{\Lambda a}{\Lambda_1 a}\right] \simeq \frac{m^2}{(\Lambda a)^2} \qquad (2.312)$$

We will discuss separately the cases $m = 0$ and $m \neq 0$. Indeed for $m = 0$ the right-hand side of eqn. 2.310 is zero and this equation is split into two independent ones, the solutions of which yield TE and TM modes. The parts of the fields due to the electric and magnetic potentials are indeed uncoupled in eqns. 2.304 and 2.309.

Case (a): $m = 0$, TE_{0n} modes
The mode equation

$$\frac{J_0'(\Lambda a)}{\Lambda a\, J_0(\Lambda a)} - \frac{H_0^{(2)'}(\Lambda_1 a)}{\Lambda_1 a H_0^{(2)}(\Lambda_1 a)} = 0 \qquad (2.313)$$

or, in approximate form

$$\frac{J_0'(\Lambda a)}{J_0(\Lambda a)} + j\frac{\Lambda a}{\Lambda_1 a} \simeq 0 \tag{2.313}$$

expresses the continuity of $E_{\phi 0}$ and H_{z0} and yields the TE modes:

$$E_{\phi 0} = j\omega\mu_0 \Lambda P_0 J_0'(\Lambda\rho) \tag{2.315}$$

$$H_{\rho 0} = -\Gamma \Lambda P_0 J_0'(\Lambda\rho) \tag{2.316}$$

$$H_{z0} = \Lambda^2 P_0 J_0(\Lambda\rho) \tag{2.317}$$

where we may use the identity $J_0'(\Lambda\rho) = -J_1(\Lambda\rho)$. We will delay the discussion of approximate solutions of eqn. 2.314.

Case (b): $m = 0$, TM_{0n} modes
The mode equation

$$\frac{J_0'(\Lambda a)}{\Lambda a J_0(\Lambda a)} - \kappa_1'\frac{H_0^{(2)'}(\Lambda_1 a)}{\Lambda_1 a H_0^{(2)}(\Lambda_1 a)} = 0 \tag{2.318}$$

or, in approximate form

$$\frac{J_0'(\Lambda a)}{J_0(\Lambda a)} + j\kappa_1'\frac{\Lambda a}{\Lambda_1 a} \simeq 0 \tag{2.319}$$

expresses the continuity of E_{z0} and $H_{\phi 0}$ and yields the TM modes:

$$E_{\rho 0} = -\Gamma \Lambda Q_0 J_0'(\Lambda\rho) \tag{2.320}$$

$$E_{z0} = \Lambda^2 Q_0 J_0(\Lambda\rho) \tag{2.321}$$

$$H_{\phi 0} = -j\omega\epsilon_0 \Lambda Q_0 J_0'(\Lambda\rho) \tag{2.322}$$

Case (c): $m \neq 0$, hydrid modes $EH_{mn}^{\pm}$
Eqn. 2.312 may be solved as a quadratic equation in $J_m'(\Lambda a)/J_m(\Lambda a)$. We will further assume that $|\kappa_1'(\Lambda a)^4| \ll |m^2(\Lambda_1 a)^2|$ to obtain

$$\frac{\Lambda a J_m'(\Lambda a)}{m J_m(\Lambda a)} \simeq \pm 1 - j\frac{\kappa_1' + 1}{2}\frac{(\Lambda a)^2}{m\Lambda_1 a} \tag{2.323}$$

Comparing this with the formulas

$$\frac{x J_m'(x)}{m J_m(x)} = \pm 1 \mp \frac{x J_{m\pm 1}(x)}{m J_m(x)} \tag{2.324}$$

it appears that we must have either

$$J_{m+1}(\Lambda a) \simeq j\frac{\kappa_1' + 1}{2}\frac{\Lambda a}{\Lambda_1 a} J_m(\Lambda a) \tag{2.325a}$$

or

$$J_{m-1}(\Lambda a) \simeq -j\frac{\kappa_1' + 1}{2}\frac{\Lambda a}{\Lambda_1 a} J_m(\Lambda a) \tag{2.325b}$$

In all these expressions, we may further use the approximation

$$\Lambda_1 a \simeq k_0 a \sqrt{\kappa_1' - 1}$$

wherein due care is taken of the determination 2.303. The coefficient of $J_m (\Lambda a)$ in eqns. 2.325a and b is small compared with unity, and consequently the roots of these equations are close to those of $J_{m+1} (\Lambda a) = 0$ and $J_{m-1} (\Lambda a) = 0$, respectively. Using small-argument approximations of Bessel's functions, it is easy to show that the roots $\Lambda a \simeq 0$ must be discarded. Denoting by $\xi_{m,n}$ the nth non-zero root of $J_m (\xi) = 0$ and using a perturbation technique, we find approximate solutions of eqns. 2.325a and b:

$$\Lambda_{mn}^+ a \simeq \xi_{m+1,n} \left[1 + j \frac{\kappa_1' + 1}{2 k_0 a \sqrt{\kappa_1' - 1}} \right] \tag{2.326a}$$

$$\Lambda_{mn}^- a \simeq \xi_{m-1,n} \left[1 + j \frac{\kappa_1' + 1}{2 k_0 a \sqrt{\kappa_1' - 1}} \right] \tag{2.326b}$$

The difference from Marcatili and Schmeltzer's (1964) result is only an apparent one. These authors used only the solution 2.326b of eqn. 2.325b but with positive and negative values of m. This is equivalent since $J_{-m-1} (x) = (-1)^{m+1} J_{m+1} (x)$. Note that the negative roots of $J_{m \pm 1} (\xi)$ must be discarded since they yield a value of Λa incompatible with the determination 2.302.[*] Using the approximation $\Lambda_{mn}^\pm a = \xi_{m \pm 1, n}$ and the boundary conditions, it is possible to show that the transverse fields may be expressed approximately by

$$E_{\rho m}^\pm \simeq J_{m \pm 1} (\xi_{m \pm 1, n} \, \rho/a) \cos [m (\phi - \phi_m)] \tag{2.327}$$

$$E_{\phi m}^\pm \simeq \pm J_{m \pm 1} (\xi_{m \pm 1, n} \, \rho/a) \sin [m (\phi - \phi_m)] \tag{2.328}$$

$$H_t \simeq \frac{k_0}{\omega \mu_0} \, u_z \times E_t \tag{2.329}$$

where we have neglected a common amplitude factor. We recall that ϕ_m is arbitrary and represents the azimuthal degeneracy of the structure.

The longitudinal components have second-order amplitudes. The polarisation of the hybrid $EH_{mn}^\pm$ modes is linear and varies with the azimuth. It is approximately constant over the whole tunnel cross-section for the $EH_{11}^\pm$ modes. Expressions 2.327 to 2.329 may be considered as valid for $m = 0$ too, instead of 2.315–2.317 and 2.320–2.322.

We may now return to the rotational-symmetric modes and solve the mode equations, using the same approximation technique. This yields

---

[*] Assuming that $\mathrm{Re} \left[(\kappa_1' + 1)/\sqrt{\kappa_1' - 1} \, \right] < 0$. This condition is not realised if $\kappa_1 < 3$ and $\mathrm{Im} \, \kappa_1' < \kappa_1^2 - 2\kappa_1 - 3$. If this was the case, only negative roots should be used. However this is immaterial since eqn. 2.310 is even in Λ and the solution remains unchanged.

$$\Lambda_{0n}a = \begin{cases} \xi_{1n}\left[1+j\,\dfrac{1}{k_0a\,\sqrt{\kappa_1'-1}}\right] & \text{for TE}_{0n} \text{ modes} \qquad (2.330) \\[3mm] \xi_{1n}\left[1+j\,\dfrac{\kappa_1'}{k_0a\,\sqrt{\kappa_1'-1}}\right] & \text{for TM}_{0n} \text{ modes} \qquad (2.331) \end{cases}$$

Hence we are able to calculate the propagation constant of all modes. Using eqn. 2.302 and the fact that $|\Lambda| \ll k_0$, we obtain

$$\Gamma \simeq jk_0 - j\,\frac{(\Lambda a)^2}{k_0 a^2} \qquad (2.332)$$

The attenuation constants are given by

$$\alpha_{0n} \simeq \frac{\xi_{1n}^2}{k_0^2 a^3} \begin{cases} \text{Re}\,\dfrac{1}{\sqrt{\kappa_1'-1}} & \text{for TE}_{0n} \text{ modes} \qquad (2.333) \\[3mm] \text{Re}\,\dfrac{\kappa_1'}{\sqrt{\kappa_1'-1}} & \text{for TM}_{0n} \text{ modes} \qquad (2.334) \end{cases}$$

$$\alpha_{mn}^{\pm} \simeq \frac{\xi_{m\pm1,n}^2}{k_0^2 a^3}\,\text{Re}\,\frac{\kappa_1'+1}{2\sqrt{\kappa_1'-1}} \quad \text{for hybrid modes} \qquad (2.335)$$

where ξ_{pq} is the qth non-zero root of $J_p(\xi) = 0$, $p \neq 0$. As for the rectangular tunnel, the attenuation constants are inversely proportional to the square of the frequency and to the cube of the tunnel size. Taking into account the values $\xi_{01} = 2\cdot405$ and $\xi_{11} = 3\cdot832$, it is seen that, for κ_1 real, the mode with the lowest attenuation is TE_{01} if $\kappa_1 < 4\cdot08$ and EH_{11}^- if $\kappa_1 > 4\cdot08$. It is interesting to collect the conditions under which this result is valid; in practical form

$$k_0a \gg 1$$
$$|k_0a\,\sqrt{\kappa_1'-1}| \gg m \pm 1 \qquad (2.336)$$
$$|k_0a\,\sqrt{\kappa_1'-1}| \gg \frac{\xi_{m\pm1,n}^2}{m}\,|\sqrt{\kappa_1'}| \qquad \text{if } m \neq 0$$

These formulas may usefully be compared with those for the rectangular tunnel, namely eqns. 2.297 and 2.301.

2.7.5 Curved tunnel

Actual tunnels are often curved either horizontally or vertically. We will not go into a study of the mode propagation inside a curved tunnel. The interested reader is referred to Mahmoud and Wait (1974b) for the rectangular tunnel and to Marcatili and Schmeltzer (1964) for the circular tunnel. In the latter case the problem was solved by writing Maxwell's equations in toroidal coordinates and using a perturbation technique valid when the radius of curvature R is large. A parameter

$$v = \frac{2a}{R} \left(\frac{k_0 a}{\xi_{m\pm 1,n}} \right)^2 \tag{2.337}$$

is defined. Subject to the condition $v \ll 1$ the following approximate expression is found for the attenuation of the modes as a function of the radius of curvature:

$$\alpha_{mn}(R) = \alpha_{mn}(\infty) + \frac{4k_0^2 a^3}{3R^2} \begin{cases} \dfrac{1}{\xi_{1n}^2} \operatorname{Re} \dfrac{1}{\sqrt{\kappa_1' - 1}} & \text{for } TE_{0n} \text{ modes} \quad (2.338) \\[2ex] \dfrac{1}{\xi_{1n}^2} \operatorname{Re} \dfrac{\kappa_1'}{\sqrt{\kappa_1' - 1}} & \text{for } TM_{0n} \text{ modes} \quad (2.339) \\[2ex] \dfrac{1}{\xi_{m\pm 1,n}^2} \left\{ \left[1 - \dfrac{m(m \pm 2)}{\xi_{m\pm 1,n}^2} \right] \operatorname{Re} \dfrac{\kappa_1' + 1}{2\sqrt{\kappa_1' - 1}} \right. \\[2ex] \left. \qquad + \tfrac{3}{8} \delta_m \cos 2\phi_m \operatorname{Re} \sqrt{\kappa_1' - 1} \right\} \\[1ex] \hspace{4cm} \text{for } EH_{mn}^{\pm} \text{ modes} \quad (2.340) \end{cases}$$

where $\delta_i = 1$ for $i = 0$ and $\delta_i = 0$ for $i \neq 0$.

In each case the specific attenuation is increased by the tunnel curvature. As the increase is proportional to $k_0^2 a^3$, the lower the attenuation of the straight tunnel the higher the loss due to bends, and vice versa. For the hybrid modes the increase is maximum when the electric field lies in the plane of curvature.

These formulas have been given for information only. Indeed, taking into account the conditions 2.336 under which the approximate results for the straight tunnel eqns. 2.333 to 2.335 are valid, it can be seen that the condition $v \ll 1$ requires extremely high values of the radius of curvature. For this reason, eqns. 2.338 to 2.340 are in general useless for the representation of high-frequency transmission in tunnels, but no better theoretical approach seems to exist at present. In spite of this, they have the merit of showing evidence that the choice of the highest frequencies, though justified for straight tunnels, is not adequate for curved ones.

2.8 Guided modes of wires and leaky feeders in a circular tunnel

2.8.1 Basic theory of guided modes

In this section, we are concerned with establishing the mode equation for the propagation of electromagnetic waves along cables parallel to the axis of a circular tunnel. By cable we mean either a conducting wire or a leaky coaxial cable. These devices may in addition be covered by a dielectric jacket and, additionally, by a lossy layer which may be representative of dust or mud accumulated on the external surface of the cable. That a unified treatment can be used for these various devices is the result of the thin-cable approximation which has been developed in

Section 2.3. Let us recall that this approximation consists of modelling the cable by eqn. 2.77 relating the rotational-symmetric part of the electric field on the external surface of the cable to the total current carried by the latter. The specific external impedance of a bare wire was obtained in Section 2.4 and that of a leaky coaxial cable in Section 2.5.6 where the technique for taking into account successive cylindrical layers has also been developed.

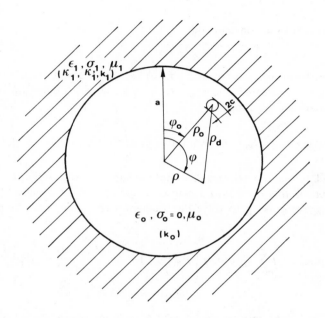

Fig. 2.18 *Geometrical and electrical parameters for a single cable in a circular tunnel*

We start our analysis by considering a circular tunnel containing a single axial cable (Fig. 2.18). The tunnel radius is a and the external medium has electrical permittivity ϵ_1, dielectric constant $\kappa_1 = \epsilon_1/\epsilon_0$, conductivity σ_1, complex dielectric constant $\kappa_1' = \kappa_1 - j\sigma_1/(\omega\epsilon_0)$, permeability μ_1 (which may differ from μ_0) and complex wave number k_1. The cable has a radius c and is located at the point (ρ_0, ϕ_0) of a system of cylindrical coordinates (ρ, ϕ, z). We are interested in finding modes with a propagation factor $\exp(-\Gamma z)$, where Γ has to be determined. The cable carries a total current $I \exp(j\omega t - \Gamma z)$. Hence the factor $\exp(j\omega t - \Gamma z)$ will be ignored in all expressions of fields and potentials. We follow the method developed by Wait and Hill (1974a). Some of the basic features of this method are also present in the paper by Fontaine, Soiron, Degauque and Gabillard (1974) and had previously been developed by one of these authors in his doctorate's thesis.

The primary fields excited by the cable current I, i.e. the field which would exist if the cable were drawn into free space, may be derived from an axial electric-type Hertz potential

$$U^{(p)} = (2\pi j\omega\epsilon_0)^{-1} I K_0 (v\rho_d) \tag{2.341}$$

where

$$v = \sqrt{-k_0^2 - \Gamma^2} \; ; \qquad \frac{\pi}{2} < \arg v \leqslant \frac{\pi}{2} \tag{2.342}$$

is the radial propagation constant and

$$\rho_d = \sqrt{\rho^2 + \rho_0^2 - 2\rho\rho_0 \cos(\phi - \phi_0)} \tag{2.343}$$

is the distance from the observation point to the cable axis. The modified Hankel function may be considered as characteristic of an outgoing cylindrical wave. The addition theorem for this function

$$K_0 (v\rho_d) = \sum_{m=-\infty}^{\infty} I_m (v\rho_<) K_m (v\rho_>) \exp\left[-jm(\phi - \phi_0)\right] \tag{2.344}$$

where $\rho_<$ and $\rho_>$ stand for the smallest and the largest of the radii ρ, ρ_0, provides a decomposition of this wave into a sum of outgoing waves emanating from the z-axis. This suggests seeking to solve the problem with an electric-type potential U given by

$$U = (2\pi j\omega\epsilon_0)^{-1} I \sum_{m=-\infty}^{\infty} I_m (v\rho_<) \left[K_m(v\rho_>) - R_m \frac{K_m (va)}{I_m (va)} I_m (v\rho_>) \right]$$

$$\exp[-jm(\phi - \phi_0)] \tag{2.345}$$

for $\rho < a$. The added terms may be seen as incoming waves to account for the reflection by the tunnel walls. The constants R_m are to be determined and play the role of reflection factors of the tunnel wall for this type of potential. However this reflection gives also rise to fields derived from a magnetic-type Hertz potential which will be written

$$V = (2\pi j\omega\epsilon_0)^{-1} I \sum_{m=-\infty}^{\infty} \Delta_m I_m (v\rho_<) I_m(v\rho_>) \exp[-jm(\phi - \phi_0)] \tag{2.346}$$

for $\rho < a$.

In the external medium, $\rho > a$, we use expressions containing only outgoing waves

$$U = (2\pi j\omega\epsilon_0)^{-1} I \sum_{m=-\infty}^{\infty} F_m K_m (u\rho) \exp[-jm(\phi - \phi_0)] \tag{2.346}$$

$$V = (2\pi j\omega\epsilon_0)^{-1} I \sum_{m=-\infty}^{\infty} G_m K_m (u\rho) \exp[-jm(\phi - \phi_0)] \tag{2.348}$$

where the radial propagation constant u is given by

$$u = \sqrt{-k_1^2 - \Gamma^2}; \qquad -\pi/2 < \arg u \leqslant \pi/2 \tag{2.349}$$

The electromagnetic fields are given by eqns. B.9 to B.14 of Appendix B. The unknown constants R_m, Δ_m, F_m and G_m can be obtained by expressing the continuity of the tangential field components at $\rho = a$. This yields a system of four linear equations in these unknowns. With a view to future approximations we will however use an alternative but equivalent method.

Starting from eqns. 2.347–2.348, we express the fields in the external medium and, by eliminating the constants F_m and G_m, we can obtain two relations between the tangential field components at $\rho = a$. These relations are written in the form

$$E_{\phi m} = \alpha_m E_{zm} + Z_m H_{zm} \tag{2.350}$$

$$H_{\phi m} = -Y_m E_{zm} + \alpha_m H_{zm} \tag{2.351}$$

for the mth harmonic of the fields. After some calculations we find

$$\alpha_m = -\frac{jm\Gamma}{u^2 a} \tag{2.352}$$

$$Z_m = -\frac{j\omega\mu_1}{u} \frac{K_m'(ua)}{K_m(ua)} \tag{2.353}$$

$$Y_m = \frac{k_1^2}{j\omega\mu_1 u} \frac{K_m'(ua)}{K_m(ua)} \tag{2.354}$$

Eqns. 2.350–2.351 depend on the external medium only and, of course, on the still unknown propagation constant Γ. Equally we could solve them for the electric field components $E_{\phi m}$ and E_{zm}, thereby defining a surface impedance matrix of the external medium. Eqns. 2.350–2.351 may thus be considered as boundary conditions imposed by the external medium and must be satisfied by the fields derived from eqns. 2.345 and 2.346 for the internal medium. Expressing this, we obtain after some calculations

$$R_m(\Gamma) = \frac{jk_0/v\, K_m'(va)/K_m(va) + \eta_0\, Y_m + \eta_0\, \delta_m}{jk_0/v I_m'(va)/I_m(va) + \eta_0\, Y_m + \eta_0\, \delta_m} \tag{2.355}$$

$$\Delta_m(\Gamma) = \frac{K_m(va)}{I_m(va)} \frac{R_m - 1}{\eta_0} \frac{\eta_0\, \delta_m}{(jm\Gamma/a)(v^{-2} - u^{-2})} \tag{2.356}$$

where $\eta_0 = \sqrt{\mu_0/\epsilon_0}$ is the intrinsic impedance of air and

$$\eta_0\, \delta_m = \frac{(jm\Gamma/a)^2\,(v^{-2} - u^{-2})^2}{jk_0/v\, I_m'(va)/I_m(va) + Z_m/\eta_0} \tag{2.357}$$

Hence, the exact expressions are known for the potentials and fields inside the tunnel, but the propagation constant Γ is not yet determined. Indeed, it remains to express the boundary condition on the external surface of the cable. As we intend to use the thin-cable approximation, we first need to write explicitly the expression for the axial field E_z. It is given by $(-v^2 U)$ from eqn. B.5 of Appendix B. Using eqn. 2.345 in which we restore the compact form eqn. 2.341 for the primary potential, we obtain

$$E_z = \frac{-v^2 I}{2\pi j\omega\epsilon_0}\left[K_0(v\rho_d) - \sum_{m=-\infty}^{\infty} R_m \frac{K_m(va)}{I_m(va)} I_m(v\rho_0) I_m(v\rho)\right.$$
$$\left. \exp[-jm(\phi - \phi_0)]\right] \tag{2.358}$$

In this expression, the sum over m represents the field reflected by the tunnel wall and in practice is assumed to be constant when the observation point moves around the cable circumference. This is the condition required for the thin-cable approximation to be valid. The boundary condition at the external surface of the cable may then be expressed at any point of this surface. Thus we choose the matching point at $\rho = \rho_0 + c$ and $\phi = \phi_0$ and obtain the mode equation

$$K_0(vc) - \sum_{m=-\infty}^{\infty} R_m \frac{K_m(va)}{I_m(va)} I_m(v\rho_0) I_m[v(\rho_0 + c)] = \frac{-2\pi j\omega\epsilon_0}{v^2} z_0(\Gamma)$$
$$\tag{2.359}$$

where $z_0(\Gamma)$ is the specific external impedance of the cable. This parameter is determined by the internal structure of the cable and may depend on the mode propagation constant Γ as we have seen in Sections 2.4 and 2.5.6.

Solving the mode equation with full generality obviously requires the use of a computer. The Newton–Raphson method for complex variables is claimed to be efficient for this purpose. The solution of the mode equation provides the propagation constant of all modes, including the waveguide modes. Of particular interest however are the modes which have a transmission line character as the frequency is lowered or as the wall conductivity becomes very high. In the case of a single monofilar wire conductor we will find one such mode, this is the monofilar mode. In the case of a leaky coaxial cable we will find the monofilar and coaxial modes described in Chapter 1. Nevertheless there is some interest in the waveguide modes when they are above cutoff, since they may then have a lower specific attenuation than the transmission line modes. It is expected that their cutoff frequencies will not be significantly modified by the presence of the cable.

It is unfortunately not possible to introduce approximations which are valid for all cases encountered in practice in subsurface radio communications. If $|k_1 a| \gg 1$, we have also $|ua| \gg 1$ and asymptotic expressions of $K_m(ua)$, $K'_m(ua)$ may be used in the calculation of Y_m and Z_m, at least for moderate values of m. However a look at Fig. 1.2 shows that the condition $|k_1 a| \gg 1$ will not be satisfied for small tunnel radii, a low wall conductivity and at low frequencies. When it is known that the mode propagation constant Γ is close to jk_0, i.e. for the monofilar mode, we have $|v| \ll k_0$ and Fig. 1.2 shows that we may have $|va| \ll 1$ if the frequency is not too high. Obviously the conditions $|ua| \gg 1$ and $|va| \ll 1$ are not met simultaneously very often. The implications of these conditions will be examined later. Only the condition $|vc| \ll 1$ is satisfied over the whole range of parameters, but this gives rise to very meagre simplification in the numerical work.

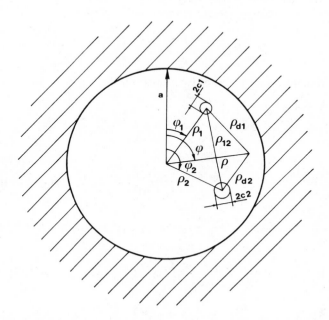

Fig. 2.19 *Geometrical parameters for a circular tunnel containing two longitudinal cables*

2.8.2 Extension to multiple wires or cables

It is now relatively easy to extend the theory to the case where the tunnel contains multiple wires or cables. We will carry out this extension for the case of two cables (Fig. 2.19). The cables are located at the points (ρ_1, ϕ_1) and (ρ_2, ϕ_2). They have radii c_1, c_2 and specific external impedances $z_1(\Gamma)$, $z_2(\Gamma)$, respectively. The method (Wait and Hill, 1974b) consists of superposing solutions of type 2.358. We write

$$
E_z = \frac{-v^2}{2\pi j \omega \epsilon_0} \left\{ I_1 \left[K_0(v\rho_{d1}) - \sum_{m=-\infty}^{\infty} R_m \frac{K_m(va)}{I_m(va)} I_m(v\rho_1) I_m(v\rho) e^{-jm(\phi-\phi_1)} \right] \right.
$$
$$
\left. + I_2 \left[K_0(v\rho_{d2}) - \sum_{m=-\infty}^{\infty} R_m \frac{K_m(va)}{I_m(va)} I_m(v\rho_2) I_m(v\rho) e^{-jm(\phi-\phi_2)} \right] \right\}
$$

$$(2.360)$$

where

$$\rho_{d1} = \sqrt{\rho^2 + \rho_1^2 - 2\rho\rho_1 \cos(\phi - \phi_1)} \tag{2.361}$$

$$\rho_{d2} = \sqrt{\rho^2 + \rho_2^2 - 2\rho\rho_2 \cos(\phi - \phi_2)} \tag{2.362}$$

and I_1, I_2 are the current carried by the cables.

The two relevant boundary conditions are now

$$E_z = z_1 I_1 \qquad \text{at} \qquad \rho = \rho_1 + c_1, \phi = \phi_1 \tag{2.363}$$

$$E_z = z_2 I_2 \qquad \text{at} \qquad \rho = \rho_2 + c_2, \phi = \phi_2 \tag{2.364}$$

This yields a homogeneous system of two equations in I_1, I_2:

$$A_{11} I_1 + A_{12} I_2 = 0 \tag{2.365}$$

$$A_{21} I_1 + A_{22} I_2 = 0 \tag{2.366}$$

with

$$A_{11} = \frac{2\pi j k_0 z_1}{v^2 \eta_0} + K_0(vc_1) - \sum_m R_m \frac{K_m(va)}{I_m(va)} I_m(v\rho_1) I_m [v(\rho_1 + c_1)] \tag{2.367}$$

$$A_{12} = K_0(v\rho_{12}) - \sum_m R_m \frac{K_m(va)}{I_m(va)} I_m(v\rho_2) I_m [v(\rho_1 + c_1)] e^{-jm(\phi_1 - \phi_2)} \tag{2.368}$$

$$A_{21} = K_0(v\rho_{12}) - \sum_m R_m \frac{K_m(va)}{I_m(va)} I_m(v\rho_1) I_m [v(\rho_2 + c_2)] e^{-jm(\phi_2 - \phi_1)} \tag{2.369}$$

$$A_{22} = \frac{2\pi j k_0 z_2}{v^2 \eta_0} + K_0(vc_2) - \sum_m R_m \frac{K_m(va)}{I_m(va)} I_m(v\rho_2) I_m [v(\rho_2 + c_2)] \tag{2.370}$$

Actually, the argument of K_0 in eqn. 2.368 should be $v\rho_{d2}$ evaluated at the matching point $\rho = \rho_1 + c_1$, $\phi = \phi_1$ but ρ_{d1} has been approximated by the distance between the cable centres:

$$\rho_{12} = \sqrt{\rho_1^2 + \rho_2^2 - 2\rho_1\rho_2 \cos(\phi_1 - \phi_2)} \tag{2.371}$$

It must indeed be stressed that the use of the thin-cable approximation in this problem requires the cable radii c_1, c_2 to be small compared with ρ_{12}. The same approximation has been made in A_{21}.

The mode equation is obtained by setting the determinant of the system eqns. 2.365–2.366 to zero. Once it has been solved the ratio I_2/I_1 is given by any one of eqns. 2.365–2.366. This ratio thus takes a well-defined value for each guided mode. For instance, in the case of a bifilar line or of a long induction loop, there are two modes having a transmission line character. These modes, which are designated the monofilar and bifilar modes, have I_2/I_1 ratios close to $+1$ and -1, respectively.

A complicated structure like the bicoaxial leaky line developed by Martin (1976) and consisting of two leaky coaxial cables will support four modes having different values of the ratio I_2/I_1. For two of them, the ratio I_2/I_1 is close to $+1$, the cables both carrying either the coaxial mode or the monofilar mode in phase. For the other two modes, the ratio is close to -1 and the cable still carry the same modes but with opposite phases.

The excitation of the modes and some approximations which can be made in the mode equation or in the expressions of the fields will be discussed later. Before doing this, it is useful to present some available numerical results.

★ 2.8.3 Results for a monofilar wire conductor

Some values of the specific attenuation were obtained by Fontaine *et al.* (1974) at a frequency of 1 MHz and for a ground conductivity $\sigma_1 = 10^{-2}$ S m^{-1}. The calculations were made for tunnel radii ranging from 1 to 10 m, and the ratio ρ_0/a was varied between 0 and 0.9. The authors used a quasi-static approximation which will be developed later in this book. Their results are shown in Fig. 2.20. For the values of frequency and conductivity used here, the skin depth in the ground is about 5 m. When the tunnel radius is smaller than this value, the specific attenuation is not greatly affected by ρ_0. This is normal, since the return current in the ground is distributed in a large area all around the tunnel, in a manner which does not depend much on the wire position. This does not remain true for larger tunnel radii.

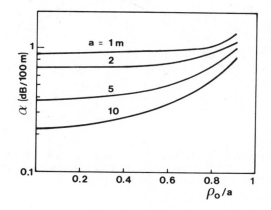

Fig. 2.20   *Specific attenuation of the monofilar mode for a bare monofilar wire in a circular tunnel at 1 MHz. The conductivity is $\sigma_1 = 10^{-2}$ S m^{-1}, the other parameters are unspecified*

Wait and Hill (1974a) have calculated the specific attenuation of the monofilar mode for the quite large frequency range 1 MHz – 1 GHz. The other parameters are given in the caption of Fig. 2.21 which shows the results. The effect of the wall proximity as described above is well confirmed at low frequencies. It increases with frequency because the skin depth in the ground decreases. The decrease of the attenuation with frequency above 100 MHz may be explained as follows. If the monofilar wire conductor were drawn in free space, it would support a Sommer-feld-type surface wave since the wire conductivity is finite (Stratton, 1941). As the frequency increases, the effective radius of this wave decreases and becomes progressively smaller than the distance from the conductor to the wall. Consequently the attenuation decreases because of a smaller loss in the wall. At frequencies of the order of 1 GHz or more, this loss becomes negligible and the attenuation curve follows that of a Sommerfeld wave, thereby increasing again with frequency. This effect, which is of course enhanced if the conductor has a dielectric coating. (Wait and Hill, 1976b), is of rather academic interest. Indeed, one is generally interested in modes which can provide radio communications to users located anywhere in the cross-section of the tunnel. Furthermore, surface waveguides are extremely sensitive to moisture and scattering loss due to obstacles located within the effective radius. Note that the dielectric coating has a negligible influence at low frequencies, as expected.

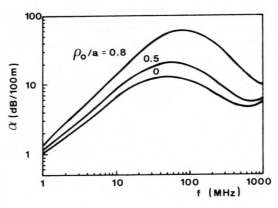

Fig. 2.21   *Specific attenuation of the monofilar mode for a single conducting wire: $a = 2\,m$, $\kappa_1 = 10$, $\sigma_1 = 10^{-2}\,S\,m^{-1}$, $c = 1\,cm$. Wire conductivity $\sigma_w = 10^5\,S\,m^{-1}$.*

Results were also obtained for the rather low frequencies 20 kHz to 1 MHz and for $\rho_0 = 0$ (Wait and Hill, 1975c). This case finds practical applications in some communications systems used in mine hoisting shafts where the shaft cable is used as a conductor. One conclusion of these calculations is that the specific attenuation depends only weakly on the ground conductivity. This is not surprising. Indeed at these low frequencies the skin depth in the ground is inversely proportional to the square root of the conductivity. Consequently the return current in the ground is distributed in an area which is roughly inversely proportional to σ_1 and the resistance of the return path is only weakly dependent on σ_1.

The influence of the distance from the conductor to the tunnel wall was investigated by Wait and Hill (1976a). The relative independence of the attenuation on this distance and on the ground conductivity (Fig. 2.22) at low frequencies is confirmed. Furthermore, the attenuation is proportional to frequency. This can easily be explained since the skin depth is proportional to $f^{-1/2}$, the effective area of the return current to f^{-1} and the resistance of the return path to f. These properties do not remain valid above 1 MHz because the skin depth in the ground is no longer large compared with the tunnel radius.

Further interesting data on the monofilar mode will be found in the next section.

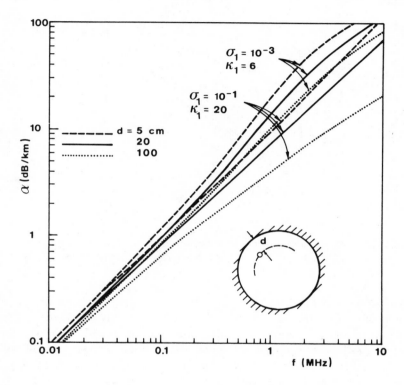

Fig. 2.22 *Influence of the distance from the conductor to the tunnel wall on the monofilar mode attenuation*

★ *2.8.4 Results for bifilar line*

In the context of this section, the term bifilar line will cover two different situations. The first occurs when the spacing between the two wires is very small compared with the distance from this line to the tunnel wall and is relevant to the use of a ribbon feeder. Exact calculations (Wait, 1975a, b) have shown that the bifilar mode does not differ significantly from the balanced mode of the same line drawn into free space, as long as the distance from the line to the wall remains large compared with the interwire spacing. This statement applies to fields (Hill and Wait, 1974a) as well as to the propagation constant. This conclusion is rather obvious but it had not been accepted by everyone at the time these papers were published, although it had been amply demonstrated by Deryck's experiments (1973).

The monofilar mode of the ribbon feeder does not differ significantly from that of a monofilar wire conductor located at the same place in the tunnel, as expected. To demonstrate this, the reader is invited to compare the attenuation curves of Fig. 2.23 (Wait, 1975b) for the ribbon feeder to those of Fig. 2.21. The difference is negligible.

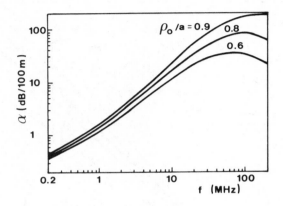

Fig. 2.23   *Specific attenuation of the monofilar mode of a ribbon feeder: $a = 2\,m$, $\kappa_1 = 10$, $\sigma_1 = 10^{-2}\,S\,m^{-1}$, $c_1 = c_2 = 1\,mm$, $\rho_{12} = 2\,cm$, wire conductivity $10^6\,S\,m^{-1}$*

The paper by Hill and Wait (1974a) is very interesting since it contains, to the author's knowledge, the only available data on the field distribution in the tunnel space. The results are given for $f = 20\,\text{MHz}$, $a = 2\,\text{m}$, $\kappa_1 = 10$, $\sigma_1 = 10^{-2}\,\text{S m}^{-1}$, which yields $\kappa_1' = 10 - j9$ and a skin depth in the rock of about $1\cdot8$ m. They provide a remarkable confirmation of the qualitative picture given by Fig. 1.5, although the skin depth is not small compared with the tunnel radius. Unjustified extrapolation of this observation should however be avoided.

The case where the interwire spacing is no longer small is very interesting since it is representative of the influence of axial conductors which may exist in a tunnel, in addition to an intentional monofilar wire conductor, and also of the long induction loop.

A very interesting exercise carried out by Fontaine *et al.* (1974) is illustrated in Fig. 2.24. The tunnel radius is $a = 4\,\text{m}$. One of the two wires is kept fixed at $\rho_1 = 3\cdot7$ m, while the other is allowed to rotate around it at a fixed distance of $0\cdot5$ m. The specific attenuation was calculated for the two modes and is showed on the figure. Also illustrated is the absolute value of the ratio I_2/I_1 for the two modes as a function of the position of the second wire. Curves giving $|V_2/V_1|$ have also been drawn; wire voltages are indeed defined in the quasi-static approach used by the authors. It appears that the ratio I_2/I_1 may differ significantly from ± 1.

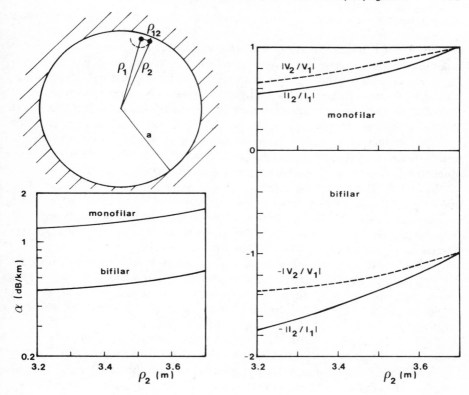

Fig. 2.24   *Characteristics of the monofilar and bifilar modes for two wires in a circular tunnel. The parameters are f = 1 MHz, $\sigma_1 = 10^{-2}$ S m^{-1}, a = 4 m, $\rho_1 = 3\cdot7$ m. Wire 1 is fixed and wire 2 is rotated around it at a fixed distance $\rho_{12} = 0\cdot5$ m*

These results have important implications for the use of a monofilar wire conductor in tunnels containing additional axial conductors such as water pipes, electric power cables and so on. These conductors participate in the propagation mechanism. They carry a fraction of the ongoing current in the monofilar mode and are part of the return path in the bifilar mode. In practice, any source will excite both modes, but after some distance the bifilar mode[*] will be dominant because of its lower attenuation. It is clear that the monofilar wire conductor should be positioned as far as possible from the other conductors if it is intended to develop intense fields in the tunnel space. Ideally, the monofilar wire and the inadvertent conductors should be located on opposite walls. In these conditions,

---

[*] Actually this mode is the intentional monofilar mode of the structure when we consider that the inadvertent conductors are an integral part of the return path. Conductors like water pipes are in general grounded at regular intervals by their fixtures.

the inadvertent conductors may have a beneficial influence on the field distribution and on the attenuation. Indeed, they reduce the resistance of the return path.

This effect was nicely illustrated by Wait and Hill (1977a) for radio frequency transmission via a trolley wire with a rail return. The theory developed in Section 2.8.2 was extended to include the case where one of the conductors is embedded in the rock medium, without being insulated. The numerical analysis shows that, at least for the investigated frequencies (less than 800 kHz), the rail drains a large part of the return current when the rock conductivity is low. In these conditions the specific attenuation is roughly independent of the rock conductivity and significantly lower than for the monofilar mode without a rail return. For rock conductivities above 1 S m^{-1}, by contrast, the rail return has negligible influence.

We will now comment on another series of results. These were obtained for a half-circular tunnel above a perfectly conducting plane (Hill and Wait, 1976d) with a single monofilar wire. The ground plane was supposed to have a similar effect to rails resting on the tunnel floor. This is probably overidealised, but we are interested in the solution of this problem since it provides the exact solution for the bifilar mode with two wires located at the same distance from the wall, i.e. $\rho_1 = \rho_2$, in a circular tunnel. This situation may be representative of the long induction loop. The calculations have been made for an angular separation $\phi_1 - \phi_2 = 90°$ on Fig. 2.19, $\kappa_1 = 10$ and for a wire radius $c = 1\cdot5 \text{ cm}$. The wire conductivity is $\sigma_w = 5\cdot7 \times 10^7 \text{ S m}^{-1}$ unless otherwise specified.

Figure 2.25 is illustrative of the influence of the distance from the wires to the wall on the attenuation. Fig. 2.25a relates to a tunnel with a fixed radius $a = 2 \text{ m}$, $\sigma_1 = 10^{-3} \text{ S m}^{-1}$; the distance $(a - \rho_0)$ from the wires to the wall is varied from 80 to 20 cm. The dashed curve shows the influence of the wire conductivity, which has been reduced to $\sigma_w = 5\cdot7 \times 10^6 \text{ S m}^{-1}$, $a - \rho_0$ being kept equal to 20 cm. Fig. 2.25b shows the attenuation for a fixed distance $a - \rho_0 = 20 \text{ cm}$, with $\sigma_1 = 10^{-3} \text{ S m}^{-1}$, when the tunnel radius is varied. It is clear that the specific attenuation is sensitive to the distance $(a - \rho_0)$ rather than to the tunnel radius. This is just the contrary of the result obtained for the monofilar mode of a single conductor (Fig. 2.22). The explanation of this difference is as follows. We explained this behaviour for the single wire by the fact that the return path for the current in the rock medium is quite broad and depends hardly at all on the wire location, since the skin depth in the rock is very large at those frequencies. The bifilar mode of the symmetrical two-wire line does not involve a *net* current in the rock. There nevertheless exists a current density in the rock. Indeed we may consider that the two wires induce in the rock two monofilar-mode-like current densities but with opposite phases. These current densities compensate each other at large distances from the two wires. Consequently the resulting current density is only important in those regions of the rock which are close to the wires. They increase as the distance $a - \rho_0$ is reduced, thereby yielding an increased attenuation.

A consequence of this discussion is that the attenuation of a long induction loop will increase substantially if the two wires are not symmetrical with respect to the tunnel. It is known that maintaining the symmetry of a long induction loop is quite

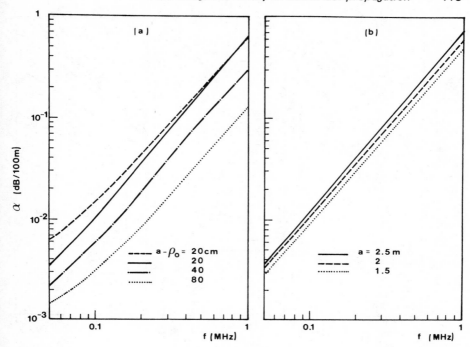

Fig. 2.25 *Influence of wire-to-wall distance ($a-\rho_0$) and of the tunnel radius a for a long induction loop*
(a) $a = 2$ m, $a-\rho_0$ variable. Dashed curve for a wire conductivity of $5 \cdot 7 \times 10^6$ S m^{-1}
(b) $a-\rho_0 = 0 \cdot 2$ m, a variable

difficult in a real environment. Other conductors present in the tunnel, rock inhomogeneities and so on may have a disastrous influence on the specific attenuation.

Figure 2.26 is taken from the same paper and illustrates the evolution of the specific attenuation of the bifilar mode as a function of the rock conductivity. The latter was varied from 10^{-4} to 10^4 S m^{-1}. As the evolution is not monotonic the results are presented in Fig. 2.26a and b. Figure 2.26c shows for comparison the attenuation for the monofilar mode of a single wire in the same conditions: $a = 2$ m, $a - \rho_0 = 20$ cm.

The evolution of the attenuation of the bifilar mode with rock conductivity is somewhat strange and needs some explanation. The field of this mode extends into the rock medium and, as the latter has a dielectric constant $\kappa_1 = 10$, the phase velocity of the wave lies somewhere between those of the air and of the rock medium. Actually, for $a - \rho_0 = 20$ cm and $\sigma_1 < 10^{-1}$ S m^{-1}, the phase constant β was found to be between $1 \cdot 2\,k_0$ and $1 \cdot 27\,k_0$. The wave is thus faster than plane waves in the external medium. For very low conductivities, e.g. $\sigma_1 = 10^{-4}$ S m^{-1},

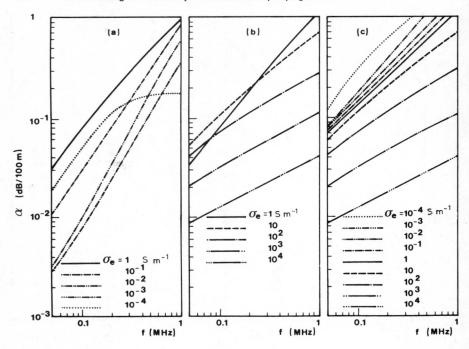

Fig. 2.26   *Dependence of the specific attenuation on rock conductivity (σ_e on the diagrams)*
a and *b* For the bifilar mode of a long induction loop
c For the monofilar mode of a single conductor

the rock medium acts as a dielectric in which the fast wave has a leaky character (Collin and Zucker, 1969). The relatively large attenuation is mainly due to radiation. This leaky behaviour decreases as σ_1 increases, since the rock progressively becomes a conductor. The attenuation reaches a minimum for σ_1 close to 10^{-2} S m^{-1}. No such effect is observed for the monofilar mode. Again, this is due to the fact that the return current and the fields of this mode extend over a very large area in the rock; the result is that the phase velocity of this mode is close to that of the rock medium and, consequently, the leaky character is not marked.

When the rock conductivity increases above 10^{-2} S m^{-1}, the attenuation of the bifilar modes increases, but it starts decreasing again above 1 S m^{-1}. This is due to the complex evolution of the current density in the rock, with the mutual cancellation of the current densities induced by the two wires as explained above. For relatively high conductivities and frequencies, the skin depth in the rock is smaller than the tunnel radius; the two current densities do not overlap. The bifilar line then behaves as two independent monofilar wire conductors excited with opposite phases, as explained in Section 1.4.

The analysis may be extended to a half-circular tunnel with perfectly conducting floor containing two non-identical wires (Hill and Wait, 1977b).

★ 2.8.5 Results for a leaky coaxial cable

The model used for calculating the specific external impedance of a leaky coaxial cable was developed in Section 2.5.6. This impedance is strongly dependent on the longitudinal propagation constant Γ and is reactive if the cable is lossless.

As a useful exercise, we will investigate the behaviour of a leaky coaxial cable drawn in free space. For greater simplicity, we will assume that the inner conductor has an infinite conductivity and that the shield is uncoated. The specific external impedance of the cable is then given by eqn. 2.189 which will be rewritten as

$$z_0 = j\omega m_t \frac{\beta^2 - \beta_{co}^2}{\beta^2 - \beta_\infty^2} \tag{2.372}$$

where $\beta_{co} = k_0 \sqrt{\kappa}$ is the phase constant of a perfectly shielded cable, κ being the dielectric constant of the cable insulation and

$$\beta_\infty = \beta_{co} \sqrt{1 + m_t/l_0} \tag{2.373}$$

In the external air medium, the transverse fields are given by

$$E_{z0} = -v^2 K_0 (v\rho) \tag{2.374}$$

$$H_{\phi 0} = -j\omega\epsilon_0 v K_0' (v\rho) \tag{2.375}$$

Using small-argument approximations for the modified Bessel function and writing $z_0 = E_{z0}/(2\pi c H_{\phi 0})$, where c is the cable radius, we obtain

$$-\ln \frac{C c \sqrt{\beta^2 - k_0^2}}{2} = \frac{2\pi k_0 \omega m_t}{\eta_0} \frac{\beta^2 - \beta_{co}^2}{(\beta^2 - k_0^2)(\beta^2 - \beta_\infty^2)} \tag{2.376}$$

The right-hand side of this equation is shown in Fig. 2.27 as a function of β/k_0 for a particular choice of parameters. The left-hand side is real for $\beta > k_0$ only. It is a relatively large positive quantity (larger than 5 on Fig. 2.27) and a slowly varying

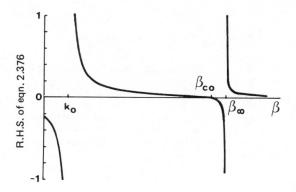

Fig. 2.27 *The right-hand side of eqn. 2.376 as a function of β. Calculations have been made for $f = 30\,MHz$, $c = 5\,mm$, $\kappa = 2·5$, $l_0 = 263·5\,nH\,m^{-1}$ (50 Ω cable) and $m_t = 20$ $nH\,m^{-1}$.*

function of β. Consequently, eqn. 2.376 has two roots β_m and β_c which are slightly higher than k_0 and β_∞, respectively. They correspond to the monofilar and coaxial modes of the structure in free space. Both waves are slow compared with the velocity of light in free space and consequently they have a surface wave character. It is remarkable that the coaxial mode is even slower than that of a perfectly shielded cable; this is due to the reactive loading of the leaky shield. Actually the term leaky is misleading since it is generally used for fast wave structures (Collin and Zucker, 1969). It is important to keep these properties in mind (cf. Table 1.2) when we try to interprete physically the behaviour of leaky coaxial cables used in tunnels.

Some results of calculations for a leaky coaxial cable drawn longitudinally inside a tunnel have been published[*] (Wait and Hill, 1975b, c; Hill and Wait, 1978b). As it does not complicate the numerical work very much, the assumption has been made that the cable shield is covered by a dielectric jacket. The latter may be covered itself by a lossy film which may represent moisture and dust accumulated on the cable. Fig. 2.28 shows the specific attenuation of the monofilar mode for the conditions specified in the caption. These conditions are identical to those of Fig. 2.21. Comparison of these results shows that the attenuation of the monofilar mode is, as expected, the same for a monofilar wire conductor and for a leaky coaxial cable. The slight difference which appears above 100 MHz is probably due to the cable jacket.

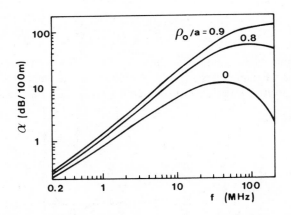

Fig. 2.28 *Specific attenuation of the monofilar mode of a leaky coaxial cable for $a = 2\,m$, $\kappa_1 = 10$, $\sigma_1 = 10^{-2}$ S m^{-1}. Inner conductor: radius $1 \cdot 5\,mm$, $\sigma = 5 \cdot 7 \times 10^7$ S m^{-1}. Shield: radius $10\,mm$, $m_t = 40$ nH m^{-1}. Plastic jacket: radius $11 \cdot 5\,mm$, $\kappa_j = 3$. No lossy film*

[*] We have some doubts about the validity of the attenuation curves contained in the third of these papers.

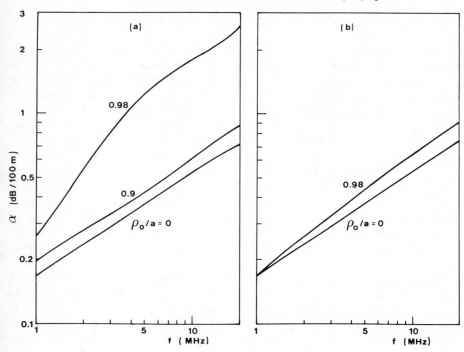

Fig. 2.29   *Effect of wall proximity on the specific attenuation of the coaxial mode for a leaky feeder*
a For $m_t = 40$ nH m^{-1}
b For $m_t = 10$ nH m^{-1}
The other parameters are $a = 2$ m, $\sigma_1 = 10^{-3}$ S m^{-1}, $\kappa_1 = 10$. Internal conductor: copper, 1·5 mm radius. Insulation: $\kappa = 2·5$. Outer conductor radius 10 mm. Jacket thickness 1·5 mm and dielectric constant 3

We have selected the curves of Fig. 2.29a and b for the attenuation of the coaxial mode. For a cable with a very high specific transfer inductance ($m_t = 40$ nH m^{-1}) the attenuation is very sensitive to close proximity of the wall. For a moderate transfer inductance (10 nH m^{-1}) this effect is rather limited. It is insignificant for well-shielded cables (2 nH m^{-1}). In any case, the attenuation is indistinguishable from its free-space value when the cable lies along the tunnel axis. It is then essentially due to the finite resistivity of the inner conductor, since the resistance of the shield was neglected in the calculations. We have some doubts about the quantitative values of attenuation for $\rho_0/a = 0·98$ since the thin-cable approximation is somewhat questionable under these conditions. The result is however qualitatively satisfactory.

2.9 Guided modes of wires and leaky feeders in a rectangular tunnel

As we observed in Section 2.7.3, analytical investigation of wave propagation in a rectangular tunnel with four imperfectly conducting walls is exceedingly complicated. This obviously remains true when the tunnel contains axial conductors or leaky feeders. Actually this problem is akin to that of dipole excitation in an empty tunnel. Indeed, the problem of an axial electric dipole of moment *Ids* located at a point with coordinates $(x_0, y_0, 0)$ may be solved by the methods used in Sections 2.6.2 and 2.7.3. Using a Fourier transform on the z-coordinate, the applied current density is decomposed into a summation of terms of the type *Ids* $\delta(x - x_0) \delta(y - y_0) \exp(-jhz)$. These are identical to the current of a guided mode with propagation constant $\Gamma = -jh$ along a thin cable located at (x_0, y_0) in the tunnel cross-section. The only difference between the two problems will arise at the final stage of the analytical study, when the boundary conditions on either the dipole or the cable are expressed. Thus we conclude that the existance of axial cables in a rectangular tunnel does not involve additional difficulties in the analytical investigations.

For this reason and also owing to lack of space, we will not go into the details of theoretical analyses, but rather refer the reader to the relevant literature. As mentioned previously, simplifying assumptions are required to make the treatment of the rectangular structure possible. The theoretical analysis for a tunnel with perfectly conducting side walls, containing a single longitudinal wire, was developed by Mahmoud and Wait (1974). Detailed numerical results at 1 GHz are available in Mahmoud (1974b) and include the waveguide modes, It appears that some waveguide modes are significantly modified by the presence of the wire, but only when the wire is located close to the horizontal wall, while other waveguide modes are only weakly modified.

The theory was extended to two-wire lines (Mahmoud 1974a; Wait 1975b) and to leaky coaxial cables (Mahmoud and Wait, 1976). The general conclusions which may be drawn from the numerical results do not differ from those we arrived at for the circular tunnel.

2.10 Some excitation problems

In previous sections we have been mainly concerned with the properties of guided modes in tunnels which may contain axial conductors or leaky cables. Some attention has already been devoted to the excitation of modes by dipoles for the cases of the low-conductivity layer (Sections 2.6.4, 2.6.5 and 2.7.2) and of the empty rectangular tunnel (Section 2.7.3). In the following sections we will concentrate exclusively on excitation problems and related areas of interest, in order finally to be able to predict the transmission loss of subsurface radio links.

2.10.1 Gap excitation of monofilar wire

Simple ways of launching guided waves along an axial conductor strung in a tunnel consist of connecting a signal generator either in series within the conductor or in parallel between the conductor and the ground. The parallel excitation is sometimes used in practice, but it is seriously impaired by the difficulty of providing a good ground connection. The same difficulty would need to be included in the modelling of this connection for a relevant theoretical analysis to result. Hence we will concentrate on the series excitation which avoids this practical and theoretical problem. Furthermore, this method is the basic ingredient of mode conversion techniques.

We consider a voltage generator inserted in a short gap made in a monofilar wire located in a circular tunnel. The geometrical and electrical parameters have been defined in Fig. 2.18 and we will use the same notation as in Section 2.8.1. The monofilar wire is interrupted in the interval $-d < z < d$ and a voltage generator is inserted there. The electric field E_z in this gap is approximately equal to $|-V_0/(2d)|$, where V_0 is the gap voltage, taken as positive when the point $z = d$ is at a higher potential than $z = -d$. We will assume that E_z and the conductor current $I(z)$ are related by

$$E_z = -[V_0/(2d)] p_d(z) + z_0 I(z) \tag{2.377}$$

where

$$P_d(z) = \begin{cases} 1 & \text{for} \quad |z| < d \\ 0 & \text{for} \quad |z| > d \end{cases} \tag{2.378}$$

and where E_z is evaluated at $\rho_d = c$ for all z. Eqn. 2.377 provides a boundary condition on the external surface of the wire. In this equation the expression $z_0 I(z)$ is symbolic and expresses that, outside the gap, the thin-wire approximation applies to all modal components of $I(z)$. Inside the gap, eqn. 2.377 is only approximate since the second term should not exist; actually this term yields only a second-order effect when d is small, but it avoids running into a difficult three-part Wiener–Hopf problem.

We will use Fourier transform pairs of type[*]

$$\mathscr{U}(\rho, \phi, h) = \int_{-\infty}^{\infty} U(\rho, \phi, z) \exp(-jhz) \, dz$$

$$U(\rho, \phi, z) = (2\pi)^{-1} \int_{-\infty}^{\infty} \mathscr{U}(\rho, \phi, h) \exp(jhz) \, dh \tag{2.379}$$

Then eqns. 2.341 to 2.358 are valid for the Fourier transforms $\mathscr{U}$, $\mathscr{I}$, $\mathscr{E}_z$, $\mathscr{J}$ of U, V, E_z, I, provided Γ is replaced by $(-jh)$. The difference with Section 2.8.1 starts when we express the boundary condition on the external surface of the wire. We must now use the Fourier transform of eqn. 2.377, i.e.

---

[*] The change in notation with respect to Hill and Wait (1974b) is justified by better compatibility with other sections of this book.

$$\mathscr{E}_z = -V_0 \frac{\sin(hd)}{hd} + z_0(-jh)\mathscr{I} \tag{2.380}$$

This yields

$$(2\pi j\omega\epsilon_0)^{-1}\mathscr{I} = \frac{\mathscr{V}_0}{D(h)}\frac{\sin(hd)}{hd} \tag{2.381}$$

where

$$D(h) = v^2\left\{K_0(vc) - \sum_m R_m(-jh)\frac{K_m(va)}{I_m(va)}I_m(v\rho_0)I_m[v(\rho_0+c)]\right\}$$

$$+ 2\pi j\omega\epsilon_0\,z_0(-jh) \tag{2.382}$$

Hence expression 2.381 for the current may be used in all the equations of Section 2.8.1.

It now remains to evaluate the inverse Fourier transforms of the potentials 2.345 and 2.346. This may be done by ordinary function-theoretic methods. The singularities of $\mathscr{U}(h)$ and $\mathscr{V}(h)$ include branch points at $h = \pm jk_0$, $h = \pm jk_1$ which correspond to unguided radiation. The poles of $\mathscr{U}(h)$ and $\mathscr{V}(h)$ are the roots of $D(h) = 0$; comparing with eqn. 2.359, it is seen that these poles are located at $h = \pm h_n$, where $h_n = -j\Gamma_n$, and Γ_n is the propagation constant of the nth guided mode. The poles denoted by h_n are located in the fourth quadrant. We will restrict our attention to the guided modes. Using the residue method requires the closing of the integration path by half-circles of infinite radius, either in the lower half-plane for $z < 0$ or in the upper half-plane for $z > 0$. Actually, because of the factor $\sin(hd)$ in eqn. 2.381, closure is permitted only for $|z| > d$, i.e. outside the gap. This yields the guided part of the Hertz potentials[*]

$$U = -j\,V_0\sum_n\frac{\sin(h_nd)}{h_nd}\frac{\exp(-jh_n|z|)}{D'(h_n)}$$

$$\sum_m\exp[-jm(\phi-\phi_0)]I_m(v_n\rho_0)\left[K_m(v_n\rho)-R_m(h_n)\frac{K_m(v_na)}{I_m(v_na)}I_m(v_n\rho)\right] \tag{2.383}$$

$$V = -j\,V_0\,\mathrm{sgn}(z)\sum_n\frac{\sin(h_nd)}{h_nd}\frac{\exp(-jh_n|z|)}{D'(h_n)}$$

$$\sum_m\exp[-jm(\phi-\phi_0)]\Delta_m(h_n)I_m(v_n\rho_0)I_m(v_n\rho) \tag{2.384}$$

The fields may be obtained by the formulas of Appendix B. The current along the wire is given by

---

[*] We have a minus sign in eqns. 2.383 to 2.386 which does not exist in the corresponding equations of Hill and Wait (1974b). This is not an error since our function $D(h)$ is actually $D(-h)$ in this paper.

$$I(z) = \frac{-2\pi k_0 \ V_0}{\eta_0} \sum_n \frac{\sin(h_n d)}{h_n d} \ \frac{\exp(-jh_n \ |z|)}{D'(h_n)}.$$

(2.385)

Hence it is possible to evaluate Poynting's vector and to calculate the power lost into the walls and into the wire by integration on these surfaces. Power is indeed lost into the wire if the latter has a finite conductivity. The ϕ and z integrations can easily be performed analytically since only exponentials are involved (Hill and Wait, 1974b). The total power delivered by the gap source is obviously the sum of these lost powers.

The total power may be obtained by a simpler method. If the gap is very short, the source may be considered as a lumped device and the active power is given by

$$P = V_0 \ \text{Re} \ [I(d+0)] = \frac{-2\pi k_0 \ V_0^2}{\eta_0} \sum_n \text{Re} \left[\frac{1}{D'(h_n)} \right]$$

(2.386)

when d tends to zero.

It is also possible to develop formulas similar to eqns. 2.383–2.384 but valid inside the gap. For this purpose we replace $\sin(hd)$ by $[\exp(jhd) - \exp(-jhd)]/(2j)$ in the Fourier transform and split the latter into two terms accordingly. For one of these terms the contour may be closed in the upper half-plane and for the other one in the lower half-plane. It is then possible to calculate the power delivered by the source as the flux of Poynting's vector across the gap. When d tends to zero we retrieve eqn. 2.386. This procedure will be detailed in Section 2.10.4 for a similar problem. Some relevant comments will be made there.

Actually, eqns. 2.384 to 2.386 are not exact since they include only the contribution of the guided modes. Furthermore, when the frequency is low enough, the waveguide modes are cutoff and attention may be restricted to the monofilar mode. The results shown in Figs. 2.30 and 2.31 have been obtained in that way (Hill and Wait, 1974b). In particular, Fig. 2.31 shows that about 20% of the power delivered by the source is dissipated by the wire resistance, the remaining 80% being lost into the tunnel wall.

It is very instructive to compare these results with those of a very rough quasi-static model. Let us assume that the wire and the tunnel wall are perfectly conducting. The characteristic impedance of the transmission line made by these conductors is given by

$$Z_0 = \frac{\eta_0}{2\pi} \ \ln \frac{a - \rho_0'}{c + \rho_0 - \rho_0'}$$

(2.387)

where

$$\rho_0' = \frac{a^2 + \rho_0^2 - c^2}{2\rho_0} - \sqrt{\left(\frac{a^2 + \rho_0^2 - c^2}{2\rho_0} \right)^2 - a^2}$$

(2.388)

In this simplified model, the gap 'sees' twice the impedance Z_0 in series and the power delivered by the source is

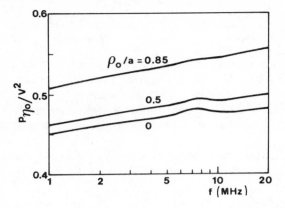

Fig. 2.30   *Normalised total power supplied to the monofilar mode as a function of frequency for three conductor locations. The other parameters are $a = 2m$, $\kappa_1 = 10$, $\sigma_1 = 10^{-2}\ S\ m^{-1}$, $c = 0.35\ cm$ and $\sigma_w = 10^5\ S\ m^{-1}$*

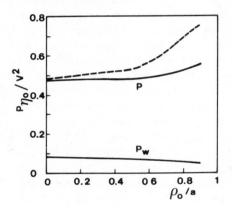

Fig. 2.31   *Normalised total power P supplied to the monofilar mode as a function of conductor location. The frequency is 7 MHz. Also shown is the power lost in the conductor P_w and the prediction P_0 of P given by eqn. 2.389*

$$P_0 = \frac{V^2}{2Z_0} \tag{2.389}$$

For comparison purposes the quantity $P_0\ \eta_0/V^2 = \eta_0/(2Z_0)$ is shown by the broken line in Fig. 2.31. It is seen that very rough approaches, such as are frequently used in engineering work, may give relatively accurate results.

2.10.2 Radiation of dipoles in tunnels

(a) Analytical solution

In this section, we will analyse the radiation of electric and magnetic Hertz dipoles located in a tunnel that may or not contain one or several axial cables. By cables we mean either conductors or leaky coaxial cables. We will restrict our attention to circular tunnels. The reader interested in rectangular tunnels is referred to Mahmoud and Wait (1974a, 1976) and Mahmoud (1974b) and to Section 2.7.3 of the present book.

The coupling of dipoles to the modes propagating inside a circular tunnel was partly investigated by Hill and Wait for several configurations of axial cables: single wire with dielectric coating (Wait and Hill, 1977b), two-wire line (Hill and Wait, 1976b, 1979). Two papers are devoted to the impedance of dipoles in a circular tunnel without (Wait and Hill, 1976c) and with (Hill and Wait, 1978a) an axial conductor. In some of these analyses, the excitation problem is solved by invoking the reciprocity theorem. This approach has already been used in Section 2.6.3 of this book and will be applied later to engineering models. A more direct attack is used in some of the papers mentioned above. The last two articles are of this type and have been used widely in the writing of the present section.

We consider a general statement of the problem allowing for the presence of any type of dipole (electric or magnetic, transverse or axial), with or without axial cables. The final results will be detailed for special cases. Cables, if any, will initially be regarded as axial current line sources with as yet unknown current distributions. The equations will be written for cables located at $\rho = \rho_c$, $\phi = \phi_c$ with current distribution $I_c(z)$. Electric dipoles with moment Ids or magnetic dipoles with moment $I_m ds$, if any, are located at a point with cylindrical coordinates $(\rho_0, \phi_0, 0)$. We will initially assume that we have simultaneously an axial electric dipole and an axial magnetic dipole at this point. The axial sources, i.e. dipoles and cables, may be dropped at some stage of the analysis. Transverse dipoles may exist but do not need to be defined yet.

We will solve the problem using axial electric and magnetic Hertz potentials $U(\rho, \phi, z)$ and $V(\rho, \phi, z)$. Since Maxwell's equations are linear, we may write

$$U = U^{(te)} + U^{(tm)} + U^{(ae)} + U^{(am)} + U^{(c)} \tag{2.390}$$

$$V = V^{(te)} + V^{(tm)} + V^{(ae)} + V^{(am)} + V^{(c)} \tag{2.391}$$

where the superscripts refer to the parts of the solution which result from transverse electric, transverse magnetic, axial electric and axial magnetic dipoles and from the cables, respectively.

We first consider the axial sources, i.e. the last three terms of eqns. 2.390 and 2.391. For this type of source, the use of axial electric and magnetic Hertz potentials has been justified in Section 2.2.2. These potentials satisfy the non-homogeneous Helmholtz equations B.1 and B.2 of Appendix B. The solution of the problem will be greatly facilitated by the use of representations of the type

$$F(\rho, \phi, z) = O_{\phi_s} F_m(\rho, z) \tag{2.392}$$

where O_{ϕ_s} is the operator

$$O_{\phi_s} F_m = \sum_{m=-\infty}^{\infty} F_m \exp\left[-jm(\phi - \phi_s)\right] \tag{2.393}$$

where ϕ_s is the azimuth of the source, and where $F_m(\rho, z)$ is given by the inverse Fourier transform

$$F_m(\rho, z) = \frac{1}{2\pi} \int_{-\infty}^{\infty} \mathscr{F}_m(\rho, h) e^{jhz} \, dz \tag{2.394}$$

For instance $U^{(ae)}$ and $V^{(ae)}$ satisfy the equations

$$(\nabla^2 + k^2) U^{(ae)} = (j\omega\epsilon_0)^{-1} Ids \frac{\delta(\rho - \rho_0)\delta(\rho - \rho_0)\delta(z)}{\rho} \tag{2.395}$$

$$(\nabla^2 + k^2) V^{(ae)} = 0 \tag{2.396}$$

for $0 < z < a$. Taking into account the Fourier series

$$\delta(\phi - \phi_0) = (2\pi)^{-1} \sum_{m=-\infty}^{\infty} \exp\left[-jm(\phi - \phi_0)\right] \tag{2.397}$$

we have

$$\left(\frac{\partial^2}{\partial\rho^2} + \frac{1}{\rho}\frac{\partial}{\partial\rho} + k_0^2 - h^2 - \frac{m^2}{\rho^2}\right) \begin{bmatrix} \mathscr{U}_m^{(ae)} \\ \mathscr{V}_m^{(ae)} \end{bmatrix} = \begin{bmatrix} (2\pi j\omega\epsilon_0)^{-1} Ids \dfrac{\delta(\phi - \phi_0)}{\rho_0} \\ 0 \end{bmatrix} \tag{2.398}$$

and similar but homogeneous equations for $\rho > a$. The general solution of equations of this type is the Green's function C.25 of Appendix C and is thus known. We have then to match the solutions at the boundary $\rho = a$.

Actually, this work has already been performed in Section (2.8.1) for axial electric sources and we may write

$$\begin{bmatrix} \mathscr{U}_m^{(ae)} \\ \mathscr{U}_m^{(c)} \end{bmatrix} = (2\pi j\omega\epsilon_0)^{-1} \begin{bmatrix} Ids \\ \mathscr{T}_c(h) \end{bmatrix} I_m(v\rho_<) \left[K_m(v\rho_>) - R_m \frac{K_m(va)}{I_m(va)} I_m(v\rho_>) \right] \tag{2.399}$$

$$\begin{bmatrix} \mathscr{V}_m^{(ae)} \\ \mathscr{V}_m^{(c)} \end{bmatrix} = (2\pi j\omega\epsilon_0)^{-1} \begin{bmatrix} Ids \\ \mathscr{T}_c(h) \end{bmatrix} \Delta_m I_m(v\rho_<) I_m(v\rho_>) \tag{2.400}$$

where $\mathscr{T}_c(h)$ is the Fourier transform of $I_c(z)$. The quantities v, R_m and Δ_m are defined in Section 2.8.1 and are to be used with Γ replaced by $(-jh)$.

A task similar to that of Section 2.8.1 has to be performed for the axial magnetic source. The result is of course the dual of the foregoing one and is written as

$$\mathcal{U}_m^{(am)} = (2\pi j\omega\mu_0)^{-1} I_m \, ds \, \Delta'_m \, I_m (v\rho_<) I_m (v\rho_>) \tag{2.401}$$

$$\mathcal{V}_m^{(am)} = (2\pi j\omega\mu_0)^{-1} I_m \, ds \, I_m (v\rho_>) \left[K_m (v\rho_<) - S_m \frac{K_m (va)}{I_m (va)} I_m (v\rho_>) \right] \tag{2.402}$$

where

$$S_m = \frac{\dfrac{jk_0}{v} \dfrac{K'_m (va)}{K_m (va)} + Z_m/\eta_0 + \delta'_m/\eta_0}{\dfrac{jk_0}{v} \dfrac{I'_m (va)}{I_m (va)} + Z_m/\eta_0 + \delta'_m/\eta_0} \tag{2.403}$$

$$\Delta'_m = \frac{- \eta_0 (S_m - 1) K_m (va)}{I_m (va)} \frac{\delta'_m/\eta_0}{\dfrac{jm\Gamma}{a} (v^{-2} - u^{-2})} \tag{2.404}$$

$$\delta'_m/\eta_0 = \frac{\left(\dfrac{jm\Gamma}{a} \right)^2 (v^{-2} - u^{-2})^2}{\dfrac{jk_0}{v} \dfrac{I'_m (va)}{I_m (va)} + \eta_0 \, Y_m} \tag{2.405}$$

where Z_m and Y_m have been defined in Section (2.8.1). It can be shown that we have

$$\eta_0 \, \Delta_m = - \Delta'_m/\eta_0 \tag{2.406}$$

We now turn to the consideration of the transverse dipoles. We feel that it is necessary to comment somewhat on the use of axial electric and magnetic potentials in this case, in conjunction with a Fourier transform on the z-coordinate and a Fourier series on ϕ. In Section 2.2.2 we have justified the use of these potentials in a sourceless region or in a region which contains only axial sources. We may thus use these potentials for the derivation of the fields of transverse dipoles provided we exclude an infinitesimal volume containing the source. In the remaining region U and V satisfy a homogeneous Helmholtz equation. It is clear that we may use an ordinary Fourier transform on the axial coordinate when $\rho \neq \rho_0, \phi \neq \phi_0$. For $\rho = \rho_0$ and $\phi = \phi_0$ however we should use one-sided Fourier transforms (Noble, 1958) to solve the Helmholtz equations. A summary of the properties of these transforms is given in Appendix D.

From the theorem of the derivative for one-sided Fourier transforms, it results that the one-sided transforms $\mathcal{U}_{m\mathrm{N}}$ and $\mathcal{U}_{m\mathrm{P}}$ satisfy the differential equations

$$\left(\frac{\partial^2}{\partial\rho^2} + \frac{1}{\rho}\frac{\partial}{\partial\rho} - v^2 - \frac{m^2}{\rho^2}\right)\mathscr{U}_{m\mathrm{N}} = \left[jh\,U_m + \frac{\partial U_m}{\partial z}\right]_{z=+0} \tag{2.407}$$

$$\left(\frac{\partial^2}{\partial\rho^2} + \frac{1}{\rho}\frac{\partial}{\partial\rho} - v^2 - \frac{m^2}{\rho^2}\right)\mathscr{U}_{m\mathrm{P}} = \left[-jh\,U_m - \frac{\partial U_m}{\partial z}\right]_{z=-0} \tag{2.408}$$

The behaviour of the potential as $|z|$ tends to infinity is such that the analytical regions of $\mathscr{U}_{m\mathrm{N}}$ and $\mathscr{U}_{m\mathrm{P}}$ overlap in a region of the h-plane containing the real axis. The ordinary two-sided Fourier transform thus exists and it satisfies

$$\left(\frac{\partial^2}{\partial\rho^2} + \frac{1}{\rho}\frac{\partial}{\partial\rho} - v^2 - \frac{m^2}{\rho^2}\right)\mathscr{U}_m = jh\,\Delta U_m + \Delta U'_m \tag{2.409}$$

where

$$\Delta U_m = U_m|_{z=+0} - U_m|_{z=-0} \tag{2.410}$$

$$\Delta U'_m = \frac{\partial U_m}{\partial z}\bigg|_{z=+0} - \frac{\partial U_m}{\partial z}\bigg|_{z=-0} \tag{2.411}$$

We have identical equations for the magnetic potential.

In order to determine ΔU_m, $\Delta U'_m$, ΔV_m and $\Delta V'_m$ we can use eqns. B.3, B.4, B.6 and B.7 to calculate the discontinuities $\Delta\,E_t$, $\Delta\,H_t$ of the transverse fields at $z = 0$. As these discontinuities are related to the dipole currents, the right-hand side of eqn. 2.409 is thus known. This equation can then be solved by a Green's function technique. This approach is lengthy and full of traps. Consequently we consider the use of one-sided Fourier transforms more as a justification for the use of axial potentials with a representation of the type eqns. 2.392–2.394 rather than as a solving technique.

An alternative and more efficient method was used by Hill and Wait in the paper mentioned previously. For an electric dipole of moment Ids located at $(\rho_0, \phi_0, 0)$ and parallel to the x-axis, the primary potential is x-oriented and given by

$$\pi'^{\mathrm{P}}_x = Ids/(4\pi j\omega\epsilon_0)R^{-1}\,\exp(-jk_0 R) \tag{2.412}$$

where R is the distance from the dipole to the observation point. On using the integral representation

$$\frac{\exp(-jk_0 R)}{R} = \frac{1}{\pi}\int_{-\infty}^{\infty} K_0\,(v\rho_\mathrm{d})\exp(jhz)\,dh \tag{2.413}$$

where $\rho_\mathrm{d} = [\rho^2 + \rho_0^2 - 2\rho\rho_0\,\cos(\phi - \phi_0)]^{1/2}$, and the addition theorem of eqn. C.23

$$K_0\,(v\rho_\mathrm{d}) = \sum_{m=-\infty}^{\infty} I_m\,(v\rho_<)K_m\,(v\rho_>)\exp\,[-jm(\phi - \phi_0)] \tag{2.414}$$

a representation of the type of eqns. 2.392–2.394 is obtained for the primary fields. The primary axial potentials $\mathscr{U}_{m,p}$ and $\mathscr{V}_{m,p}$ are obtained by writing

$$- v^2 \mathscr{U}_{m,p} = \mathscr{E}_{zm,p} \tag{2.415}$$

$$- v^2 \mathscr{V}_{m,p} = \mathscr{H}_{zm,p} \tag{2.416}$$

according to eqns. B.5, B.8. Note that these equations are valid for $\rho \neq \rho_0$, $\phi \neq \phi_0$ only, since we used the theorem of the derivative for one-sided Fourier transforms. This restriction does not influence the solution since the remainder of the derivation consists of adding reflected cylindrical harmonics and expressing the boundary conditions 2.350 and 2.351 at $\rho = a$. The derivation is quite similar to the one carried out in Section 2.8.1 and can be found in the referenced papers. We will restrict ourselves to giving the final result.

The general form which is arrived at is, for $\rho_0 < \rho < a$:

$$\mathscr{U}_m = A_m K_m (v\rho) + P_m I_m (v\rho) \tag{2.417}$$

$$\mathscr{V}_m = B_m K_m (v\rho) + Q_m I_m (v\rho) \tag{2.418}$$

where A_m, B_m, P_m and Q_m are functions of h related by

$$P_m = - R_m \frac{K_m (va)}{I_m (va)} A_m + \Delta'_m B_m \tag{2.419}$$

$$Q_m = \Delta_m A_m - S_m \frac{K_m (va)}{I_m (va)} B_m \tag{2.420}$$

Specifically, we have for the transverse (x-oriented) electric dipole

$$A_m^{(te)} = \frac{h \, Ids}{4\pi\omega\epsilon_0 v} \, [I_{m-1} (v\rho_0) e^{-j\phi_0} + I_{m+1} (v\rho_0) e^{j\phi_0}] \tag{2.421}$$

$$B_m^{(te)} = \frac{j \, Ids}{4\pi v} \, [I_{m-1} (v\rho_0) e^{-j\phi_0} - I_{m+1} (v\rho_0) e^{j\phi_0}] \tag{2.422}$$

and for the transverse magnetic dipole

$$A_m^{(tm)} = \frac{-j \, I_m ds}{4\pi v} \, [I_{m-1} (v\rho_0) e^{-j\phi_0} - I_{m+1} (v\rho_0) e^{j\phi_0}] \tag{2.423}$$

$$B_m^{(tm)} = \frac{-h \, I_m ds}{4\pi\omega\mu_0 v} \, [I_{m-1} (v\rho_0) e^{-j\phi_0} - I_{m-1} (v\rho_0) e^{j\phi_0}] \tag{2.424}$$

It is quite remarkable, but not surprising, that the five terms of the potentials 2.390–2.391 all have the general form of eqns. 2.417 to 2.420 in which the A_m and B_m terms give the primary potentials and the P_m and Q_m are due to reflection on the tunnel wall. Actually, R_m and S_m are the reflection factors of cylindrical harmonics on the tunnel wall for the electric and magnetic potentials, respectively. This reflection however includes a conversion from one potential into the other.

The conversion factors are Δ_m and Δ'_m [actually $\Delta_m K_m(va)/I_m(va)$ and $\Delta'_m K_m (va)/I_m(va)$]. They are related by eqn. 2.406 which express reciprocity.

Hence we may consider that the exact expressions of the five terms of the potentials in eqns. 2.390–2.391 are known, except that the cable currents, if any, are still unknown. They may be obtained in a straightforward manner by expressing the boundary condition on the cables, as was done in Sections 2.8.1 and 2.8.2, but now including the dipole terms. For instance, we have for a single cable

$$\mathscr{T}_c(h) = \frac{-4\pi^2 j\omega\epsilon_0 v}{D(h)} \sum_{m=-\infty}^{\infty} [A_m(h) K_m(v\rho_c) + P_m(h) I_m(v\rho_c)]$$

$$\exp[-jm(\phi_c - \phi_0)] \tag{2.425}$$

where

$$D(h) = 2\pi j\omega\epsilon_0 z_0(-jh) + v^2 \left\{ K_0(vc) - \sum_{n=-\infty}^{\infty} R_n(h) \frac{K_n(va)}{I_n(va)} I_n(v\rho_c) \right.$$

$$\left. I_n[v(\rho_c + c)] \right\} \tag{2.426}$$

and where we obviously have to use the relevant values for A_m and P_m.

We may now discuss some features and applications of the solution. In the first instance we consider the excitation of modes. In view of the application of the residue method, we need to identify the poles of the Fourier transforms. If no axial cable is present, the poles are those of R_m, S_m, Δ_m and Δ'_m. These quantities have a common denominator

$$D_m(h) = \left(\frac{hm}{a}\right)^2 (v^{-2} - u^{-2}) + \left[\frac{jk_0 I'_m(va)}{v I_m(va)} + \eta_0 Y_m\right] \left[\frac{jk_0 I_m(va)}{v I_m(va)} + Z_m/\eta_0\right]$$

$$\tag{2.427}$$

It is easy to show that $D_m(h) = 0$ is just the mode equation 2.310 of the empty tunnel, but now expressed with modified Bessel functions. Similarly, for a tunnel containing a single cable, $D(h)$ is identified as the mode equation 2.359. Calculating the guided modes by the residue method is a straightforward task which will not be detailed here.

★ *(b) Dipole impedance*
As mentioned at the beginning of this section, the foregoing solution has been used to obtain the radiation resistance of dipoles in a circular tunnel which may or not contain an axial cable. The approach which is used is known as the e.m.f. method. For an electric dipole, it consists of calculating the impedance as the limit

$$Z = \lim_{s \to 0} \frac{-E \cdot ds}{I} \tag{2.428}$$

where s is the separation of the observation point at which E is calculated, and of the dipole. Similarly, for a magnetic dipole consisting of an infinitesimal loop of area dA and unit normal n, carrying an electric current I, one has

$$Z = \lim_{s \to 0} \frac{j\omega\mu_0 H \cdot (n dA)}{I} \tag{2.429}$$

We recall that $I dA$ and $I_m ds$ are related by eqn. 2.6.

Since the potentials have been obtained above as the sum of a primary part and of a reflected part, the application of the e.m.f. method provides the result in the form

$$Z = Z_0 + \Delta Z$$

where Z_0 is the free-space value and ΔZ is the perturbation due to the presence of the tunnel and of the possible cables. The reader interested in the derivation of the results is directed to the referenced papers. We will present and comment on some of the numerical results. We have selected the results for the transverse dipoles since they are more important than those for the axial dipoles, at least from the point of view of applications to communication systems.

The calculations have been made for a tunnel radius of 2 m. The conductivity of the external medium σ_1 is taken to be 10^{-2} S m^{-1}, and the dielectric constant κ_1 is taken to be 10. It is also assumed that $\mu_1 = \mu_0$. The results presented here concern the input resistance R. The quantity shown is the ratio R/R_0, where R_0 is the free-space value. Actually this ratio is independent of the length or area of the antenna, as long as the latter remains small compared with the geometrical size of the structure. The results are thus not restricted to infinitesimal dipoles. Of course, at low frequencies, the dipoles will in practice be much smaller than the wavelength. In this case, we have for the electric dipole

$$R_0 = \frac{\eta_0 (k_0 ds)^2}{6\pi} \tag{2.430}$$

and for the magnetic dipole

$$R_0 = \frac{\eta_0 k_0^4 (dA)^2}{6\pi} \tag{2.431}$$

Figs. 2.32 and 2.33 concern the circumferential electric dipole (c.e.d.) and the radial magnetic dipole (r.m.d.) in an empty tunnel. These results apply also to the case where the tunnel contains a cable located at $\phi_c = \phi_0$ for any $\rho_c \neq \rho_0$, since there is obviously no coupling between the dipole and the cable in this situation.

Figs. 2.34 and 2.35 are for the cases of the radial electric dipole (r.e.d.) and of the circumferential magnetic dipole (c.m.d.). The continuous lines are for the empty tunnel and the dotted lines are for the case where the tunnel contains a single wire conductor located at $\phi_c = \phi_0$ and $\rho_c = 0.9$. For these locations, this orientation of the dipoles yields maximum coupling to the cable.

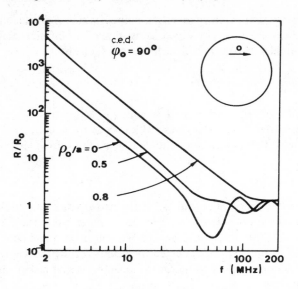

Fig. 2.32   *Relative input resistance of circumferencial electric dipole in empty tunnel or in tunnel containing a cable located at $\phi_c = \phi_0$ for any $\rho_c \neq \rho_0$*

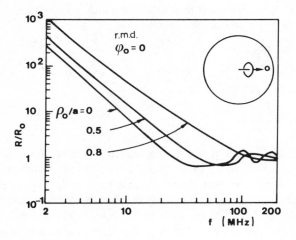

Fig. 2.33   *Relative input resistance of radial magnetic dipole in empty tunnel or in tunnel containing a cable located at $\phi_c = \phi_0$ for any $\rho_c = \rho_0$*

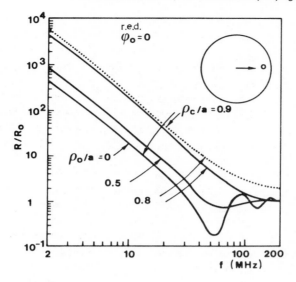

Fig. 2.34   *Relative input resistance of radial electric dipole. Continuous curves are for empty tunnel. Dotted curves for tunnel containing a cable located at $\phi_c = \phi_0$ and $\rho_c/a = 0.9$, for the cases where $\rho_0/a = 0$ and 0.8*

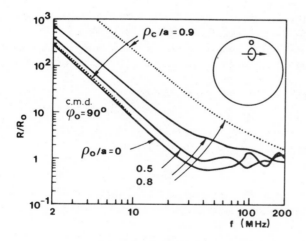

Fig. 2.35   *Relative input resistance of circumferencial magnetic dipole. Continuous curves are for empty tunnel. Dotted curves for tunnel containing a cable located at $\phi_c = \phi_0$ and $\rho_c/a = 0.9$, for the cases where $\rho_0/a = 0$ and 0.8*

Common to all cases is the fact that the ratio R/R_0 is significantly increased at low frequencies. At high frequencies however the radiation resistance oscillates about its free-space value. The increase at low frequencies is the result of losses in the tunnel wall, due to a strong interaction between the wall and the near-field region of the dipole. It is thus not surprising that this effect is more marked when the dipole is located close to the wall, nor that it ceases when the radius of the near-field region, roughly given by k_0^{-1}, becomes smaller than the distance from the dipole to the wall. The increase of R/R_0 at low frequencies is proportional to ω^{-2} and, because of the frequency behaviour of R_0 in eqns. 2.430–2.431, we may conclude that R is independent of the frequency for the electric dipole and proportional to ω^2 for the magnetic dipole.

In the cases where there is a coupling between the dipole and a cable, a further increase in R/R_0 is observed. This effect is barely noticeable when the dipole is located in the centre of the tunnel (with the cable at $\rho_c = 0\cdot9a$), but is significant for $\rho_0 = 0\cdot8a$. It is more marked for the magnetic dipole than for an electric dipole. This is an indication of the better efficiency of the magnetic dipole for the excitation of the monofilar mode, but also of a different distribution of the electric and magnetic fields of this mode in the tunnel cross-section. More on this subject will be found in Section 2.12.2. This kind of conclusion should however be used with much caution, since it may happen that the presence of a cable reduces the input resistance of a dipole, as was found for the axial electric dipole.

The reactance of the dipole is also modified by the tunnel. When expressed relative to R_0, i.e. in the form $|\Delta X|/R_0$, the reactance change is extremely high at low frequencies, being of the order of $|\Delta X|/R_0 \simeq 4.10^6/f^2$ where f is in MHz. However it should be observed that the Q-factor of the dipole in free space $Q_0 = |X_0|/R_0$ is itself extremely high. A lower bound for Q_0 is given by $\lambda^3/(\pi ds)^3$, where ds is the largest size of the dipole (Chu, 1948). For $ds = 0\cdot2$ m this yields $|X_0|/R_0 > 10^8/f^3$. Consequently the relative change $|\Delta X/X_0|$ is rather limited. It is more pronounced for the electric than for the magnetic dipole.

2.10.3 Coaxial cable with interrupted shield

★ *(a) General considerations*
The idea of creating energy exchanges between the interior of a non-leaky coaxial cable and the external space through an annular slot consisting of a local interruption of the outer conductor was originated by the author in 1969. The intention was to convey communication signals with a low specific attenuation inside the cable, and to release the minimum required energy in the tunnel space by a few slots located at discrete places along the cable, in order to provide communication to mobile receivers located therein. As the radiation process involved is reciprocal, two-way communications can be established by this method between a base station connected to the cable and mobile transceivers, and also between the latter. It had been observed in previous work (Delogne, 1969a, b) that any discontinuity along a multimode transmission line causes mode conversion and, if the structure is open,

radiation. Furthermore, it was known that the radiation tends to be concentrated in directions close to the tunnel axis for open structures which are able to guide waves with a velocity slightly lower than light in free space. It is thus not surprising that this idea has been at the source of the mode conversion techniques now used in subsurface radio communication systems.

The outer surface of an ordinary non-leaky coaxial cable, provided with a dielectric coating, appeared to be an adequate open structure since it can convey a surface wave known as a Goubau wave. Actually, if there were no coating, it would be necessary to take into account the finite conductivity of the external conductor to calculate the guided wave. The latter is called a Sommerfeld wave in this case.

In the initial analysis that was undertaken (Delogne and Liégeois, 1971), it was supposed that the cable was strung in free space.[*] In order to simplify the problem, it was also assumed that the external surface of the coaxial cable could be modelled by a surface impedance. Later on, the model was refined to include an exact description of the dielectric coating (Wait and Hill, 1975a, d e) as well as losses (Hill and Wait, 1977a), but this yields an insignificant change in the results. The coupling between the slot and a dipole located in its near-field region (Hill and Wait, 1975b) and the radiation of a slot in an imperfectly shielded cable (Wait, 1975c; Hill and Wait, 1976a) were also analysed. We will not go into this additional complication, since the radiation processes through the slot and the imperfect shield are only weakly coupled. A more useful extension was to consider the cable drawn in a circular tunnel (Hill and Wait, 1975a). In this case the surface wave guided by the external surface of the cable becomes the monofilar mode. This extension will not be explained here since it can easily be obtained by combining the results of the present section with those of Section 2.10.1.

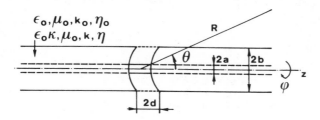

Fig. 2.36   *Annular slot in a coaxial cable*

Thus we keep the simple model used in the initial paper. The coaxial cable (Fig. 2.36) is located in free space along the z-axis. We denote by a the radius of the inner conductor, b the external radius of the cable, κ the dielectric constant of the insulation, $k_0 = \omega \sqrt{\epsilon_0 \mu_0}$, $k_1 = k_0 \sqrt{\kappa}$, $\eta_0 = \sqrt{\mu_0/\epsilon_0}$, $\eta = \eta_0/\sqrt{\kappa}$ and

[*]To be complete, we should mention a short paper by Beal and Dewar (1968), completed by a more extensive analysis (Dewar and Beal, 1970), in which annular slots were contemplated as the core of a surface-wave launcher. The further papers have shown that this principle could not work.

$$Y_0 = 2\pi/[\eta \ln(b/a)] \tag{2.432}$$

the characteristic admittance of the cable. The annular slot extends over the interval $-d < z < d$. The external surface of the cable is described by a reactive surface impedance

$$(E_z/H_\phi)_{\rho=b+0} = Z_s = j X_s \tag{2.433}$$

For a dielectric coated conductor, the dispersive surface impedance can be calculated using eqn. 2.182 when we assume that the cable radius is small compared with the transverse wave number. This yields

$$Z_s(\Gamma) = \frac{-\rho_2 (k_0^2 \kappa_2 + \Gamma^2)}{j\omega\epsilon_0 \kappa_2} \ln(\rho_2/\rho_1) \tag{2.434}$$

where ρ_1 and ρ_2 are the internal and external radii of the dielectric coating, κ_2 is the dielectric constant of the coating and Γ is the axial propagation constant.

The Goubau wave which can be guided by the external surface of the cable may be derived from an axial electric potential with rotational symmetry. Using the formulas of Appendix B, we obtain the fields of this mode in the form

$$E_\rho(\rho, z) = jh_1 u_1 B_0 K_1(u_1\rho) \exp(-jh_1 z) \tag{2.435}$$

$$E_z(\rho, z) = -u_1^2 B_0 K_0(u_1\rho) \exp(-jh_1 z) \tag{2.436}$$

$$H_\phi(\rho, z) = j\omega\epsilon_0 u_1 B_0 K_1(u_1\rho) \exp(-jh_1 z) \tag{2.437}$$

where jh_1 is the mode propagation constant, $u_1 = \sqrt{h_1^2 - k_0^2}$, and B_0 is an arbitrary amplitude constant. Actually the mode propagation constant is obtained by expressing the surface impedance condition at $\rho = \rho_2$. This yields

$$\frac{u_1 K_0(u_1\rho_2)}{K_1(u_1\rho_2)} = \frac{\rho_2 (k_0^2 \kappa_2 - h_1^2)}{\kappa_2} \ln(\rho_2/\rho_1) \tag{2.438}$$

Assuming $u_1\rho_2 \ll 1$ and using asymptotic approximations for the modified Bessel's functions, we obtain the approximate mode equation

$$-\ln \frac{1\cdot7810\, \rho_2 \sqrt{h_1^2 - k_0^2}}{2} = \frac{k_0^2 \kappa_2 - h_1^2}{\kappa_2 (h_1^2 - k_0^2)} \ln(\rho_2/\rho_1) \tag{2.439}$$

In this form there exists a powerful iteration method to solve this equation (Stratton, 1941): at the kth step of the iteration the approximation $h_1^{(k)}$ is substituted in the right-hand side of the equation and the left-hand side is solved to obtain $h_1^{(k+1)}$. Convergence is achieved in a few steps starting from an initial approximation such as $h_1^{(0)} = 1\cdot01 k_0$.

In the present case, ρ_2 stands for b, while ρ_1 is the radius of the external conductor. Fig. 2.37 is an abacus for the calculation of the ratio h_1/k as a function of the ratio $\rho_2/\rho_1 = b/\rho_1$ and of the parameter.

$$x = k_0 b \tag{2.440}$$

which is proportional to frequency.

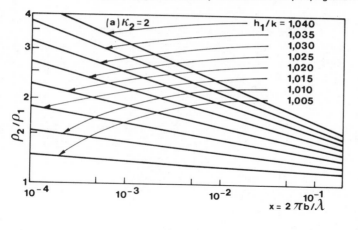

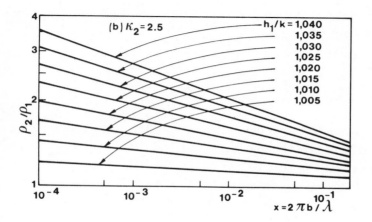

Fig. 2.37   *Abacus for the solution of the mode equation of a G-line*

The approximation which is used here consists of taking for $Z_s(jh)$ a constant value equal to $Z_s(jh_1)$. The problem which will be solved is to calculate the interesting quantities when the dominant TEM mode of the coaxial cable

$$H_i = \frac{e^{-jkz}}{\rho} u_\phi \tag{2.441}$$

$$E_i = \eta \frac{e^{-jkz}}{\rho} u_\rho \tag{2.442}$$

propagating in the positive direction of the z-axis, is incident on the annular slot. With the normalisation used, the incident field carries a power

$$P_i = 2\pi^2/Y_0 \tag{2.443}$$

Instead of using an axial electric potential, as is frequently done in this book, we will work directly with the azimuthal magnetic field $H(\rho, z)$ satisfying

$$\left(\frac{\partial^2}{\partial\rho^2} + \frac{1}{\rho}\frac{\partial}{\partial\rho} + \frac{\partial^2}{\partial z^2} + k^2 - \frac{1}{\rho^2}\right) H = 0 \tag{2.444}$$

where the correct wave number has to be used. The electric field may be obtained by

$$E_\rho = \frac{-1}{j\omega\epsilon}\frac{\partial H}{\partial z} \tag{2.445}$$

$$E_z = \frac{1}{j\omega\epsilon}\left(\frac{\partial}{\partial\rho} + \frac{1}{\rho}\right) H \tag{2.446}$$

One unknown of the problem is the axial electric field in the slot. This quantity will be written in the form

$$E_z(b, z) = -Vf(z) \tag{2.447}$$

where the unknown function $f(z)$ is such that

$$\int_{-d}^{d} f(z)\,dz = 1 \tag{2.448}$$

With this normalisation, the slot voltage is V, which is also unknown. On assuming that E_z is known we are able to calculate the fields inside and outside the cable.

(b) External fields
Using a Fourier transform representation of the type

$$H(\rho, z) = \frac{1}{2\pi}\int_{-\infty}^{\infty} \mathcal{H}(\rho, h)\,e^{jhz}\,dh \tag{2.449}$$

it is elementary to see that the solution of eqn. 2.444 is given by

$$\mathcal{H}(\rho, h) = A(h) H_1^{(2)}(\Lambda\rho) \tag{2.450}$$

where

$$\Lambda = \sqrt{k^2 - h^2}; \qquad -\pi < \arg\Lambda \leqslant 0 \tag{2.451}$$

Here $A(h)$ is an arbitrary function which can be determined by expressing the boundary condition on the external surface of the cable. Proceeding as in Section 2.10.1, we write this condition in the form

$$\frac{1}{j\omega\epsilon_0}\left(\frac{\partial}{\partial\rho} + \frac{1}{\rho}\right) H(\rho, z) = -Vf(z) + Z_s H(\rho, z) \tag{2.452}$$

After Fourier transformation we easily obtain $A(h)$ and finally

$$\mathscr{H}(\rho, h) = \frac{-j \, x \, V}{\eta_0} \frac{F(h) \, H_1^{(2)} \, (\Lambda\rho)}{\Lambda b \, H_0^{(2)} \, (\Lambda b) + C \, H_1^{(2)} \, (\Lambda b)} \tag{2.453}$$

where C is the positive constant

$$C = -j\omega\epsilon_0 \, b \, Z_s = x \, X_s/\eta_0 \tag{2.454}$$

and $F(h)$ is the Fourier transform of $f(z)$.

Actually this expression would be exact if we used the dispersive model eqn. 2.434 with Γ replaced by $(-jh)$.

It remains to calculate the inverse Fourier transform by

$$H(\rho, z) = \frac{-j \, x \, V}{2\pi \, \eta_0} \int_{-\infty}^{\infty} \frac{F(h) \, H_1^{(2)} \, (\Lambda\rho) \, e^{jhz}}{\Lambda b \, H_0^{(2)} \, (\Lambda b) + C \, H_1^{(2)} \, (\Lambda b)} \, dh \tag{2.455}$$

It is impossible to do this exactly since $F(h)$ is an unknown function. However we may observe that, according to the known properties of Fourier transforms, $F(h)$ is a slowly varying function of h if the gap is short. Furthermore the normalisation eqn. 2.448 implies that $F(0) = 1$. Consequently we may use the approximation $F(h) \simeq 1$ when $|hd| \ll 1$. For instance, if we assume that $f(z)$ is constant for $|z| < d$, we have

$$F(h) = \sin(hd)/(hd) \tag{2.456}$$

At a large distance from the slot, we can evaluate eqn. 2.455 by the method of steepest descent (Morse and Feshbach, 1953). The saddle-point is at $h = -k \cos \theta$, where θ is the colatitude in the system of spherical coordinates (R, θ, ϕ) defined in Fig. 2.36. Using the approximation $F(-k \cos \theta) \simeq 1$, we find the radiated field

$$H_r(R, \theta) \simeq \frac{-j \, x \, V}{\pi\eta_0} \frac{e^{-jkR}}{R} \frac{1}{x \sin \theta \, H_0^{(2)} \, (x \sin \theta) + C \, H_1^{(2)} \, (x \sin \theta)} \tag{2.457}$$

The external field however does not only include radiation, but also surface waves travelling away from the slot along the external surface of the cable. It may indeed be observed that the equation $\Delta(h) = 0$, where

$$\Delta(h) = \Lambda b \, H_0^{(2)} \, (\Lambda b) + C \, H_1^{(2)} \, (\Lambda b) \tag{2.458}$$

is just the mode equation 2.438. The integrand of eqn. 2.455 thus has poles at $h = \pm \, h_1$. The surface waves can be obtained by the residue method. This requires the integration contour to be closed by large half-circles in either the upper or the lower h-plane. Since $f(z) = 0$ for $|z| > d$, $F(h)$ has the exponential behaviour exp (jhd) in the lower half-plane and $\exp(-jhd)$ in the upper half-plane. Taking account of the factor $\exp(jhz)$ in eqn. 2.455, closure is allowed in the lower half-plane for $z < -d$, and in the upper half-plane for $z > d$. The pole at h_1, though on the real axis for a lossless structure, is to be considered as lying in the fourth quadrant. Calculation of the residues is straightforward and yields the surface waves

$$H_{\pm} \simeq \frac{k_0 \, c_1 \, V}{\eta_0} \, \frac{K_1 \, (u_1 \rho)}{K_1 \, (u_1 b)} \, e^{\mp jhz} \tag{2.459}$$

where the upper and lower signs are valid for $z > d$ and $z < -d$, respectively, and where we used the approximation $F(\pm H_1) \simeq 1$. The coefficient c_1 is defined by

$$c_1 = \frac{(u_1 b)^2}{bh_1 \, [C^2 + 2C - (u_1 b)^2]} \tag{2.460}$$

A more exact expression, taking the dispersive character of Z_s into account is (Wait and Hill, 1975e)

$$c_1' = \frac{(u_1 b)^2}{bh_1 \left[C^2 + 2C \dfrac{k_0^2 \, (\kappa_1 - 1)}{k_0^2 \, \kappa_1 - h_1^2} - (u_1 b)^2 \right]} \tag{2.461}$$

but the difference is negligible since h_1 is very close to k_0.

The radial electric field of the guided waves may be calculated using eqn. 2.445. The power carried by each guided wave can be obtained by integrating Poynting's vector from $\rho = b$ to infinity. Using the integral (Erdélyi *et al.*, 1953)

$$\int \rho \, K_{\nu}^2 \, (u\rho) \, d\rho = -1/2 \, \rho^2 \, \{ [K_{\nu}' \, (u\rho)]^2 - [K_{\nu} \, (u\rho)]^2 \, (1 + u^{-2} \, \rho^{-2} \cdot \nu^2) \} \tag{2.462}$$

we obtain

$$P_{1\pm} \simeq \frac{\pi x c_1}{2\eta_0} \tag{2.463}$$

(c) External impedance

If we intended to solve the problem exactly we would also need to calculate an exact expression for the magnetic field $H(b + 0, z)$ on the external face of the slot. Indeed, V and $F(h)$ must be determined by equating $H(b + 0, z)$ and $H(b - 0, z)$ on the interval $|z| < d$. We thus run into a Wiener–Hopf problem (Noble, 1958). This complication may be avoided and an approximate solution will be found by equating only some kind of average value of $H(b \pm 0, z)$. For this purpose, we consider the external space as a load fed through the slot, which acts as a voltage generator. The complex power delivered to the external space is given by $Y_e \, |V|^2$, where Y_e is defined as the external admittance of the slot.

But we can also calculate this power as the flux of Poynting's vector through the slot. This yields

$$P = Y_e \, |V|^2 = -2\pi b \int_{-d}^{d} E_z^* \, (b, z) \, H(b + 0, z) \, dz$$

$$= 2\pi b \, V \int_{-\infty}^{\infty} f^* \, (z) \, H(b + 0, z) \, dz \tag{2.464}$$

where we consider $f(z)$ to be zero for $|z| > d$. Using Parseval's identity and eqn. 2.453, we obtain

$$Y_e = \frac{2\pi jbx}{\eta_0} \int_{-\infty}^{\infty} \frac{|F(h)|^2 \, H_1^{(2)} (\Lambda b)}{\Lambda b \, H_0^{(2)} (\Lambda b) + C H_1^{(2)} (\Lambda b)} \, dh \tag{2.465}$$

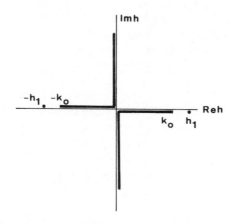

Fig. 2.38 *The complex h-plane showing the poles at $\pm h_1$ and the branch cuts*

In order to evaluate this integral, we consider the complex h-plane (Fig. 2.38). Apart from the poles at $\pm h_1$, the other singularities of the integrand in eqn. 2.465 are branch cuts issuing from the points $\pm k_0$ and due to eqn. 2.451. Since $F(h)$ behaves as the exponentials $e^{\pm jhd}$, we may not close the contour without taking some precautions. For this purpose, using the notation of Appendix B, we split $f(z)$ into two functions

$$f(z) = f_+(z) + f_-(z) \tag{2.466}$$

where $f_+(z)$ and $f_-(z)$ are non-zero only on $0 < z < d$ and $-d < z < 0$, respectively. Accordingly, we have for the Fourier transforms

$$F(h) = F_N(h) + F_P(h)$$

$$|F(h)|^2 = \{|F_N(h)|^2 + F_N(h) F_P^*(h)\} + \{|F_P(h)|^2 + F_P(h) F_N^*(h)\} \tag{2.467}$$

Now, at infinity, $F_N(h)$ behaves as $\exp(-jhd/2)$ and $F_P(h)$ as $\exp(jhd/2)$, and closure may be carried out in the lower half-plane for the first term of eqn. 2.467 and in the upper half-plane for the second term. For the reasons mentioned previously, the two terms of eqn. 2.467 are slowly varying functions of h. They are approximately equal to $1/2$ when $|hd| \ll 2$. This approximation will be used here. The integral 2.465 may now be written in the form

$$Y_e \simeq Y_1 + Y_r \tag{2.468}$$

where

$$Y_1 = G_1 \triangleq \frac{2\pi x c_1}{\eta_0} \tag{2.469}$$

is the pole contribution and Y_r is obtained as the result of the integration enlacing the branch cuts. Actually the term Y_1 is responsible for the power transferred to the guided waves, as may be checked by a comparison with eqn. 2.463, whereas Y_r corresponds to the radiated waves. Note that Y_1 is real.

Since Λ changes sign when crossing the branch cuts, the integrals along the two sides of the branch cuts may be combined into a single integral. Assuming that the two terms of eqn. 2.467 are equal to 1/2, it appears that Y_r is real and due to the horizontal part of the branch cut. This is not surprising since the branch cut defines a continuous spectrum of cylindrical modes. Only the non-evanescent modes, i.e. those with real h, contribute active power. It is found that

$$Y_r = G_r \triangleq \frac{4x}{\pi} \int_0^x \frac{1}{|\xi H_0^{(2)}(\xi) + C H_1^{(2)}(\xi)|^2} \, dt \tag{2.470}$$

where $\xi = \sqrt{x^2 - t^2}$.

With the approximation made for the two terms of eqn. 2.467, the external impedance is purely real. If the exact but as yet unknown expressions were used, we would find additional susceptances B_1 and B_r. They were calculated by Delogne and Liégeois (1971), assuming that $f(z)$ is constant. Actually the external susceptance remains small, unless the ratio d/b becomes very small. The results eqns. 2.468 to 2.470 remain accurate as long as the slot width $(2d)$ remains small compared with the wavelength, not compared with the cable radius as is sometimes believed.

(d) Internal fields
The total fields at the interior of the cable are given by the sum of the incident fields eqns. 2.441–2.440 and of the scattered fields. The latter are due to the axial electric field eqn. 2.447 considered as given. Taking the Fourier transform of the scattered field and expressing the boundary conditions $E_{zs} = 0$ for $\rho = a$ and $E_{zs} = V f(z)$ for $\rho = b$ is a straightforward calculation. We find

$$\mathcal{H}_s(\rho, h) = \frac{jx\kappa V}{\eta_0} \frac{F(h) \, [J_1(\Lambda'\rho) N_0(\Lambda'a) - J_0(\Lambda'a) N_1(\Lambda'\rho)]}{\Lambda'b \, [J_0(\Lambda'a) N_0(\Lambda'b) - J_0(\Lambda'b) N_0(\Lambda'a)]} \tag{2.471}$$

where

$$\Lambda' = \sqrt{k_0^2 \kappa - h^2} \tag{2.472}$$

It is not necessary to specify the determination of the square root, since eqn. 2.471 is an even function of Λ' and there is no branch cut. This function has poles at $\pm h_0 = \pm k_0 \sqrt{\kappa}$, due to $\Lambda' = 0$ and corresponding to the dominant TEM mode of the coaxial cable, and at $\pm h_n$, due to the zero of the bracketed factor in the denominator and corresponding to the evanescent modes. The h_n are imaginary.

A procedure similar to the one used for the exterior of the cable may be set up for the calculation of the inverse Fourier transform. The slot excites forward and backward TEM waves with equal amplitudes, provided we assume that $F(k_0 \sqrt{\kappa})$ $\simeq 1$. The backward wave is the reflected TEM wave and is given by

$$H_r \simeq \frac{-V\sqrt{\kappa}}{2\eta_0 \ln(b/a)} \frac{e^{jkz}}{\rho} \qquad \text{for} \qquad z < -d \tag{2.473}$$

while the sum of the forward and incident waves is the TEM wave transmitted beyond the slot

$$H_t \simeq \left(1 + \frac{xV}{2\eta_0 \ln(b/a)}\right) \frac{e^{-jkz}}{\rho} \qquad \text{for} \qquad z > d \tag{2.474}$$

The slot also excites evanescent modes which are responsible for reactive power accumulated in the vicinity of the slot.

We can also calculate the flux of Poynting's vector through the slot, given by

$$P = -2\pi b \int_{-d}^{d} E_z^*(b, z)\, [H_i(b, z) + H_s(b, z)]\, dz \tag{2.475}$$

We find

$$P \simeq 2\pi V - \left(\frac{Y_0}{2} + jB_i\right) V^2 \tag{2.476}$$

where the first term is due to the incident wave and the second term to the scattered field. Actually the term $1/2\, Y_0\, V^2$ is due to the poles at $\pm jk_0 \sqrt{\kappa}$, i.e. to the scattered TEM mode, and the term $B_i V^2$ to the evanescent modes. The exact evaluation of the susceptance B_i requires a knowledge of $f(z)$. Calculations based on the approximation $f(z) = (2d)^{-1}$ yielded

$$B_i = \frac{-\pi\sqrt{\kappa}}{\eta_0 \ln(b/a)} \frac{k_1 d - \sin(k_1 d)\cos(k_1 d)}{(k_1 d)^2}$$

$$+ \sum_{n=1}^{\infty} \frac{\pi\kappa k_0 d}{\eta_0} \frac{[J_0(\Lambda_n' a)]^2}{[J_0(\Lambda_n' a)]^2 - [J_0(\Lambda_n' b)]^2} \frac{2|h_n|d - 1 + \exp(-2|h_n|d)}{|h_n d|^3}$$

$$\tag{2.477}$$

The second term of eqn. 2.476 may be interpreted as follows. If the slot were fed externally by a voltage generator, the interior of the cable would act as an admittance

$$Y_i = Y_0/2 + jB_i \tag{2.478}$$

It is not surprising that the real part is $Y_0/2$ since the slot 'sees' two coaxial cables connected in series.

★ *(e) Equivalent circuit for the slot*

The slot voltage is still unknown. As previously stated, we will not calculate $f(z)$ exactly but restrict outself to evaluating the slot voltage. We are equating the expressions 2.464 and 2.475 of the complex power flowing through the slot. This yields

$$V = \frac{2\pi}{(Y_i + Y_e)}$$

$$= \frac{2\pi}{Y_e/2 + G_1 + G_r + j(B_i + B_e)} \tag{2.479}$$

where we assume that an external susceptance has also been calculated. The transmission line voltage for the incident wave eqn. 2.442 is given by

$$V_i = -\int_a^b \frac{\eta}{\rho}\, d\rho = \frac{-2\pi}{Y_0} \tag{2.480}$$

Now it appears that the slot voltage and all parameters pertaining to the reflection and transmission of the TEM mode inside the coaxial cable may be obtained from the lumped equivalent circuit shown in Fig. 2.39.

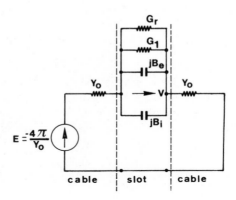

Fig. 2.39   *Equivalent circuit of the annular slot*

The slot may be considered as a two-port network with a scattering matrix

$$S = \begin{vmatrix} S_{11} & S_{12} \\ S_{12} & S_{11} \end{vmatrix} \tag{2.481}$$

where

$$S_{11} = \frac{1}{1 + 2y_t} \tag{2.482}$$

$$S_{12} = 1 - S_{11} = \frac{2y_t}{1 + 2y_t} \tag{2.483}$$

and

$$y_t = g_t + jb_t = \frac{G_1 + G_r + j(B_i + B_e)}{Y_0} \tag{2.484}$$

is the normalised slot admittance. The relative reflected power is given by

$$p_{ref} = \frac{P_{ref}}{P_i} = |S_{11}|^2 \tag{2.485}$$

the relative transmitted power by

$$p_t = \frac{P_t}{P_i} = |S_{12}|^2 \tag{2.486}$$

The relative power $(1 - p_{ref} - p_t)$ is transferred through the external medium. It is the sum of the relative power of the surface waves

$$p_1 = \frac{P_1}{P_i} = \frac{4g_1}{(1 + 2g_t)^2 + 4b_t^2} \tag{2.487}$$

and of the relative radiated power

$$p_r = \frac{P_r}{P_i} = \frac{4g_r}{(1 + 2g_t)^2 + 4b_t^2} \tag{2.488}$$

k *(f) Radiation pattern*
We may consider the slotted cable as an antenna which is fed a power P_i equal to that of the incident wave. Using the far-field expression 2.457, we may calculate Poynting's vector N of the radiated wave and define the antenna gain

$$G(\theta) = \frac{4\pi R^2 N}{P_i} \tag{2.489}$$

This yields

$$G(\theta) = \frac{16x^2 \sqrt{\kappa}}{\pi \eta \, Y_0 \, |1 + 2y_t|^2} \frac{1}{|x \sin\theta \, H_0^{(2)} (x \sin\theta) + C H_1^{(2)} (x \sin\theta)|^2} \tag{2.490}$$

The gain function is symmetrical about $\theta = 90°$ and needs only to be calculated for $0 < \theta < 90°$. Because of reciprocity, the gain, though defined for the emission by the cable, may also be used to calculate the power transferred inside the cable when a plane wave is incident from the exterior. Such a wave excites two waves of the TEM mode, travelling away from the slot inside the cable. It must be stressed that the use of classical receiving antenna formulas will give the power of each of these waves, not the total power.

We can also define a directive gain by

$$D(\theta) = \frac{4\pi R^2 N}{P_r} \tag{2.491}$$

where we use the radiated power instead of the incident power. The directive gain is normalised in such a way that its integral over all directions is (4π). It is obviously related to the gain by

$$G(\theta) = p_r D(\theta) \tag{2.492}$$

★ *(g) Numerical results*

Some numerical results will not be presented. They were obtained by assuming that the axial electric field E_z is constant in the slot, i.e. $f(z) = (2d)^{-1}$. A more exact model would only modify the slot susceptance, but no fundamental change is expected.

In the first instance, the effect of the slot width was investigated. The parameters $b = 5$ mm, $a = 1 \cdot 34$ mm, $\kappa = 2 \cdot 5$, $x = 0 \cdot 0314$, $h_1/k_0 = 1 \cdot 01$ were selected. They correspond to a typical 50 Ω cable, with a 2 mm thick PVC jacket ($\kappa_1 = 2 \cdot 5$). Table 2.2 shows the results obtained when the slot width is varied from 1% of the cable diameter to $(2b)$. The slot conductance is not affected by the slot width. The susceptance varies with the ratio (d/b). This variation is very slow and seems to be logarithmic, as expected from similar work on the impedance of gap fed antennas (Collin and Zucker, 1969). About 60% of the power is reflected, 8% transmitted, 20% converted into surface waves and 12% radiated. Trying to reduce the insertion loss $(- 10 \log p_t)$ by reducing the slot width would be based on a serious delusion, since the performance parameters are almost unaffected by this quantity.

Table 2.2   *Influence of the slot width on the performance parameters*

d/b	y_t	p_{ref}	p_t	p_1	p_r
		dB	dB	dB	dB
$0 \cdot 01$	$0 \cdot 127 + j\, 0 \cdot 178$	$-2 \cdot 31$	$- 9 \cdot 4$	$-7 \cdot 15$	$-9 \cdot 76$
$0 \cdot 05$	$0 \cdot 127 + j\, 0 \cdot 140$	$-2 \cdot 18$	$-10 \cdot 6$	$-7 \cdot 02$	$-9 \cdot 64$
$0 \cdot 1$	$0 \cdot 127 + j\, 0 \cdot 119$	$-2 \cdot 12$	$-11 \cdot 2$	$-6 \cdot 97$	$-9 \cdot 58$
$0 \cdot 5$	$0 \cdot 127 + j\, 0 \cdot 076$	$-2 \cdot 04$	$-12 \cdot 5$	$-6 \cdot 86$	$-9 \cdot 49$
1	$0 \cdot 127 + j\, 0 \cdot 054$	$-2 \cdot 00$	$-13 \cdot 1$	$-6 \cdot 85$	$-9 \cdot 46$

The next investigation consisted of varying the ratio h_1/k_0, keeping d/b and all other parameters constant (Table 2.3). A change in h_1/k_0 from $1 \cdot 002$ to $1 \cdot 05$ could be obtained by changing the PVC jacket thickness from 4 to 100% of the outer conductor radius. It is seen that the total gap admittance, namely the sum $(G_1 + G_r)$, remains practically constant. However the ratio G_1/G_r increases with h_1/k_0, which means that more power is transferred to the surface waves and less to the radiated   field. The directive gain (Fig. 2.40) $D(\theta)$ increases significantly when the jacket thickness decreases. The influence is even more marked for the gain

Table 2.3   *Influence of the velocity ratio h_1/k_0 of the surface waves on the performance parameters of the slot*

h_1/k_0	y_t	G_1	G_r	p_1	p_r
		mS m^{-1}	mS m^{-1}	dB	dB
1·002	0·125 + j 0·120	1·42	1·07	−7·5	− 7·6
1·005	0·127 + j 0·120	1·54	0·99	−7·2	− 9·1
1·010	0·127 + j 0·119	1·63	0·89	−7·0	− 9·6
1·020	0·125 + j 0·118	1·76	0·76	−6·7	−10·2
1·050	0·122 + j 0·116	1·88	0·57	−6·3	−11·4

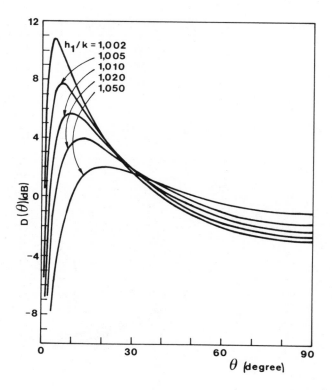

Fig. 2.40   *Influence of the velocity ratio h_1/k_0 of the surface wave on the directive gain*

$G(\theta)$, because the decrease in directivity with increasing h_1/k is enhanced by the variation of p_r. Actually, the maximum gain $G(\theta)$ varies from $2·9$ dB ($\theta = 5°$) for $h_1/k_0 = 1·002$ to $−9·5$ dB ($\theta = 20°$) for $h_1/k_0 = 1·05$. This type of radiation pattern is extremely useful when the slotted cable is used to provide radio communications in tunnels, since it favours radiation in the axial directions rather than toward the lateral walls. Thin dielectric jackets appear to be interesting. Obviously, radiation here means excitation of the waveguide modes and is useful only above the cutoff frequency of the tunnel. However below this cutoff the excitation of the monofilar mode can easily be studied by considering the impedance calculated in Section 2.10.1 as the slot load.

A final series of calculations was made for a given cable ($b = 5$ mm, $a = 1 \cdot 34$ mm, $\kappa = 2 \cdot 5$, $d/b = 0 \cdot 1$ and 2 m thick PVC jacket), and now varying the frequency. The results are shown in Table 2.4 and Fig. 2.41. The variation of the slot admittance is mainly due to that of the frequency parameter x, rather than to that of h_1/k. Actually, the increase in h_1/k is partly compensated by that of x, with the result that p_1 and p_r both increase. The directive gain $D(\theta)$ decreases with

Table 2.4   *Influence of frequency on the performance parameters of the slot*

f	h_1/k	y_t	p_{ref}	p_t	p_1	p_r
MHz			dB	dB	dB	dB
30	$1 \cdot 0068$	$0 \cdot 0765 + j\,0 \cdot 0285$	$-1 \cdot 24$	$-17 \cdot 0$	$-7 \cdot 8$	$-11 \cdot 9$
60	$1 \cdot 0075$	$0 \cdot 0871 + j\,0 \cdot 0443$	$-1 \cdot 42$	$-15 \cdot 6$	$-7 \cdot 5$	$-11 \cdot 2$
300	$1 \cdot 0100$	$0 \cdot 127\ \ + j\,0 \cdot 0119$	$-2 \cdot 12$	$-11 \cdot 3$	$-6 \cdot 9$	$-\,9 \cdot 6$
600	$1 \cdot 0120$	$0 \cdot 156\ \ + j\,0 \cdot 204$	$-2 \cdot 78$	$-\,8 \cdot 5$	$-6 \cdot 9$	$-\,9 \cdot 0$

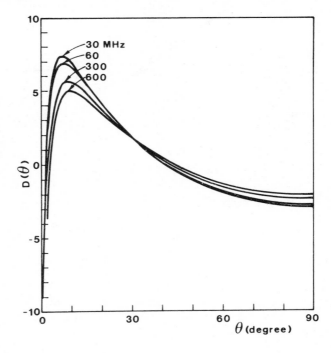

Fig. 2.41   *Influence of frequency on the directive gain*

frequency, but the gain $G(\theta)$ remains practically constant and close to $-4\,\text{dB}$. Some calculations and verification measurements were also made at $3\,\text{GHz}$ where the insertion loss $(-10\log p_t)$ drops to about $3\,\text{dB}$.

2.11 Transmission loss

The dipole excitation of electromagnetic waves inside a tunnel which may contain either a monofilar wire conductor or a leaky coaxial cable was analysed in Section 2.10.2. Hence we are able to calculate the excited fields for either the monofilar or the coaxial mode. The radiation resistance of electric and magnetic dipoles was also calculated.

Let us now consider receiving dipoles. For an electric dipole of effective vectorial length l_{re}, the received electromotive force is given by

$$v_r = -l_{re} \cdot E \tag{2.493}$$

For a receiving magnetic dipole (loop of effective area A_{re}) with unit normal n, we have

$$v_r = j\omega\mu_0 A_{re}\, n \cdot H \tag{2.494}$$

Consequently, if we have calculated the fields excited by the transmitting dipole, we are able to calculate the mutual impedance between the transmitting and receiving dipoles by

$$Z_m = \frac{v_r}{I} \tag{2.495}$$

where I is the current fed into the transmitting dipole.

When the axial separation z of the dipoles is small, the evaluation of Z_m is as complicated as that of the self-impedances carried out in Section 2.10.2. The task is considerably simpler when z is large, since the dominant contribution is from the mode with the lowest attenuation. The mutual impedance is then given by a formula of the type $Z_m = Z_{0m}\exp(-\Gamma z)$, where Γ is the propagation constant of the dominant mode and Z_{0m} is a function of the orientation and location of the dipoles in the tunnel cross-section. In the leaky feeder technique the dominant mode is the coaxial (or bifilar) mode.

If we assume impedance matching at the transmitting and receiving ends, the transmission loss is given by

$$L = 10\log\frac{P_t}{P_r} = 10\log\frac{4\,R_t R_r}{|Z_m|^2} \tag{2.496}$$

where P_t is the available transmitter power, P_r is the received power and R_t, R_r are the transmitting and receiving dipole resistances.

Several calculations of transmission loss in the presence of a leaky coaxial cable have been published for rectangular (Mahmoud and Wait, 1976) and circular tunnels (Hill and Wait, 1976b, 1979). We will not go into the calculations, but rather comment on some typical results. These are for a circular tunnel with $a = 2$ m, $\kappa_1 = 10$, $\sigma_1 = 10^{-3}$ S m^{-1}. The leaky coaxial cable is located at $\rho_c/a = 0.9$ and has the following parameters: inner conductor radius 1·5 mm, outer conductor radius 10 mm, dielectric constant 1·5, conductivity of the inner conductor 5.7×10^7 S m^{-1}. The dipoles are located at the same azimuth as the cable ($\phi_0 = \phi_c$) and oriented for maximum coupling. The effective length of electric dipoles is taken to be 0·5 m, which corresponds to an actual length of 1 m. The loop area of magnetic dipoles is taken to be 1 m^2.

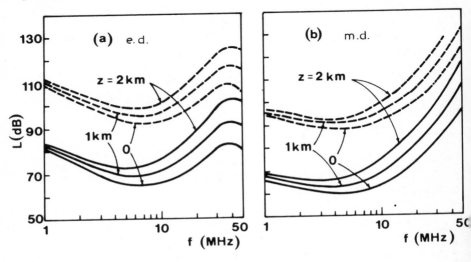

Fig. 2.42   *Transmission loss for electric and magnetic dipoles. Continuous curves for $m_t = 10$ nH m^{-1}. Broken curves for $m_t = 2$ nH m^{-1}. Dipoles located at $\rho_0/a = 0$*

We are particularly interested in the frequency behaviour of the transmission loss and in a comparison between electric and magnetic dipoles. Figure 2.42 has been arranged to facilitate this comparison. The curves for $z = 0$ are shown to illustrate the two-way coupling loss to and from the cable. The curves show a broad minimum between 5 and 10 MHz and appear to be slightly favourable to the magnetic dipole. A relatively large value of the specific transfer inductance is of paramount importance. However an overlarge value should be avoided, particularly for long distance transmission and when the cable has to be strung very close to the wall, since it results in a considerable increase in the coaxial mode attenuation.

The discussion may usefully be clarified by consideration of Fig. 2.43, which shows the self- and mutual impedances for axial separation of 1 km, for a specific transfer inductance of 40 nH m^{-1} and for another value of ρ_0/a. The mutual impedance of electric dipoles decreases with frequency. This is the result of the increase of the cable specific attenuation with frequency. At low frequencies however this effect is offset by the frequency behaviour of the electric field distribution in the tunnel cross-section. For the magnetic dipole, these effects are largely compensated by the ω factor in eqn. 2.494. The input resistances follow closely the frequency behaviour mentioned in Section 2.10.2: roughly constant for the electric dipole and proportional to ω^2 for the magnetic dipole.

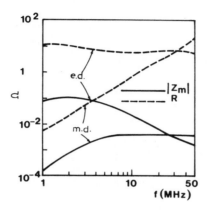

Fig. 2.43 *Mutual and self-impedances for electric and magnetic dipoles: $z = 1\,km$, $m_t = 40$ nH m^{-1}. Dipoles located at $\rho_0/a = 0\cdot5$*

These results, though very instructive, are somewhat unrealistic since they do not take account of the practical difficulties in realising impedance matching (Delogne, 1976b). In the frequency range of interest, actual antennas are electrically short and exhibit a nearly reactive impedance, having a very high Q-factor. As we saw in Section 2.10.2, the reactance of the electric dipole is more dependent on the dipole location in the tunnel than that of the magnetic dipole. Furthermore, the electric dipole has a higher Q-factor than the magnetic dipole. Impedance matching is consequently considered impossible for the electric dipole in the tunnel environment. The slight advantage in favour of the magnetic dipole exhibited by the transmission curves reinforces this conclusion. This however does not remain true above a few tens of megahertz.

2.12 Quasi-static limits

2.12.1 Electric and magnetic potentials

In this section, we will investigate the low-frequency behaviour of waves guided by axial cables inside tunnels. Here again the term cable is used for a single-wire conductor as well as for a leaky coaxial cable. The analysis will be carried out in detail for a single cable located in a circular tunnel, for which the basic equations and notation were given in Section 2.8.1. The extension to multiple cables and to other tunnel shapes will be investigated briefly in Section 2.12.5.

The electric Hertz potential for the case considered is given by any one of eqns. 2.341, 2.344 and 2.345 and the magnetic Hertz potential by 2.346, where R_m and Δ_m are defined by eqns. 2.355 and 2.356. Since the Bessel functions I_m and K_m are even functions of m, we have $R_m = R_{-m}$ and $\Delta_m = -\Delta_{-m}$. Hence we will write U as a cosine series and V as a sine series, with positive values of m. As our interest is in low frequencies, where the tunnel wall may be considered as a relatively good conductor, we observe that $R_m = 1$, $\Delta_m = 0$ for a perfectly conducting wall and we recast the potentials into the form (Wait, 1977)

$$U = U_{\text{pc}} + \Delta U \tag{2.497}$$

$$V = \Delta V \tag{2.498}$$

where

$$U_{\text{pc}} = (2\pi j\omega\epsilon_0)^{-1} I \Lambda(\rho,\phi) \tag{2.499}$$

$$\Delta U = (2\pi j\omega\epsilon_0)^{-1} I \left(\frac{jk_0}{v}\right)^2 \Omega(\rho,\phi) \tag{2.500}$$

$$\Lambda(\rho,\phi) = \left[K_0(v\rho_>) - \frac{K_0(va)}{I_0(va)} I_0(v\rho_>) \right] I_0(v\rho_<)$$

$$+ 2 \sum_{m=1}^{\infty} \left[K_m(v\rho_>) - \frac{K_m(va)}{I_m(va)} I_m(v\rho_>) \right] I_m(v\rho_<) \cos[m(\phi-\phi_0)] \tag{2.501}$$

$$\frac{k_0}{v^2} \Omega(\rho,\phi) = (R_0 - 1) \frac{K_0(va)}{I_0(va)} I_0(v\rho) I_0(v\rho_0)$$

$$+ 2 \sum_{m=1}^{\infty} (R_m - 1) \frac{K_m(va)}{I_m(va)} I_m(v\rho) I_m(v\rho_0) \cos[m(\phi-\phi_0)] \tag{2.502}$$

Clearly, U_{pc} is the potential for the case of a perfectly conducting wall, while ΔU and $\Delta V = V$ are the perturbations due to the finite conductivity of the wall (but assuming that Γ takes on its actual value).

Two assumptions will be required to develop quasi-static approximations. In the first instance, we suppose that $|va| \ll 1$, which obviously entails $|v\rho| \ll 1$ and $|v\rho_0| \ll 1$. Since $v = \sqrt{-k_0^2 - \Gamma^2}$ and Γ does not differ from (jk_0) by more than 50%, even for the coaxial mode of a leaky coaxial cable, it is seen that this condition is well fulfilled up to several tens of megahertz for current tunnel sizes. The second assumption is that the complex dielectric constant κ_1' of the wall is large compared with unity. Since $\kappa_1' = \kappa_1 - j\sigma_1/(\omega\epsilon_0)$, this condition is in general well satisfied, particularly below about 10 MHz, for current wall conductivities.

The first assumption only is required to approximate eqn. 2.501. On using the small-argument approximations of Bessel's functions C.19 and C.20, it is not difficult to obtain

$$\Lambda(\rho,\phi) = \ln \frac{a}{\rho_d} - \sum_{m=1}^{\infty} \frac{1}{m} \left(\frac{\rho\rho_0}{a^2}\right)^m \cos[m(\phi - \phi_0)] \tag{2.503}$$

The summation may be reduced using the identity

$$\sum_{m=1}^{\infty} \frac{1}{m} p^m \cos(m\psi) = -\frac{1}{2} \ln(1 - 2p\cos\psi + p^2) \tag{2.504}$$

if $p < 1$. Referring to Fig. 2.44, the cable located at C has a (negative) electrostatic image I located at a radius $\rho_i = OI$ given by

$$\rho_i = \frac{a^2}{\rho_0} \tag{2.505}$$

In this way, we get

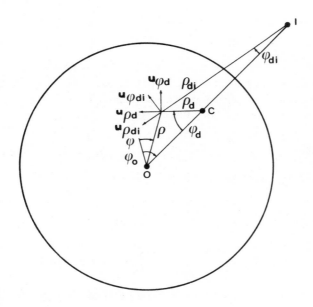

Fig. 2.44   *Electrostatic image I of a cable C in a circular tunnel. The figure defines notation used in Section 2.12*

$$\Lambda(\rho, \phi) = \ln \frac{a}{\rho_d} - \ln \frac{\rho_i}{\rho_{di}} \qquad (2.506)$$

where ρ_d and ρ_{di} are the distances from the observation point P to the cable and to its image. We have thus

$$U_{pc} = \frac{1}{2\pi\epsilon_0} \frac{I}{j\omega} \left(\ln \frac{a}{\rho_d} - \ln \frac{\rho_i}{\rho_{di}} \right) \qquad (2.507)$$

Since the electric charge per unit length along the cable is $\Gamma I/(j\omega)$, it is seen that U_{pc} is just the electrostatic potential of the cable and of its image, divided by Γ.

The reduction of $\Omega(\rho, \phi)$ is somewhat more complicated. We have, using the Wronskian relation C. 16,

$$R_m - 1 = \frac{-1}{K_m(va) I_m(va)} \left\{ \frac{jk_0}{v^2 a} \left[\frac{jk_0}{v} \frac{I'_m(va)}{I_m(va)} + \frac{Z_m}{\eta_0} \right] \right\} \Big/$$

$$\left\{ \left[\frac{jk_0}{v} \frac{I'_m(va)}{I_m(va)} + \eta_0 Y_m \right] \left[\frac{jk_0}{v} \frac{I'_m(va)}{I_m(va)} + \frac{Z_m}{\eta_0} \right] + \left(\frac{jm\Gamma}{a} \right)^2 (v^{-2} - u^{-2})^2 \right\} \qquad (2.508)$$

Remembering that $v^2 = -k_0^2 - \Gamma^2$, $u^2 = -k_0^2 \kappa'_1 - \Gamma^2$, and that Γ does not differ much from (jk_0), the assumption $|\kappa_1| \gg 1$ entails $|u^2| \gg |v^2|$ and

$$u \simeq jk_1 = \gamma_1 \qquad (2.509)$$

the propagation constant of the external medium. We assume that this medium has a permeability $\mu_1 = \mu_0$. We may then neglect Z_m/η_0 with respect to $\eta_0 Y_m$, since $Z_m/\eta_0 = \eta_0 Y_m/\kappa'_1$. We further use small-argument approximations for the Bessel's functions of argument (va), yielding for $m > 0$

$$\frac{v^2}{k_0^2} (R_m - 1) \frac{K_m(va)}{I_m(va)} \simeq \frac{1}{[I_m(va)]^2} \frac{1}{m + jk_0 \eta_0 Y_m} \qquad (2.510)$$

On using eqns. 2.354, 2.509 and C.18 we finally obtain

$$\Delta U = (2\pi j\omega\epsilon_0)^{-1} I \left(\frac{jk_0}{v} \right)^2 \left\{ \frac{K_0(\gamma_1 a)}{\gamma_1 a K_1(\gamma_1 a)} \right.$$

$$\left. + \sum_{m=1}^{\infty} \frac{2K_m(\gamma_1 a)}{\gamma_1 a K_{m+1}(\gamma_1 a)} \left(\frac{\rho\rho_0}{a^2} \right)^m \cos[m(\phi - \phi_0)] \right\} \qquad (2.551)$$

wherein the curled bracket is $\Omega(\rho, \phi)$. It is also not difficult to see that we have

$$\Delta_m \simeq \frac{\Gamma}{\omega\mu_0} (R_m - 1) \frac{K_m(va)}{I_m(va)} \qquad (2.512)$$

and consequently

$$\Delta V \simeq (2\pi j \omega \epsilon_0)^{-1} \frac{-\Gamma}{j\omega\mu_0} \left(\frac{jk_0}{v}\right)^2 \sum_{m=1}^{\infty} \frac{2K_m(\gamma_1 a)}{\gamma_1 a\, K_{m+1}(\gamma_1 a)} \left(\frac{\rho\rho_0}{a}\right)^m \sin\left[m(\phi - \phi_0)\right]$$

$$(2.513)$$

As yet no assumption has been made about the value $\gamma_1 a$. One extreme case is for $|\gamma_1 a| \ll 1$, i.e. when the skin depth in the wall is much larger than the tunnel radius. We have then, using small-argument approximations for $K_m(\gamma_1 a)$ and $K_{m+1}(\gamma_1 a)$

$$\Delta U \simeq (2\pi j \omega \epsilon_0)^{-1} I \left(\frac{jk_0}{v}\right)^2 \left\{ -\ln(0\cdot89\,\gamma_1 a) + \sum_{m=1}^{\infty} \frac{1}{m} \left(\frac{\rho\rho_0}{a}\right) \cos\left[m(\phi - \phi_0)\right] \right\}$$

$$\text{for } |\gamma_1 a| \ll 1 \quad (2.514)$$

$$\Delta V \simeq (2\pi j \omega \epsilon_0)^{-1} I \frac{-\Gamma}{j\omega\mu_0} \left(\frac{jk_0}{v}\right)^2 \sum_{m=1}^{\infty} \frac{1}{m} \left(\frac{\rho\rho_0}{a}\right) \sin\left[m(\phi - \phi_0)\right]$$

$$\text{for } |\gamma_1 a| \ll 1 \quad (2.515)$$

On using eqn. 2.504 and the similar identity

$$\sum_{m=1}^{\infty} \frac{1}{m} p^m \sin(m\psi) = \text{arctg} \frac{p \sin \psi}{1 - p \cos \psi} \quad (2.516)$$

we obtain the closed forms

$$\Delta U \simeq (2\pi j \omega \epsilon_0)^{-1} I \left(\frac{jk_0}{v}\right)^2 \left[-\ln(0\cdot89\,\gamma_1 a) + \ln \frac{\rho_i}{\rho_{di}} \right] \text{ for } |\gamma_1 a| \ll 1$$

$$(2.517)$$

$$\Delta V \simeq (2\pi j \omega \epsilon_0)^{-1} I \frac{-\Gamma}{j\omega\mu_0} \left(\frac{jk_0}{v}\right)^2 (-\phi_{di}) \qquad \text{for } |\gamma_1 a| \ll 1 \quad (2.518)$$

where ϕ_{di} is the angle shown in Fig. 2.44.

The other extreme case is when $|\gamma_1 a| \gg 1$. For the terms such that $|\gamma_1 a| \gg m$ we may use the semi-convergent large-argument approximation (Erdéleyi *et al.*, 1953)

$$K_m(\gamma_1 a) \simeq \sqrt{\frac{\pi}{2\gamma_1 a}}\, e^{-\gamma_1 a} \left[1 + \frac{4m^2 - 1}{8\gamma_1 a} + \frac{(4m^2 - 1)(4m^2 - 9)}{2!\,(8\gamma_1 a)^2} + \dots \right]$$

$$(2.519)$$

to obtain (Pogorzelski, 1979)

$$\frac{K_m(\gamma_1 a)}{\gamma_1 a K_{m+1}(\gamma_1 a)} = \frac{1}{\gamma_1 a} - \frac{2m+1}{2(\gamma_1 a)^2} + \frac{(2m+1)(2m+3)}{8(\gamma_1 a)^3} + 0\left[\frac{1}{(\gamma_1 a)^4}\right]$$

$$\text{for } |\gamma_1 a| \gg m \qquad (2.520)$$

The condition $|\gamma_1 a| \gg m$ can obviously not be satisfied for all m, but if we assume that it is fulfilled by all terms which are significant in the representation of the potentials, we may use eqn. 2.520 for all m and attempt to obtain closed-form expressions for the development of ΔU and ΔV in an inverse power series of $(\gamma_1 a)$. We will at present restrict ourselves to the first term of the series. This yields for eqns. 2.511 and 2.513

$$\Delta U \simeq (2\pi j\omega\epsilon_0)^{-1} I\left(\frac{jk_0}{v}\right)^2 \frac{1}{\gamma_1 a} \frac{\rho_i^2 - \rho^2}{\rho_{di}^2} \qquad \text{if } |\gamma_1 a| \gg 1 \qquad (2.521)$$

$$\Delta V = (2\pi j\omega\epsilon_0)^{-1} I\frac{-\Gamma}{j\omega\mu_0}\left(\frac{jk_0}{v}\right)^2 \frac{1}{\gamma_1 a} \frac{\rho\rho_i \sin(\phi - \phi_0)}{\rho_{di}^2}$$

$$\text{if } |\gamma_1 a| \gg 1 \qquad (2.522)$$

This is the first-order correction term to take account of the finite wall conductivity, when the skin depth in the wall is small compared with the tunnel radius.

★ *2.12.2 Electromagnetic fields in the tunnel space*

The electromagnetic fields in the tunnel space are obviously of primary importance for transmission loss calculations. Following eqns. 2.497 and 2.498, we will write

$$E = E_{pc} + \Delta E \qquad (2.523)$$

$$H = H_{pc} + \Delta H \qquad (2.524)$$

where E_{pc}, H_{pc} are derived from U_{pc}. Under the assumptions $|va| \gg 1$ and $|\kappa_1'| \gg 1$, U_{pc} is given by eqn. 2.507. Referring to eqns. B.9 and B.10 for the calculation of the transverse electric field, we see that

$$(E_{pc})_t = -Q \text{ grad } \psi \qquad (2.525)$$

where

$$Q = \frac{\Gamma I}{j\omega} \qquad (2.526)$$

$$\psi = \frac{1}{2\pi\epsilon_0}\left[\ln\frac{a}{\rho_d} - \ln\frac{\rho_i}{\rho_{di}}\right] \qquad (2.527)$$

According to the continuity theorem of electric charge, Q is the charge per unit length along the cable. The function ψ is the electrostatic potential for charges per unit length equal to 1 C m^{-1} for the cable and -1 C m^{-1} for its image, with the tunnel wall taken as the zero potential reference. The gradient in eqn. 2.525 can most conveniently be calculated by referring to the notation of Fig. 2.44. We have

$$(E_{pc})_t = \frac{\Gamma I}{2\pi j \omega \epsilon_0} \left(\frac{1}{\rho_d} u_{pd} - \frac{1}{\rho_{di}} u_{\rho di} \right) \tag{2.528}$$

It is also easily shown, using the formulas of Appendix B, that

$$(H_{pc})_t = \frac{1}{\eta_0} \frac{jk_0}{\Gamma} u_z \times (E_{pc})_t \tag{2.529}$$

$$= \frac{I}{2\pi \rho_d} u_{\phi d} - \frac{I}{2\pi \rho_{di}} u_{\phi di} \tag{2.530}$$

Our conclusion is that $(E_{pc})_t$ – apart from a factor Γ/jk_0 – and $(H_{pc})_t$ have the same values as for the TEM mode.

We now consider the perturbations ΔE and ΔH due to the finite wall conductivity starting from eqns. 2.511 and 2.513. We obtain for the perturbation of the transverse electric field

$$\Delta E_t \simeq 0 \tag{2.531}$$

which is a direct consequence of eqn. 2.512. This result is valid whatever the value of $\gamma_1 a$. We may thus draw an important conclusion: *the transverse electric field distribution in the quasi-static approximation* $|va| \ll 1, |\kappa_1'| \gg 1$, *is the same as for the TEM mode with a perfectly conducting wall, even if the skin depth in the wall is larger than the tunnel radius.* It is given by eqn. 2.528.

Next we consider the perturbation of the magnetic field. Using the formulas of Appendix B and eqns. 2.511–2.513, we obtain

$$\begin{bmatrix} \Delta H_\rho \\ \Delta H_\phi \end{bmatrix} = \frac{-I}{2\pi \rho} \sum_{m=1}^{\infty} \frac{2m \, K_m (\gamma_1 a)}{\gamma_1 a \, K_{m+1} (\gamma_1 a)} \left(\frac{\rho \rho_0}{a^2} \right)^m \begin{bmatrix} \sin \\ \cos \end{bmatrix} [m(\phi - \phi_0)]$$

$$\tag{2.532}$$

There is thus a major difference with that of the electric field. The transverse magnetic field is perturbed by the finite wall conductivity. The governing parameter is $\gamma_1 a$, i.e. the ratio of the tunnel radius to the skin depth in the wall, and also the ratio ρ_0/a. It is quite important to observe that, *though frequency dependent, the magnetic field does not depend on the mode propagation constant. It has, for instance, the same distribution for the coaxial and monofilar modes of a leaky feeder.*

We will now consider two extreme cases. If $|\gamma_1 a| \ll 1$, we may use small-argument approximations of the modified Bessel functions or, equivalently, start from eqns. 2.517 and 2.518. The calculations are most conveniently made in the system of cylindrical coordinates $(\rho_{di}, \phi_{di}, z)$ shown in Fig. 2.44. It is found that ΔH_t just cancels the second term of eqn. 2.530, and we are left with

$$H_t \simeq \frac{I}{2\pi\rho_d} u_{\phi d} \qquad \text{if } |\gamma_1 a| \ll 1 \qquad (2.533)$$

In other words, *the magnetic field can be calculated from the total cable current by the Biot–Savart law without taking account of the return current in the wall, when the skin depth in the latter material is much larger than the tunnel radius.* This conclusion was already drawn by Pogorzelski (1979).

The other extreme case is for $|\gamma_1 a| \gg 1$, i.e. *when the skin depth in the wall material is small. To zero order, the transverse magnetic field is given by eqn. 2.530, i.e. by Biot–Savart law applied to the cable and to its negative image.* The first-order perturbation may be obtained from eqns. 2.521 and 2.522, using the formulas of Appendix B.

No attention has been devoted to the axial fields E_z, H_z since they are very simply related to U and V, for which approximate expressions have been developed in Section 2.12.2.

★ *2.12.3 Transmission line equations*

In the chapters to follow we will make intensive use of transmission line equations to describe the propagation of modes along axial cables inside tunnels. Consequently, it is very important to justify this approach and at the same time to give explicit definitions of the transmission line parameters. In any case we assume that the conditions $|va| \ll 1$ and $|\kappa_1'| \gg 1$ are fulfilled. As we have seen in the previous section, the transverse electric field then has the electrostatic behaviour described by eqns. 2.525 to 2.527. As a result, the line integral of the electric field along any path joining the cable to the tunnel wall in a transverse plane is independent of the choice of this path. As a matter of convenience, we choose the radial path for $\phi = \phi_0$ and define the transmission line voltage by[*]

$$V_m = \int_{\rho_0+c}^{a} E_\rho \, d\rho \qquad (2.534)$$

The subscript m refers to the word 'monofilar' but this calls for some clarification. For a leaky cable we found two eigenmodes which were called monofilar and coaxial. When coupled-line theory is applied to leaky cables, the terms monofilar and coaxial are used for the cable exterior and interior, respectively. This misleading terminology has however become standard practice. The monofilar voltage, as defined by eqn. 2.534, includes the external fields of the coaxial eigenmode as well as those of the monofilar eigenmode. Using eqns. 2.525 to 2.527, we obtain

$$-\Gamma I = j\omega c_m V_m \qquad (2.535)$$

where

$$c_m = \frac{2\pi\epsilon_0}{\ln\left[(a^2 - \rho_0^2)/(ac)\right]} \qquad (2.536)$$

[*] The sign convention used here considers the tunnel wall as the outward conductor and the cable as the return conductor, as has already been done in Section 2.5.1.

is the specific capacity of the cable w.r.t. the wall. The transmission line current is defined as[*] $I_m = I$ and, coming back to z-dependent variables, we may write eqn. 2.535 as

$$-\frac{dI_m}{dz} = j\omega c_m V_m \tag{2.537}$$

This is the first transmission line equation.

In order to establish the second transmission line equation, we start from the integral form of Maxwell's first equation

$$\int_C E \cdot dl = -j\omega\mu_0 \int\int_S H \cdot dS$$

The path C is chosen in the $\phi = \phi_0$ plane, running as shown in Fig. 2.45. This yields

$$\frac{dV_m}{dz} = +j\omega\mu_0 \int_{\rho_0+c}^{a} H_\phi \, d\rho - E_z(a, \phi_0) + E_z(\rho_0 + c, \phi_0) \tag{2.538}$$

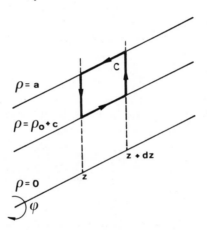

Fig. 2.45 *Integration path used in eqn. 2.538*

The first and second terms may be evaluated using expressions given in Sections 2.12.1 and 2.12.2, while the third term is just $z_0(\Gamma)I$, as results from the boundary condition at the external surface of the cable. On using the decomposition $H_\phi = (H_\phi)_{pc} + \Delta H_\phi$, we obtain the second transmission line equation

$$-\frac{dV_m}{dz} = (z_{m0} + z_0) I_m \tag{2.539}$$

---

[*] The sign convention used here considers the tunnel wall as the ongoing conductor and the cable as the return conductor, as has already been done in Section 2.5.1.

where z_{m0} is given by

$$z_{m0} = j\omega l_{pc} + z_e \tag{2.540}$$

Here

$$l_{pc} = \frac{\mu_0}{2\pi} \ln \frac{a^2 - \rho_0^2}{ac} \tag{2.541}$$

is the specific inductance for the case of perfectly conducting cable and tunnel wall (TEM case), while

$$z_e = \frac{j\omega\mu_0}{2\pi} \Omega(a, \phi_0)$$

$$= \frac{j\omega\mu_0}{2\pi} \left[\frac{K_0(\gamma_1 a)}{\gamma_1 a K_1(\gamma_1 a)} + \sum_{m=1}^{\infty} \frac{2K_m(\gamma_1 a)}{\gamma_1 a K_{m+1}(\gamma_1 a)} \left(\frac{\rho_0}{a}\right)^{2m} \right] \tag{2.542}$$

may be considered as the specific impedance of the tunnel wall, and z_0 is the specific external impedance of the cable. Actually z_{m0} is the specific series impedance for the case where the cable is perfectly conducting.

In the decomposition eqn. 2.540, $j\omega l_{pc}$ comes from the integration of $(H_\phi)_{pc}$ in eqn. 2.538, while z_e is due to ΔH_ϕ and $E_z(a, \phi_0)$. As expected, z_e will be small compared with $(j\omega l_{pc})$ when $|\gamma_1 a| \gg 1$, i.e. when the skin depth in the wall material is very small. In the opposite extreme case where $|\gamma_1 a| \ll 1$, i.e. when the tunnel radius is small compared with the skin depth in the wall material, it is better to evaluate eqn. 2.538 from eqn. 2.533. This yields

$$z_{m0} \simeq \frac{j\omega\mu_0}{2\pi} \left[\ln \frac{a - \rho_0}{c} + \frac{K_0(\gamma_1 a)}{\gamma_1 a K_1(\gamma_1 a)} + \sum_{m=1}^{\infty} \frac{2K_m(\gamma_1 a)}{\gamma_1 a K_{m+1}(\gamma_1 a)} \left(\frac{\rho_0}{a}\right)^m \right] + z_0 \tag{2.543}$$

The characteristic impedance of the monofilar mode should of course be calculated as $Z_m = [z_m/(j\omega c_m)]^{1/2}$.

★ *2.12.4 Modal equation*

According to transmission line theory, the mode propagation constant is given by

$$\Gamma^2 = j\omega c_m(z_{m0} + z_0) \tag{2.544}$$

where z_{m0} is given by either eqn. 2.540 or eqn. 2.543. As a check, it may be verified that the same result can be obtained by making approximations resulting from $|va| \ll 1$, $|\kappa_1'| \gg 1$ directly in the mode equation 2.359. In so far as z_0 depends on Γ however eqn. 2.544 is only an implicit equation in Γ.

Further approximations can be made if $\gamma_1 a$ is either very small or very large. For $|\gamma_1 a| \ll 1$ we can use the small-argument approximation of Bessel's functions in eqn. 2.543 to obtain

$$z_{m0} \simeq \frac{j\omega\mu_0}{2\pi} \left[\ln\frac{a}{c} + \frac{K_0(\gamma_1 a)}{\gamma_1 a K_1(\gamma_1 a)} \right] + z_0$$

$$\simeq -\frac{j\omega\mu_0}{2\pi} \ln(0\cdot89 \, \gamma_1 c) + z_0 \qquad \text{for } |\gamma_1 a| \ll 1 \qquad (2.545)$$

It is quite remarkable that this result is independent of the tunnel radius and wire location. This is not surprising since it is the result of the spreading of the return current in a quite large area of the rock medium.

On the other hand, when $\gamma_1 a$ is large, eqn. 2.540 is more suitable. As long as $|\gamma_1 a| \gg m$ for all cylindrical harmonics which are significant in representing the fields at the tunnel wall, we may use the approximation eqn. 2.520 and obtain eqn. 2.542 in closed form. The result is (Pogorzelski, 1979)

$$z_e = \frac{j\omega\mu_0}{2\pi} \left\{ \left[\frac{1}{\gamma_1 a} - \frac{1}{2(\gamma_1 a)^2} + \frac{3}{8(\gamma_1 a)^3} \right] \frac{1+p}{1-p} \right.$$

$$\left. \left[\frac{-2}{(\gamma_1 a)^2} + \frac{2}{(\gamma_1 a)^3} \right] \frac{p}{(1-p)^2} + \frac{1}{(\gamma_1 a)^3} \frac{p(1+p)}{(1-p)^3} \right\}$$

$$\text{for } |\gamma_1 a| \to \infty \qquad (2.546)$$

where $p = (\rho_0/a)^2$.

The approximate formulas 2.545 and 2.546 may be used to solve the mode equation without resorting to complicated computer programs involving the use of Bessel functions with complex arguments. A brute force attack on the problem consists of using eqn. 2.545 for $|\gamma_1 a| < 1$ and eqn. 2.546 for $|\gamma_1 a| > 1$. Pogorzelski (1979) obtained some relatively good results by this method for $\rho_0/a = 0$ and $0\cdot5$, but we observed strong oscillations in the approximate attenuation curves for larger values of (ρ_0/a). This is due not only to the fact that $|\gamma_1 a| \gg m$ is not verified for all significant cylindrical harmonics, but also to the semi-convergent character of eqn. 2.519. These oscillations may be damped somewhat by using only the first term of eqn. 2.546. We are unfortunately forced to the conclusion that quantitative results cannot be obtained without some complicated calculations.

★ *2.12.5 Extensions of quasi-static analysis*

The quasi-static approach developed above can easily be extended to multiple axial wires in a circular tunnel. The electrostatic character of the transverse electric field will be kept, provided the conditions $|va| \ll 1$, $|\kappa_1'| \gg 1$ are satisfied. The first transmission line equation may now be written in matrix form

$$-\frac{d\,I}{dz} = j\omega\,c\,V \qquad (2.547)$$

where I, V are the column matrices of the wire currents and voltages, and c is the matrix of specific capacity coefficients calculated from electrostatic theory.

Things are unfortunately more complicated for the magnetic field. The procedure that needs to be followed is however straightforward. Referring to Section 2.8.2 we simply need to split the potentials in the form $U = U_{pc} + \Delta U$, $V = \Delta V$, as we did for the single-conductor case, and then calculate the approximations due to $|va| \ll 1$ and $|\kappa_1'| \gg 1$. One is led to the second transmission line equation

$$- \frac{dV}{dz} = [j\omega l_{pc} + z_e + z_0] I \tag{2.548}$$

where l_{pc} is the matrix of specific induction coefficients for the TEM case, z_e is a matrix of specific impedances for the external medium and z_0 is the diagonal matrix of the specific external impedances of the cables. Matrix l_{pc} is related to c by the standard relation

$$l_{pc}c = \frac{1}{\sqrt{\epsilon_0 \mu_0}} E \tag{2.549}$$

where E is the unit matrix. l_{pc} and c may be obtained by standard electrostatic methods. Matrix z_e expresses the effect of the finite conductivity of the medium and includes self and mutual impedances. The elements of z_e can be approximated by

$$(z_e)_{ik} = \frac{j\omega\mu_0}{2\pi} \left\{ \frac{K_0(\gamma_1 a)}{\gamma_1 a K_1(\gamma_1 a)} + \sum_{m=1}^{\infty} \frac{2K_m(\gamma_1 a)}{\gamma_1 a K_{m+1}(\gamma_1 a)} \left(\frac{\rho_i \rho_k}{a^2} \right)^m \cos[m(\phi_i - \phi_k)] \right\} \tag{2.550}$$

where (ρ_i, ϕ_i) and (ρ_k, ϕ_k) indicate the locations of the ith and kth wires. This equation provides the extension of eqn. 2.542. As in the single-cable case, the difficulty lies in evaluating z_e.

Calculations using the quasi-static approximation were made for a semi-circular tunnel with a perfectly conducting ground plane, containing either one (Wait and Hill, 1977a) or two (Hill and Wait, 1977b) axial conductors. Some results of this investigation have been presented in Section 2.8.4, where it was observed that the semi-circular tunnel with a perfectly conducting ground plane reduces to the full circular tunnel containing negative images of the conductors. Seidel and Wait (1978a) considered the case of a braided coaxial cable in a semi-circular tunnel.

The transmission line model provides a method for treating axial discontinuities and will be widely used in subsequent chapters. For the time being, we will restrict ourself to mentioning some applications available in the literature. Wait and Hill (1978) considered a semi-circular tunnel containing two axial conductors, one of which has a shunt load. A periodic variation of the conductor impedance was examined by Wait (1978) and this was applied (Seidel and Wait, 1978b) to the periodic insertion of leaky sections in a non-leaky cable. Mode conversions due to non-uniformities of the tunnel wall were also analysed (Seidel and Wait, 1979a).

Apart from the previously mentioned work on the rectangular tunnel (Section

2.9), far less literature is available for non-circular tunnels containing axial cables. The elliptical cross-section was investigated by Seidel and Wait (1979b). Their conclusion was that, for frequencies up to 20 MHz, the specific attenuation is relatively insensitive to the ellipticity if the cable-to-wall separation and the tunnel cross-sectional area are kept constant. A very interesting application of Katsenelenbaum's integral equation to low-frequency propagation along a thin wire in an arbitrarily shaped tunnel was made by Kuester and Seidel (1979). By using the electrostatic field distribution as a trial function for a variational expression, these authors arrive at an equation for the propagation constant and, by making low-frequency approximations, they demonstrate our eqn. 2.545. They have thus shown that the latter is valid at low frequencies, irrespective of the tunnel shape and wire location. This is not unexpected since, when the skin depth in the rock is very large, the tunnel then just plays the role of a thin insulation of the conducting wire located in a conducting medium. The mode eqn. 2.544 in which eqn. 2.545 is used together with the specific capacity c_m for the actual geometrical structure can provide useful results up to a few megahertz.

Finally, before closing this chapter, we would like to mention some applications of the approximate method of complex images (Wait and Spies, 1969; Wait, 1969) to propagation along a thin wire located above a conducting half-space. This method has been applied to the calculation of propagation along a monofilar conductor located in a tunnel bored in a low-conductivity coal seam (Lagace and Emslie, 1978a; Emslie and Lagace, 1978). Although this application may be suspect from a theoretical viewpoint, it gave a fairly good agreement with experimental results. Some tentative extensions were proposed by Bannister and Dube (1978).

Chapter 3

Modes and mode conversion

3.1 Modes and transmission line models

3.1.1 Introduction

In this chapter we shall consider the problem of wave propagation along axial cables strung inside a tunnel within the framework of transmission line theory. By cable we mean either a single-wire conductor or a leaky coaxial cable. The same problem was analysed from the point of view of electromagnetic theory in the previous chapter: this is obviously the only rigorous approach. We shall summarise here the main conclusions which were established for a circular tunnel, but are conjectured to remain valid for other tunnel shapes. Our main interest has been and still is in modes characterised by a propagation factor $\exp(-\Gamma x)$, where

$$\Gamma = \alpha + j\beta \tag{3.1}$$

is the axial complex wave number, and x is the axial coordinate. This notation has been used instead of z as in the previous chapter, because z is preferred for impedances.

A quasi-static approach including a transmission line model can be set up provided at least two conditions are met. Defining a transverse wave number $v = (k_0^2 - \Gamma^2)^{1/2}$, where $k_0 = \omega(\epsilon_0\mu_0)^{1/2} = 2\pi/\lambda_0$ is the free-space wave number and λ_0 the free-space wavelength, it is required that the tunnel size should be significantly larger than $|v^{-1}|$. Since in practice Γ does not frequently differ by more than 50% from k_0, this condition is met for frequencies less than several tens of megahertz for most tunnel sizes. The second condition requires that $|\kappa_1'| \gg 1$, where κ_1' is the complex dielectric constant of the tunnel wall, given by $\kappa_1' = \kappa_1 - j\sigma_1/(\omega\epsilon_0)$, where κ_1 is the dielectric constant and σ_1 is the conductivity. This condition is undoubtedly well satisfied below that transmition frequency, for which the two terms of κ_1' are of equal magnitude. Above it, it requires the dielectric constant of the tunnel wall to be significantly larger than unity. Although the second condition is only weakly satisfied for most media above a few megahertz, the resulting errors are undoubtedly negligible when compared to those due to insufficient knowledge of the electrical parameters of the ground, or to drastic modelling of the tunnel environment.

It has been shown that, under these conditions, the distribution of the transverse electric field in the tunnel space follows the laws of electrostatics, i.e. may be calculated as if the cable and the tunnel wall were perfect conductors. The axial component of the electric field is considered as a second-order effect. As a result, a transmission line voltage V_m may be defined in the usual way by considering the external surface of the cable and the tunnel wall as the transmission line conductors. Denoting by I_m the total current carried by the cable, we have the first transmission line equation

$$\frac{dI_m}{dx} = -j\omega c_m V_m = -y_m V_m \tag{3.2}$$

where c_m is the specific capacity. This parameter, as well as the transverse electric field, can be obtained by standard methods of electrostatic theory, including conformal mapping, image theory, rheographic and other methods. It may be useful to note that the charge per unit length on the cable is given by eqn. 3.2.

The situation is unfortunately not so simple for the magnetic field. A governing parameter for the structure of this field is the propagation constant of the external medium

$$\gamma_1 = [j\omega\mu_0(j\omega\epsilon_0\kappa_1 + \sigma_1)]^{1/2} \tag{3.3}$$

the imaginary part of which is the inverse of the skin depth in this medium. When $|\gamma_1^{-1}|$ is much smaller than the geometrical parameters of the structure cross-section, like the tunnel size and the distance from the cable to the wall, the transverse magnetic field distribution is close to that of the TEM mode with perfect conductors. As may be seen from Fig. 1.2, this condition is sometimes met for highly conducting tunnel walls but only above a few tens of megahertz. The other and opposite case is when $|\gamma_1^{-1}|$ is large compared with the tunnel size. This occurs at low frequencies, i.e. up to a few megahertz, and means that the skin depth in the external medium is several times larger than the tunnel size. We may then consider that the magnetic field induced in the tunnel space by the current flowing in the external medium is negligible. Consequently, the magnetic field may be considered as due to the cable only and may be calculated using the Biot–Savart law.

In spite of this complex behaviour, we may write the second transmission line equation

$$\frac{dV_m}{dx} = -z_{m0}I_m + E_x \tag{3.4}$$

where E_x is the axial electric field at the external surface of the cable. The sign convention which is used here considers the current as positive when it flows along the ground and returns along the conductor. The specific impedance z_{m0} reflects the complex behaviour of the magnetic field and cannot be obtained by simple means. It may be cast into the form

$$z_{m0} = j\omega l_{pc} + z_e \tag{3.5}$$

where l_{pc} is the specific inductance of the monofilar mode for the TEM case, i.e. if we assume that the cable and the tunnel wall are perfect conductors, whereas z_e is called the specific impedance of the external medium considered as a transmission line conductor. The real part of z_e accounts for ohmic losses in the external medium. Its imaginary part expresses the change in specific inductance due to the departure of the magnetic field from the TEM approximation in the tunnel space and in the external medium. Actually, for reasons explained above, z_e is expected to be essentially resistive at high frequencies. It will also be much smaller than ωl_{pc} provided that $\alpha \ll \beta$ in eqn. 3.1.

The decomposition 3.5 of z_{mo} is not adequate at low frequencies since it departs greatly from $j\omega l_{pc}$. It has been shown that, provided the skin depth in the external medium is significantly larger than the tunnel size, z_m becomes independent of the tunnel shape and size, and of the cable location. It is then given by

$$z_{mo} \simeq \frac{\omega \mu_0}{2\pi} \arg \gamma_1 + \frac{j\omega \mu_0}{2\pi} \ln \frac{1}{0\cdot89|\gamma_1 c|} \tag{3.6}$$

where c is the cable radius.

A fundamental property of the quasi-static approximation is that the magnetic field, though having a complicated dependence on frequency, does not vary with the mode propagation constant Γ. This property applies also to the current distribution in the external medium and to z_m. This actually is required to validate the transmission line model.

3.1.2 Application to a monofilar wire conductor

We consider a monofilar wire conductor of radius c, conductivity σ_w and permeability μ_w, located in a tunnel. It is assumed that the wire is made of a good conductor ($\sigma_w \gg \omega\epsilon_0$). As shown in Section 2.4, E_x in eqn. 3.4 is related to the monofilar current by an impedance condition

$$E_x = -z_0 I_m \tag{3.7}$$

where z_0, the specific impedance of the wire, is given by eqns. 2.76 and 2.90. These equations may be approximated by

$$z_0 \simeq \frac{1}{\pi c^2 \sigma_w} \tag{3.8}$$

if the skin depth in the conductor is much larger than the wire radius, and by

$$z_0 \simeq \frac{1}{2\pi c}\sqrt{\frac{j\omega\mu_w}{\sigma_w}} \tag{3.9}$$

when the inverse occurs. It is easily recognised that eqn. 3.8 is the d.c. resistance of the wire, while eqn. 3.9 takes account of the skin effect. As is obvious from eqn. 3.4 this impedance must be added to the specific impedance of the wall z_{mo} to obtain the specific series impedance of the transmission line.

Propagation along a single monofilar wire conductor has thus been reduced to a classical transmission line model. Since transmission line theory is assumed to be known by the reader, we will restrict our attention to some consideration of the values of the transmission line parameters in the present case. the operational parameters are the characteristic impedance $Z_m = \sqrt{(z_{m0} + z_0)/(j\omega c_m)}$, the specific attenuation α and the relative phase velocity $v_r = v/(3 \times 10^8) = \omega \sqrt{\epsilon_0 \mu_0}/\beta$. They depend on the numerous geometrical (tunnel shape and size, wire radius and location) and electrical (σ_w, μ_w, σ_1, κ_1) parameters that describe the structure. The number of these parameters is quite large and we are not able to give tables or diagrams of the operational parameters as functions of the frequency, but we can comment on the values attained.

Fortunately it turns out that the characteristic impedance does not vary over a large range with the structure parameters and with the frequency. This can be explained as follows. The specific capacity is only a logarithmic function of the geometrical parameters (see e.g. eqn. 2.356 for the circular tunnel). Except at very low frequencies, where the wire resistance may dominate, the series impedance ($z_{m0} + z_0$) is essentially a reactance. At low frequencies, the latter is given by the second term of eqn. 3.6: the specific inductance may then be approximated by

$$l_{mo} \simeq \frac{\mu_0}{2\pi} \ln \frac{1}{0 \cdot 89 \, c \sqrt{\omega \mu_0 \sigma_1}} \tag{3.10}$$

It varies inversely with c, ω and σ_1, but the variation is logarithmic. At high frequencies, the specific inductance tends to its TEM approximation, which is again only a logarithmic function of the geometrical parameters (see e.g. eqn. 2.541 for the circular tunnel). Finally it turns out that Z_m is nearly real and decreases smoothly with frequency, but remains confined to a rather limited range of values. Typically it lies in the range 400–600 Ω at 100 kHz and 200–400 Ω at 50 MHz. A more accurate knowledge of Z_m is not required for engineering purposes.

For the same reasons, the relative phase velocity is a smoothly increasing function of frequency. Typical values are 0·4–0·85 at 100 kHz, 0·6–0·9 at 1 MHz, 0·80–0·98 at 10 MHz and 0·9–1·0 at 50 MHz. The upper limit of these ranges is reached for the most conductive tunnel walls.

It is far more difficult to evaluate the specific attenuation, and one has frequently to resort to experimental (Fig. 1.6) or theoretical (Sections 2.8 and 2.9) data, but the latter are only available for a limited range of geometrical structures and electrical parameters. At low frequencies however eqn. 3.6 may be used to estimate the attenuation, provided the ground conductivity is known.

3.1.3 Application to leaky coaxial cables

The transfer of electromagnetic energy through small apertures made in the shield of a coaxial cable was approached briefly in Section 1.5.1 and in some detail in Section 2.5. The reader who skipped Chapter 2 is nevertheless invited to read Sections 2.5.1 and 2.5.5, in which the concept of specific transfer impedance is explained without resorting to advanced electromagnetic theory.

In the present chapter, propagation along a leaky coaxial cable located in a tunnel will be analysed by means of coupled line theory. We consider the exterior and the interior of the cable as two transmission lines, whose currents flow along the tunnel wall and the inner conductor, respectively, and whose common return is considered to be the cable shield. It is assumed that the shield is electrically thin, i.e. that the axial field E_x is continuous across the shield. The shield is modelled by a specific transfer impedance z_t which relates E_x to the shield current I_s by

$$E_x = z_t I_s \tag{3.11}$$

We shall frequently suppose that z_t is an inductance

$$z_t = j\omega m_t \tag{3.12}$$

where the specific transfer inductance m_t does not depend on the frequency or on the mode propagation constant. In more refined analyses, z_t may have a resistive part and may be spatially dispersive (i.e. Γ-dependent).

The voltage V_m and the current I_m of the external transmission line satisfy the transmission line equations 3.2 and 3.4. The subscript m is intended to refer to the term 'monofilar', since it has been shown that, in the limits of the quasistatic approximation, the associated fields have the same distribution as those of a mono-filar wire conductor. The voltage and current inside the coaxial cable are denoted by V_c, I_c and satisfy similar equations. Since $I_s = -(I_c + I_m)$, we may combine eqns. 3.2, 3.4 and 3.11 into

$$\frac{dI_m}{dx} = -y_m V_m \tag{3.13}$$

$$\frac{dV_m}{dx} = -z_m I_m - z_t I_c \tag{3.14}$$

where

$$z_m = z_{m0} - z_t \tag{3.15}$$

Similarly we have

$$\frac{dI_c}{dx} = -y_c V_c \tag{3.16}$$

$$\frac{dV_c}{dx} = -z_c I_c - z_t I_m \tag{3.17}$$

where

$$z_c = z_{c0} - z_t \tag{3.18}$$

It should be observed that z_{m0} and z_{c0} are the specific series impedances for the case of a perfect shield. The equivalent circuit of an elementary length of transmission line is shown in Fig. 3.1.

Before leaving this subject for the time being, we shall discuss some questions of terminology. In coupled line theory the terms monofilar mode and coaxial mode are used for the pairs (V_m, I_m) and (V_c, I_c), and for related quantities like fields, impedances, admittances and so on. In Chapter 2 however we used the same names for solutions characterised by a propagation factor $\exp(-\Gamma x)$. The terminology is thus misleading, since the existence of a propagation factor $\exp(-\Gamma x)$ necessarily implies that of both pairs. In order to avoid any confusion in the remainder of this book, we will refer to the pairs (V_m, I_m) and (V_c, I_c) as the coupled modes and to the exponential solutions as eigenmodes.

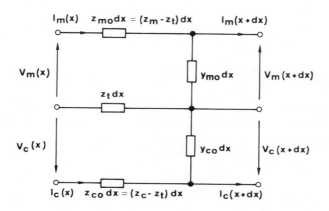

Fig. 3.1 *Equivalent circuit of an elementary length dx of a leaky coaxial cable inside a tunnel*

3.1.4 Extension to multiple wires

In this section we consider the case where several conducting wires are strung parallel to the axis of a tunnel. This includes the long induction loop, the two-wire ribbon feeder, but also the case where a monofilar wire conductor or a leaky feeder are strung in a tunnel which contains other axial conductors like power lines, trolley wires, water and air pipes and so on. These inadvertent conductors will obviously play a role in the propagation of electromagnetic waves along the tunnel. More attention will be devoted to this question in Sections 3.2.4, 3.2.5 and 3.7.

Matrix transmission line equations have been developed in Section 2.12.5 and will not be repeated here. This is a straightforward extension of the models developed in Sections 3.1.1 and 3.1.2. In particular, eqn. 2.547 is the natural extension of eqn. 3.2; we now have to use a matrix of capacity coefficients calculated from electrostatic theory, i.e. obtained by considering the wires and the tunnel wall as perfect conductors. Similarly eqn. 2.548 provides the extension of eqns. 3.4, 3.5 and 3.7. This decomposition of the specific impedance matrix is particularly suitable for the high-frequency case, where the skin depth in the tunnel wall is small.

At low frequencies, when the skin depth in the wall is large compared with the tunnel size, it seems preferable to seek for an extension of eqn. 3.6. For this purpose, let us imagine that only the kth wire carries current and let us write the equations for the kth and ith wires:

$$\frac{dV_k}{dx} = -[(z_{m0})_{kk} + z_{0k}]I_k \tag{3.19}$$

$$\frac{dV_i}{dx} = -(z_{m0})_{ik}I_k \tag{3.20}$$

For this thypothetic situation, the magnetic field is due to I_k only and it is clear that $(z_{m0})_{kk}$ may be calculated as if only the kth wire existed. Consequently, $(z_{m0})_{kk}$ is given by eqn. 3.6, i.e.

$$(z_{m0})_{kk} \simeq \frac{\omega\mu_0}{2\pi}\arg\gamma_1 + \frac{j\omega\mu_0}{2\pi}\ln\frac{1}{0\cdot89|\gamma_1 c_k|} \tag{3.21}$$

where c_k is the radius of the kth wire. On the other hand, if we subtract eqn. 3.20 from eqn. 3.19, the right-hand side must be equal to $j\omega$ times the flux of magnetic induction due to I_k across a unit length of the surface between the two wires. This flux can easily be calculated since we know that the magnetic field may be approximated by the Biot–Savart law. This yields

$$(z_{m0})_{kk} \simeq \frac{\omega\mu_0}{2\pi}\arg\gamma_1 + \frac{j\omega\mu_0}{2\pi}\ln\frac{1}{0\cdot89|\gamma_1 d_{ik}|} \tag{3.22}$$

where d_{ik} is the distance between the ith and kth wires. Eqns. 3.21 and 3.22 apply when the frequency is below a few hundreds of kilohertz; that is, they are applicable to the long induction loop, to trolley-wire systems and so on.

3.2 Coupled-line theory

3.2.1 General theory

We consider the transmission line equations in matrix form

$$\frac{d\boldsymbol{I}}{dx} = -y\boldsymbol{V} \tag{3.23}$$

$$\frac{d\boldsymbol{V}}{dx} = -z\boldsymbol{I} \tag{3.24}$$

where $\boldsymbol{I}$ and $\boldsymbol{V}$ are column vectors with n elements, and y and z are square matrices of order n. This model applies to a system of n conductors strung parallel to the axis of a tunnel, including the leaky coaxial cable ($n = 2$) described by eqns. 3.13, 3.14, 3.16 and 3.17. After derivation and elimination, these equations transform into

$$\frac{d^2 I}{dx^2} = yzI \tag{3.25}$$

$$\frac{d^2 V}{dx^2} = zyV \tag{3.26}$$

These equations have a fundamental set of n linearly independent solutions of the form

$$I(x) = I_0 \exp(\pm \Gamma x) \tag{3.27}$$

$$V(x) = V_0 \exp(\pm \Gamma x) \tag{3.28}$$

where I_0, V_0 and Γ are solutions of the eigenvalue problems

$$yzI_0 = \Gamma^2 I_0 \tag{3.29}$$

$$zyV_0 = \Gamma^2 V_0 \tag{3.30}$$

The fundamental solutions are called eigenmodes. We have implicitly assumed that the eigenvalues are non-degenerate.

As results from reciprocity, y and z are symmetric matrices and, consequently, yz and zy are the transposes of each other. They thus have the same eigenvalues Γ^2, while the eigenvectors I_0 and V_0 are left-hand and right-hand eigenvectors of yz, respectively. Constructing two modal matrices J and U, the columns of which are the eigenvectors I_0 and V_0, respectively, eqns. 3.29 and 3.30 may be written

$$yzJ = J \, Diag \, \Gamma^2 \tag{3.31}$$

$$zyU = U \, Diag \, \Gamma^2 \tag{3.32}$$

where $Diag \, \Gamma^2$ denotes the diagonal matrix of eigenvalues.

However J and U may not be chosen independently since the voltages and currents are related by eqns. 3.23 and 3.24. This imposes the constraint

$$J \, Diag \, \Gamma = yU \tag{3.33}$$

or, equivalently

$$U \, Diag \, \Gamma = zJ \tag{3.34}$$

Finally the general solution of the coupled line equations 3.23–3.24 may be written as

$$V(x) = U \, [Diag \, e^{-\Gamma x} A + Diag \, e^{+\Gamma x} B] \tag{3.35}$$

$$I(x) = J \, [Diag \, e^{-\Gamma x} A - Diag \, e^{+\Gamma x} B] \tag{3.36}$$

where A and B are two column matrices of arbitrary constants to be determined by the boundary conditions. This very compact form of the solutions shows progressive and regressive eigenmodes with generalised matrix amplitudes A and B.

It is important to understand the meaning of the various elements of these solutions. Let us for instance assume that the boundary conditions are such that $A_k \neq 0$, all other elements of A and all elements of B being zero. One has then

$$V(x) = U_k A_k \exp(-\Gamma_k x) \tag{3.37}$$

$$I(x) = J_k A_k \exp(-\Gamma_k x) \tag{3.38}$$

It is seen that A_k is a generalised amplitude factor for a progressive wave of the kth eigenmode. The wire voltages and currents for this mode are A_k times the elements of the kth columns of U and J. A given eigenmode is thus characterised by a well-defined distribution of the line voltages and currents. As results from eqns. 3.23 and 3.24, we may define the characteristic impedance matrix z/Γ_k and the characteristic admittance matrix y/Γ_k of the kth eigenmode.

The eigenvectors are not yet completely defined by eqns. 3.31 to 3.34. Since eqns. 3.31 and 3.32 are homogeneous, we are free to choose $(2n)$ normalising constants for the $(2n)$ eigenvectors. But as J and U are related by eqns. 3.33 or 3.34, n of these constants are determined. It can easily be shown that the eigenvectors U_i and J_j corresponding to different eigenvalues Γ_i and Γ_j are mutually orthogonal. Consequently the matrix $U^T J$, where T indicates transposition, is diagonal and the n remaining constants may be chosen so as to impose the constraint

$$U^T J = E \tag{3.39}$$

where E is the unit matrix. This is the usual scattering matrix normalisation. It is such that the scalars $A^T A/2$ and $B^T B/2$ give the complex power carried through the transverse plane $x = 0$ by the progressive and regressive waves, respectively. Thus $A_i^2/2$ and $B_i^2/2$ are relative to the ith eigenmode.

3.2.2 Application to leaky coaxial cables

A first application of the general coupled-line theory developed above is to leaky coaxial cables. Before starting the analysis, we first consider a perfectly shielded cable with exactly the same internal parameters as the leaky cable under study and located at the same place in the same tunnel. This rather trivial case will be referred to by a subscript 0, as was already done in Section 3.1.3. Matrices z and y are now given by

$$z_0 = \begin{bmatrix} z_{c0} & 0 \\ 0 & z_{m0} \end{bmatrix} \qquad y_0 = \begin{bmatrix} y_{c0} & 0 \\ 0 & y_{m0} \end{bmatrix} \tag{3.40}$$

It is easy to verify that the modal matrices are given by

$$U_0 = \begin{bmatrix} Z_{c0}^{1/2} & 0 \\ 0 & Z_{m0}^{1/2} \end{bmatrix} \qquad J_0 = \begin{bmatrix} Z_{c0}^{-1/2} & 0 \\ 0 & Z_{m0}^{-1/2} \end{bmatrix} \tag{3.41}$$

where

$$Z_{c0} = \sqrt{z_{c0}/y_{c0}} \qquad Z_{m0} = \sqrt{z_{m0}/y_{m0}} \tag{3.42}$$

are the characteristic impedances of the — obviously uncoupled — coaxial and monofilar modes, whereas

$$\Gamma_{c0} = \sqrt{z_{c0}y_{c0}} \qquad \Gamma_{m0} = \sqrt{z_{m0}y_{m0}} \tag{3.43}$$

are their propagation constants.

As was explained previously, $y_{m0} = j\omega c_{m0}$ where the specific capacity may be evaluated from electrostatics. The series impedance

$$z_{m0} = j\omega l_{m0} + r_{m0} \tag{3.44}$$

is nearly reactive ($r_{m0} \ll \omega l_{m0}$), but r_{m0} and l_{m0} are functions of frequency. Similar considerations apply to the coaxial mode, except that l_{c0} may be considered as constant. We may thus use classical approximations for low-loss transmission lines:

$$Z \simeq (l/c)^{1/2}$$

$$\Gamma \simeq \alpha + j\beta; \qquad \alpha \simeq r/(2Z), \qquad \beta \simeq \omega\sqrt{lc} \tag{3.45}$$

$$\beta \gg \alpha$$

with adequate subscripts, either m0 or c0. We define the velocity ratio of the two modes by

$$\rho = \frac{\beta_{c0}}{\beta_{m0}} = \frac{v_{m0}}{v_{c0}} \tag{3.46}$$

Actually $v_{c0} = 3 \times 10^8 \kappa^{-1/2}\,\mathrm{m\,s^{-1}}$, where κ is the dielectric constant of the cable insulation, whereas v_{m0} is a slowly increasing function of frequency. Above a few megahertz however v_{m0} is close to $3 \times 10^8\,\mathrm{m\,s^{-1}}$ and one has

$$\rho \simeq \sqrt{\kappa} \tag{3.47}$$

Hence we consider the leaky coaxial cable. As seen previously, we have

$$y = y_0 \tag{3.48}$$

$$z = \begin{bmatrix} z_c & z_t \\ z_t & z_m \end{bmatrix} = \begin{bmatrix} z_{c0}-z_t & z_t \\ z_t & z_{m0}-z_t \end{bmatrix} \tag{3.49}$$

where $z_t = j\omega m_t$ is the specific transfer impedance of the shield. Once the numerical values of y and z are available, it is an elementary task to compute the eigenvalues and the modal matrices. Instead, we will concentrate on approximate formulas valid for a weak coupling. Derivation of these approximations is a straightforward task which is left to the reader. We will restrict ourself to the presentation of the results.

We define two coupling coefficients[*]

$$C_1 = \frac{m_t}{2\sqrt{l_m l_c}}\,\frac{\sqrt{\rho}}{\rho - 1} \tag{3.50}$$

---

[*]The factor 2 in the denominator of these expressions is missing in Delogne (1976a).

$$C_2 = \frac{m_t}{2\sqrt{l_m l_c}} \frac{\sqrt{\rho}}{\rho + 1} \tag{3.51}$$

where $l_m = l_{mo} - m_t$ and $l_c = l_{co} - m_t$ are the specific inductances of the monofilar and coaxial modes. In the weak-coupling assumption, we suppose that

$$C_1^2, C_2^2 \quad \text{and} \quad C_1 C_2 \ll 1 \tag{3.52}$$

In practice, the values of l_{co} range from 250 to 400 nH m^{-1}, those of l_{mo} from 800 to 4000 nH m^{-1}, while $m_t = 40$ nH m^{-1} must be considered as an upper limit for specific transfer inductances. The weak-coupling assumption may thus be considered as generally valid, excepted when the velocity ratio ρ is close to unity.

The approximations which are found for the propagation contants of the two eigenmodes are then

$$\Gamma_1 = \alpha_1 + j\beta_1; \qquad \alpha_1 \simeq \alpha_{co} + (C_1 - C_2)^2 \alpha_{mo} \tag{3.53}$$

$$\beta_1 \simeq \beta_{co} \tag{3.54}$$

$$\Gamma_2 = \alpha_2 + j\beta_2; \qquad \alpha_2 \simeq \alpha_{mo} \tag{3.55}$$

$$\beta_2 \simeq \beta_{mo} \tag{3.56}$$

It should already be clear that the subscripts 1 and 2 refer to the coaxial and monofilar eigenmodes, respectively, whereas the subscripts c and m are used for the coupled modes, i.e. for signals propagating inside and outside the cable. These signals are of course a linear combination of the two eigenmodes.

The modal matrices are approximately given by

$$U \simeq \begin{bmatrix} Z_{co}^{1/2} & -Z_{co}^{1/2}(C_1 - C_2) \\ Z_{mo}^{1/2}(C_1 + C_2) & Z_{mo}^{1/2} \end{bmatrix} \tag{3.57}$$

$$J \simeq \begin{bmatrix} Z_{co}^{-1/2} & Z_{co}^{-1/2}(C_1 + C_2) \\ Z_{mo}^{-1/2}(C_1 - C_2) & Z_{mo}^{-1/2} \end{bmatrix} \tag{3.58}$$

The meaning of the modal matrix elements given by eqns. 3.37 and 3.38 should be kept in mind when interpreting these results. For instance, it is seen that when the coaxial eigenmode propagates, for a unit voltage inside the cable, there is a wall voltage equal to $(C_1 + C_2)\sqrt{Z_{mo}/Z_{co}}$, and for a unit current flowing along the inner conductor, there is a current $(C_1 - C_2)\sqrt{Z_{co}/Z_{mo}}$ flowing along the wall. Of course, the current I_m flowing along the wall is equal to the opposite of the total current carried by the cable, as is obvious from Fig. 3.1.

The coaxial eigenmode obviously has the main part of its power propagating inside the cable. Some power is however also carried by the leakage fields outside the cable. These powers are given by

$$\tfrac{1}{2} V_c I_c^* = \tfrac{1}{2} |A_1^2|$$

$$\tfrac{1}{2} V_m I_m^* = \tfrac{1}{2} |A_1^2|(C_1^2 - C_2^2) \tag{3.59}$$

where A_1 is the normalised wave amplitude of the coaxial eigenmode. Similar considerations and formulas apply to the monofilar eigenmode. It thus appears that $(C_1^2 - C_2^2)$ is the relative power of the leakage fields. This explains also in a very simple way the increase of specific attenuation for the coaxial eigenmode given by eqn. 3.53. A similar peturbation exists for α_m but it has been neglected since it has been assumed that α_{mo} is significantly higher than α_{co}. Otherwise a leaky coaxial cable would not perform better than a simple monofilar wire conductor.

We may further comment on the coupling coefficients C_1 and C_2. The specific transfer inductance m_t is only a shield parameter and it does not by itself give full information on the intensity of the leakage fields. The relevant parameters are the coupling coefficients C_1, C_2. They also depend on the internal and external parameters through l_c, l_m and the velocity ratio ρ as shown by eqns. 3.50 and 3.51. In particular, a value of ρ close to unity may have more influence on the leakage intensity than a high transfer inductance. At high frequencies, where eqn. 3.47 is valid, this result is obtained by using a cable insulation with a small dielectric constant. For $\rho \simeq 1$, we have $C_1 \gg C_2$, but the weak-coupling assumption eqn. 3.52 does not necessarily hold any longer. Finally it should be noted that too high a value of C_1 may have drawbacks in continuous leaky feeders, since it may result in an important increase of α_1, as shown by eqn. 3.53.

3.2.3 Discontinuities along a leaky feeder: end effects

The theory developed in the previous section applies to a homogeneous leaky feeder in the sense given to this term in transmission line theory, i.e. when the transmission line parameters are independent of the axial coordinate x. Inhomogeneities can be treated by several methods which are standard practice in transmission line theory. In particular, local inhomogeneities or discontinuities may be represented by lumped circuit elements inserted in the line and give rise to boundary conditions to be used in the general solution eqns. 3.35 and 3.36. Quite generally, a discontinuity acts on both eigenmodes. This is always true for a discontinuity which acts on only one of the coupled modes, either on (V_m, I_m) or on (V_c, I_c) Consequently, if one eigenmode is incident on the discontinuity, the latter will act as a source for the other eigenmode. This is the base of intentional and inadvertent mode conversions which will be studied in subsequent sections.

A particular type of discontinuity is a radio transmitter located in the tunnel space. The transmitter is coupled to the electromagnetic wave associated with (V_m, I_m) and is thereby a source for both eigenmodes. Similarly a radio receiver is coupled to the electromagnetic waves of both eigenmodes. The modelling of radio transceivers by active lumped sources for the transmission line description will be undertaken in Chapter 4.

Circuits connected at the ends of the leaky feeder are also in general coupled to both eigenmodes. A particular example of this is when a base station is connected to the cable. The task of writing the analytical solution for this case is rather simple and is left to the reader (Degauque *et al.*, 1976). Instead of proceeding to this calculation, we will discuss qualitatively the solution with a view to obtaining a good understanding of the physics of leaky feeders. Let us suppose that a generator is connected to the input of a semi-infinite cable located along the positive half of the x-axis. We assume that the generator has no connection to the ground, i.e. to the tunnel wall, and that no cable exists for $x < 0$. We have thus $I_m(0) = 0$. Progressive waves of both eigenmodes are excited by the generator, but the monofilar current is therefore necessarily of the form

$$I_m(x) = I_{m0}[\exp(-\Gamma_1 x) - \exp(-\Gamma_2 x)] \tag{3.60}$$

This behaviour of the monofilar current is illustrated in Fig. 3.2 where logarithmic units are used in the ordinate. Straight lines 1 and 2 show the amplitudes of the two terms of eqn. 3.60 and have slopes α_c and α_m, respectively. Curve 3 exhibits the total current $I_m(x)$. Apparent standing waves are due to the fact that the two terms of eqn. 3.60 travel with different phase velocities and are successively in phase and in antiphase. It can easily be shown that the distance between two nodes is equal to $\lambda_1 \lambda_2 / (\lambda_2 - \lambda_1)$ where λ_1, λ_2 are the wavelengths of the eigenmodes. The envelope of the pseudomaxima starts 6 dB above I_{m0}. Note that curve 3 is also representative of the magnetic field in the tunnel space along a path parallel to the x-axis.

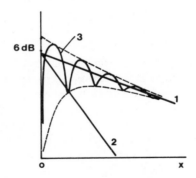

Fig. 3.2   *Leakage fields of a leaky feeder fed by a generator without connection to the ground*

A fundamental difference with classical standing waves is worth mentioning. In the common situation where standing waves are due to two waves travelling with the same velocity but in opposite directions, the standing wave patterns for current and voltage are offset by a quarter of a wavelength. This does not occur here. On the contrary the standing wave patterns for $V_m(x)$ and $I_m(x)$ are coincident. The reason for this is that the ratio V/I is positive for progressive waves and negative for regressive waves. Consequently, curve 3 is also valid for $V_m(x)$, i.e. for the electric field along the tunnel, and finally also for the power flow in the tunnel space. Actually the power of the leakage fields jumps alternately in and out of the cable. Similar phenomena occur near the end of the cable when the latter is terminated in a matched load for the coupled modes. Things are then even more complicated since we have to deal with one progressive and two regressive eigenmodes.

These end effects are rather curious, but they do not lead to useful applications. We will now examine with some detail another type of discontinuity that will find application in Section 3.5. We consider the transition from a non-leaky coaxial cable to a leaky one. The transition is located at $x = 0$: waves $(A_{c0}, A_{m0}, B_{c0}, B_{m0})$ of the perfectly shielded cable are flowing for $x < 0$ and waves (A_1, A_2, B_1, B_2) of the leaky cable are flowing for $x > 0$. The problem can be solved by using eqns. 3.35 and 3.36, with suitable subscripts, on either side of the transition and expressing the continuity of the voltages and currents at the transition. The resultant maxtrix equations

$$U_0(A_0 + B_0) = U(A + B) \tag{3.61}$$

$$J_0(A_0 - B_0) = J(A - B) \tag{3.62}$$

can be solved for B_0 and A to obtain the scattering matrix of the transition. Under the weak-coupling assumption eqn. 3.52, this yields

$$
\begin{bmatrix} B_{c0} \\ B_{m0} \\ A_1 \\ A_2 \end{bmatrix}
=
\begin{bmatrix}
-C_1 C_2 & C_2 & \sqrt{1-C_1^2-C_2^2} & -C_1 \\
C_2 & C_1 C_2 & C_1 & \sqrt{1-C_1^2-C_2^2} \\
\sqrt{1-C_1^2-C_2^2} & C_1 & -C_1 C_2 & -C_2 \\
-C_1 & \sqrt{1-C_1^2-C_2^2} & -C_2 & C_1 C_2
\end{bmatrix}
\begin{bmatrix} A_{c0} \\ A_{m0} \\ B_1 \\ B_2 \end{bmatrix}
\tag{3.63}
$$

It can be seen that an eigenmode incident on the transition is:

Reflected with a small reflection coefficient $\pm C_1 C_2$;

Reflected into the other eigenmode with a reflection coefficient $\pm C_2$;

Transmitted through the transition in the corresponding mode with a very small loss; and

Transmitted through the transition into the other mode with a transmission coefficient $\pm C_1$.

3.2.4 Application to multiple-wire systems

Apart from the long induction loop, multiple wires are not intentionally used to guide electromagnetic waves in tunnels. However, particularly in mine tunnels, inadvertent axial conductors may exist and will perturb the intended propagation modes. The problem can be reated by the general theory outlined in Section 3.2.1. The main difficulties encountered in the calculations are listed below with indications about methods which have been found efficient in solving them. Some computing facility is necessary when more than two wires are considered.

(*a*) *Evaluating matrix y*

If the conductor radii are much smaller than the distances between them and with respect to the wall, the matrix of capacity coefficients can be found by using a specific thin-wire approximation (Frankel, 1977). In the particular case of a circular tunnel, one may calculate a matrix S with elements

$$S_{kk} = \ln \frac{a^2 - \rho_k^2}{ac_k} \tag{3.64}$$

$$S_{ki} = \tfrac{1}{2} \ln \frac{\rho_i^2 \rho_k^2 + a^4 - 2a^2 \rho_i \rho_k \cos(\phi_i - \phi_k)}{a^2 [\rho_i^2 + \rho_k^2 - 2\rho_i \rho_k \cos(\phi_i - \phi_k)]} \tag{3.65}$$

where a is the tunnel radius, c_k is the radius and (ρ_k, ϕ_k) are the polar coordinates of the kth wire, and then obtain the matrix of capacity coefficients by $c = 2\pi\epsilon_0 S^{-1}$.

(*b*) *Evaluating matrix z*

This is undoubtedly the most difficult part of the problem. Methods for solving it have been discussed in Section 3.1.4.

(*c*) *Solving the mode equation*

Consider the complex matrices $P = yz$ and $Q(\lambda) = \lambda E - P$, where E is the unit matrix of order n. The characteristic polynomial is the complex polynomial

$$f(\lambda) = \mathrm{Det}\, Q(\lambda) = \lambda^n + a_1 \lambda^{n-1} + \ldots + a_n$$

and the mode equation is $f(\lambda) = 0$, where λ stands for Γ^2. The following procedure for computing and solving the mode equation was found efficient:

(i) Select n values x_i of λ and calculate $f_i = f(x_i)$. For ease and efficiency, one may use real values equispaced in the interval $(-2k_0^2, 0)$ where k_0 is the free-space wave number. Calculating f_i as the determinant of $Q(x_i)$ may be performed by first diagonalising this matrix by Gauss's pivotal-condensation method (Korn and Korn, 1961).

(ii) Obtain the coefficients a_i by solving the system of equations

$$a_1 x_i^{n-1} + \ldots + a_n = f_i - x_i^n; \qquad i = 1, \ldots, n$$

This task is facilitated by the choice of real values x_i. Gauss's method may again be used here.

(iii) Calculate the first root of $f(\lambda) = 0$ by Newton–Raphson's iteration method

$$\lambda^{(n+1)} = \lambda^{(n)} - f(\lambda^{(n)})/f'(\lambda^{(n)})$$

using $-k_0^2$ as a trial value. The polynomial and its derivative are calculated by the process of successive multiplications

$$f(\lambda) = \{[(\lambda + a_1)\lambda + a_2]\lambda + \ldots \}\lambda + a_n$$

$$f'(\lambda) = \{[n\lambda + (n-1)a_1]\lambda + (n-2)a_2 \ldots \}\lambda + a_{n-1}$$

(iv) To find the next root, carry out the long division

$$g(\lambda) = f(\lambda)/(\lambda - \lambda_1)$$

where λ_1 is the first root, using the recursion

$$b_1 = \lambda_1 + a_1$$

$$b_k = \lambda_1 b_{k-1} + a_k; \qquad k = 1, \ldots, n-1$$

to calculate the coefficients of $g(\lambda)$. Repeat steps (iii) and (iv), reducing the degree of the polynomial by one unit at each step.

(*d*) *Obtaining the eigenvectors*

The elements of the ith eigenvector J_i are calculated as the cofactors of the elements of an arbitrary line of $Q(\lambda_i)$. When the modal matrix J is complete, U is obtained from eqn. 3.33. It remains to normalise the columns of these matrices according to eqn. 3.39.

As a useful exercise, we shall now discuss some results concerning the long induction loop at 150 kHz in a circular tunnel. The tunnel parameters are: radius 2 m, conductivity $0.01\,\mathrm{S\,m^{-1}}$, dielectric constant 10. The long induction loop is made up of two wires with radius 1 mm, conductivity $5 \times 10^7\,\mathrm{S\,m^{-1}}$ and relative permeability equal to unity. Initially we consider the symmetrical location $\rho_1 = \rho_2 = 1.9\,\mathrm{m}$, $\phi_1 = 0°$ and $\phi_2 = 180°$. Two eigenmodes are found:

Mode 1: attenuation 0.02 dB/100 m.
relative velocity $k_0/\beta = 0.741$
Mode 2: attenuation 0.29 dB/100 m
relative velocity $k_0/\beta = 0.757$

The modal matrices are

$$J = \begin{bmatrix} 0.0367 & 0.0325 \\ & \exp(j2°) \\ 0.0367 & 0.9325 \\ \exp(180°) & \exp(j2°) \end{bmatrix} \qquad U = \begin{bmatrix} 13.6 & 15.5 \\ & \exp(-j2°) \\ 13.6 & 15.4 \\ \exp(j180°) & \exp(-j2°) \end{bmatrix}$$

Mode 1 is obviously the balanced or bifilar mode, whereas mode 2 is the monofilar mode.

The next step of the exercise is to consider asymmetrical locations of the two wires. Leaving wire 1 unchanged, we move wire 2 to the location $\rho_2 = 1.8\,\mathrm{m}$, $\phi_2 = 180°$. The results of the calculations are now:

Mode 1: attenuation 0.07 dB/100 m
relative velocity $k_0/\beta = 0.766$
Mode 2: attenuation 0.23 dB/100 m
relative velocity $k_0/\beta = 0.778$

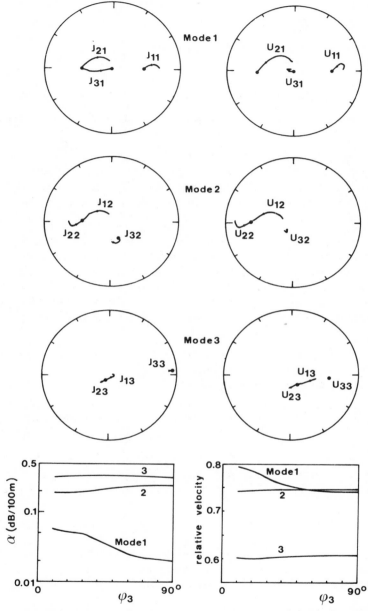

Fig. 3.3   *Eigenvectors, specific attenuation and relative velocities of the three modes for a circular tunnel containing a long induction loop and an iron tube. The arrows indicate the evolution of the eigenvector elements in the complex plane when the azimuth ϕ_3 of the iron tube is varied from 10° to 90°. The dot is for 90°. The circles indicate the scale and have a radius of 0·08 for J and of 25 for U.*

$$J = \begin{bmatrix} 0.0488 & 0.0395 \\ \exp(-j18°) & \exp(j36°) \\ 0.0404 & 0.0433 \\ \exp(-j152°) & \exp(-j21°) \end{bmatrix} \quad U = \begin{bmatrix} 18.7 & 17.4 \\ \exp(-j34°) & \exp(j26°) \\ 17.0 & 21.1 \\ \exp(-j156°) & \exp(-j20°) \end{bmatrix}.$$

Amplitude equality and phase opposition are seriously perturbed for mode 1. Some current is now flowing along the tunnel wall and, consequently, the attenuation increases by a factor of more than three. The inverse occurs for mode 2 because less current is flowing along the tunnel wall.

The exercise is continued by returning to the initial symmetrical wire location, but now adding a third conductor. The latter is an iron tube with radius 50 mm, conductivity $10^6\,S\,m^{-1}$ and relative permeability 200. We assume $\rho_3 = 1.8\,m$, while ϕ_3 is varied from $10°$ to $90°$. The results are shown in Fig. 3.3. The figure exhibits the attenuation and relative velocity of the three modes and, in the complex plane, the elements of the modal matrices as ϕ_3 varies.

The modes are most easily characterised by their properties in the symmetric case ($\phi_3 = 90°$). Mode 1 is the bifilar mode: current flows along wire 1 and returns through wire 2, while no net current flows in the tube and in the tunnel wall. Not surprisingly the attenuation is not influenced by the iron tube, but the relative velocity is increased. The reason for this is that the tube to some extent plays the role of an electrostatic screen between the transmission line and the ground. In the non-symmetrical case however the currents are seriously perturbed. The sum of the three currents is non-zero, which means that a net current is flowing in the tunnel wall. The specific attenuation is thereby increased. For small angles, a large part of the current flowing along conductor 1 returns through the iron tube, because of the relatively large value of the partial capacity coefficient C_{13}.

Similar considerations apply to modes 2 and 3. In mode 2, current flows along the two wires in parallel and returns through the ground and the tube. In mode 3, current flows along the tube and returns mainly through the ground.

3.2.5 The dedicated-wire technique

Electric traction is frequently used in mines and the trolley wire is sometimes used as a monofilar wire conductor to provide radio communications at a few hundreds of kilohertz in the haulageways. Actually the rails act as a return conductor; they may be modelled as a metallic ground plane (Wait and Hill, 1977a). The trolley wire is primarily designed for power transmission and it is not a very efficient monofilar wire for radio transmission. Devices like locomotives, rectifiers and so on act as shunt loads with rather unpredictable impedance at radio frequencies. Attempts have been made to use radio frequency chokes in series with these loads. Alternatively an additional monofilar wire is strung in the tunnel to guide electromagnetic waves. This method was proposed in the United States under the name of the dedicated-wire technique (Emslie et al., 1978).

Although there is no reason why it should have more influence than any inadvertent conductor that may exist in the tunnel, particular attention was devoted to the trolley wire, and specifically to the effect of its shunt loading on the propagation of the low-loss mode guided by the dedicated wire (Hill and Wait, 1977b; Wait and Hill, 1978). The problem can be treated by multiconductor transmission line theory. In the first instance the metallic ground plane modelling the rails is taken into account by considering a doubled-size tunnel containing image wires, as suggested by Fig. 3.4. Attention should obviously be restricted to odd modes. Solving the problem of a shunt load then reduces to a rather elementary problem.

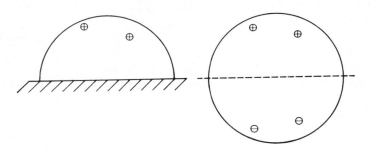

Fig. 3.4   *Equivalence of tunnel with metallic ground plane and tunnel with doubled sized and image wires*

Let us assume that the shunt load is located at $x = 0$ and that eigenmodes with matrix wave amplitudes A^- and B^+ are incident on the load for $x < 0$ and $x > 0$, respectively. Using eqns. 3.35 and 3.36, we may write the solution

$$V(x) = \begin{cases} U\left[Diag\ e^{-\Gamma x}A^- + Diag\ e^{\Gamma x}B^-\right] \\ U\left[Diag\ e^{-\Gamma x}A^+ + Diag\ e^{\Gamma x}B^+\right] \end{cases} \tag{3.66}$$

$$I(x) = \begin{cases} J\left[Diag\ e^{-\Gamma x}A^- - Diag\ e^{\Gamma x}B^-\right] \\ J\left[Diag\ e^{-\Gamma x}A^+ - Diag\ e^{\Gamma x}B^+\right] \end{cases} \tag{3.67}$$

where B^- and A^+ are the still unkown matrix amplitudes of the reflected and transmitted waves. Quite generally, the shunt load may be characterised by an admittance matrix Y and we have the continuity equations

$$V(-0) = V(+0)$$
$$I(-0) = YV(+0) + I(+0) \tag{3.68}$$

Using these as boundary conditions for eqns. 3.66 and 3.67, we obtain the scattering matrix of the shunt load

$$\begin{bmatrix} B^- \\ A^+ \end{bmatrix} = \begin{bmatrix} K & T \\ T & K \end{bmatrix} \begin{bmatrix} A^- \\ B^+ \end{bmatrix} \tag{3.69}$$

where

$$T = [E + \tfrac{1}{2} J^{-1} YU]^{-1} \tag{3.70}$$

$$K = T - E \tag{3.71}$$

The submatrices K and T are of order n if there are n wires and are generalised reflection and transmission matrices. They can easily be calculated if the modal matrices J and U are known. In general, mode conversion occurs in reflection as well as in transmission, in the sense that, if one eigenmode is incident on the shunt load, all eigenmodes are generated in the two directions away from the discontinuity. This conclusion is of practical importance and provides a good introduction to mode conversion techniques. Note that reciprocity requires K and T to be symmetric.

3.3 General properties of mode converters

By comparison with the simple monofilar-wire technique, the main idea in the use of continuous leaky coaxial cables is to benefit by the low specific attenuation of the coaxial eigenmode. This attenuation remains small because only a small part of the electromagnetic energy is released in the tunnel space in the form of leakage fields. An alternative solution to the continuous leaky feeder is to use a well-shielded coaxial cable in which mode converters are inserted at discrete places. This principle can also be used with a two-wire line like a ribbon feeder. The function of the mode converters is to convert a small part of the coaxial or bifilar mode power into the monofilar mode supported by the non-leaky transmission line. An obvious requirement for the mode converters is to provide a good impedance match and a low insertion loss for the coaxial or bifilar mode: ideally the loss should be due to the mode conversion only.

This method provides excellent flexibility in the design of a subsurface communication system because the converter parameters and spacing can be varied along the path as functions of the tunnel cross-section, acceptable cable location, distance to the base station, and so on. Mode converters are now widely used in subsurface radio communications (Delogne, 1971; De Keyser, 1972; Delogne, 1972; Deryck, 1972; Delogne, 1973; De Keyser, 1973; Delogne *et al.*, 1973; Delogne, 1974; De Keyser, 1974; Deryck, 1975; De Keyser *et al.*, 1978; Siedel and Wait, 1978c; Delogne, 1979).

A detailed description of several types of mode converter will be given in subsequent sections. Shortly we will show that the operational parameters of mode converters are subject to some limitations. The analyses carried out in the present chapter are restricted to frequencies where only transmission line type modes are significant, but it will be seen in Chapter 5 that the same devices can be used as efficient launchers for waveguide modes.

When the existence of two modes on either side is considered, a mode converter is seen as a four-port device. We will use the subscripts 1 and 2 to denote the two sides of the converter and the subscripts c and m to indicate the coaxial (or bifilar) and monofilar modes, respectively. We use the normalised waves A and B which are flowing into and out of the device, respectively. By reciprocity, the scattering matrix which relates them should be symmetric. It has thus the following most general form

$$
\begin{bmatrix} B_{c1} \\ B_{m1} \\ B_{c2} \\ B_{m2} \end{bmatrix} = \begin{bmatrix} K_{c1} & K_{cm1} & T_c & T_{c1m2} \\ K_{cm1} & K_{m1} & T_{c2m1} & T_m \\ T_c & T_{c2m1} & K_{c2} & K_{cm2} \\ T_{c1m2} & T_m & K_{cm2} & K_{m2} \end{bmatrix} \begin{bmatrix} A_{c1} \\ A_{m1} \\ A_{c2} \\ A_{m2} \end{bmatrix}
\tag{3.72}
$$

The meaning of the scattering matrix elements is as follows and is illustrated in Fig. 3.5:

K_{c1}, K_{c2}: reflection factors of the coaxial mode at sides 1 and 2;
K_{m1}, K_{m2}: reflection factors of the monofilar mode at sides 1 and 2;
K_{mc1}, K_{mc2}: mode conversion factors in reflection at sides 1 and 2;
T_{c1m2}: mode conversion factor in transmission between the coaxial mode at side 1 and the monofilar mode at side 2;
T_{c2m1}: same, between the coaxial mode at side 2 and the monofilar mode at side 1.

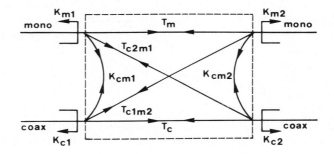

Fig. 3.5 *Flow graph to illustrate the meaning of the scattering matrix elements of a mode converter*

The mode converters are most frequently lossless devices and the scattering matrix must be self-adjoint. Moreover most types have a transverse symmetry plane for the monofilar mode and either a symmetry or an antisymmetry plane for the coaxial mode. When combined, these properties yield:

$$
|K_{c1}| = |K_{c2}| = K_c
$$
$$
|K_{m1}| = |K_{m2}| = K_m
$$

$$|K_{cm1}| = |K_{cm2}| = K_{cm}$$
$$|T_{c1m2}| = |T_{c2m1}| = T_{cm} \tag{3.73}$$

$$|K_c^2| + |T_c^2| + |K_{cm}^2| + |T_{cm}^2| = 1 \tag{3.74}$$

$$|K_m^2| + |T_m^2| + |K_{cm}^2| + |T_{cm}^2| = 1 \tag{3.75}$$

and some phase relationships. The transverse symmetry plane thus does not necessarily imply equality of the mode conversion factors in reflection and in transmission, or equality of the reflection factors or of the transmission factors for the two modes.

For obvious practical reasons, mode converters should have no connection to the tunnel wall. If the mode converter is made with lumped elements, this implies $I_{m1} + I_{m2} = 0$, or

$$A_{m1} - B_{m1} = A_{m2} - B_{m2} \tag{3.76}$$

As this condition must be satisfied for all combinations of incident waves, one has

$$K_{cm1} = -T_{c1m2} \tag{3.77}$$

$$K_{cm2} = -T_{c2m1} \tag{3.78}$$

$$K_{m1} + T_m = K_{m2} + T_m = 1 \tag{3.79}$$

In particular, it is seen that a coaxial wave incident on the converter always excites two equal monofilar waves in opposite directions. From this we conclude that a directional mode conversion can only be realised by mode converters with a non-zero electrical length.

Finally, if the mode converter is lossless, has a transverse symmetry plane, no ground connection and a zero electrical length, it is described by reflection factors K_c, K_m, transmission factors T_c, T_m and a single mode conversion factor S, with the property

$$K_c^2 + T_c^2 + 2S^2 = K_m^2 + T_m^2 + 2S^2 = 1 \tag{3.80}$$

3.4 Annular-slot mode converters

3.4.1 Principle

A short annular slot realising a complete interruption of the outer conductor of a non-leaky coaxial cable has been proposed as the main ingredient of a mode converter. The relevant electromagnetic aspects have been treated in Section 2.10.3. A simplified quasi-static analysis can be obtained by considering the cable and the tunnel as two transmission lines having a common conductor. The latter is interrupted over a short length. The problem thus reduces to an elementary circuit calculation suggested by Figs. 3.6a and b, where Z_m and Z_c are the characteristic

impedances of the monofilar and coaxial modes, respectively. The only difficult
point in this respect is to estimate the value of Z_m, but we have seen that this
quantity fortunately varies over a very limited range with the electrical parameters
of the ground, with the geometrical parameters such as tunnel cross-section and
cable location, and with the frequency. A comparison with the few exact electromagnetic solutions available between 1 and 50 MHz shows that calculations based
on the value $Z_m = 377\,\Omega$ never yield an error larger than 1 dB for the mode conversion factor. This value has been selected since it allows the simplification of
some calculations in Chapter 4.

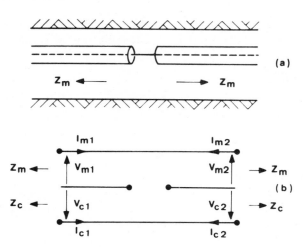

Fig. 3.6   *Principle and equivalent circuit for the quasi-static analysis of the annular-slot mode
converter (shown without lumped circuit elements)*

A naked slot does not provide a good impedance match and a low insertion loss
for the coaxial mode. The external load impedance 'seen' by the slot $(2Z_m)$ is
indeed rather large, so that most power flowing inside the coaxial cable is reflected
back inside the cable. It is thus necessary to add some lumped circuit elements in
order to improve the impedance match and to lower the insertion loss.

3.4.2 Resonant matching

In this type of mode converter, the slot impedance is deliberately lowered by connecting a reactance across it. The residual reactive effect can further be compensated
at the design frequency by inserting a dual reactance in the inner conductor. Figs.
3.7a and b show two such designs. Eqns. 3.72 to 3.80 all apply to this circuit. The
elements of the scattering matrix can easily be obtained from the circuit equations,
but the calculations are somewhat tedious. The results are shown in Fig. 3.8 for a
typical value of the ratio Z_c/Z_m. The curves are drawn as functions of a reduced

frequency v and of a Q-factor. These parameters are defined as follows:

$$\omega_0 = (LC)^{-1/2}$$

and

Fig. 3.7a: $Q = 2Z_m/(\omega_0 L);$ $v = \omega_0/\omega$

Fig. 3.7b: $Q = 2Z_m\omega_0 C;$ $v = \omega/\omega_0$

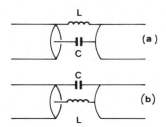

Fig. 3.7   *Resonant matching of an annular-slot mode converter*

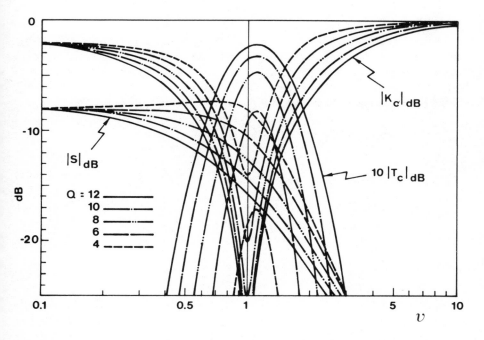

Fig. 3.8   *Performance curves of selective annular-slot mode converter*

The choice of the circuit elements allows some control over the converter performance. Typical values for Q are 6 to 8. The circuit provides a moderate bandwidth. The parameters K_m and T_m have not been shown because they are not very important in most applications, but it may be useful to know that K_m is very small and T_m close to unity.

3.4.3 Wideband matching

When mode converters have to work over very wide frequency bands, a transformer may be integrated in the annular slot. In these types of converter the bandwidth is only limited by the transformer frequency response.

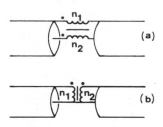

Fig. 3.9   *Wideband annular-slot mode converters*

A first example is shown in Fig. 3.9a. The winding direction is such that the coaxial mode would not develop any voltage across the slot if n_1 and n_2 were equal: hence no mode conversion would occur. A small imbalance allows the creation of some mode conversion. The device has all the properties considered in the last paragraph of Section 3.3. It is easy to show that

$$K_c = T_m = \frac{\xi}{2 + \xi}$$

$$K_m = T_c = \frac{2}{2 + \xi} \tag{3.81}$$

$$S = \frac{(2\xi)^{1/2}}{2 + \xi}$$

where

$$\xi = \frac{2Z_m}{Z_c}(1 - n_1/n_2)^2 \tag{3.82}$$

A small imbalance will thus yield a low value for the reflection factor and the insertion loss of the coaxial mode, but a high value for the same parameters of the monofilar mode.

When this is not desirable, the converter of Fig. 3.9b may be used. In spite of the apparent simplicity of the circuit, the calculation of the scattering matrix elements is extremely tedious, for there is no transverse symmetry plane. The general form eqn. 3.72 together with the properties eqns. 3.77 to 3.79 apply. One finds:

$$K_{c1} = [2(n_1^2 - n_2^2)x - (n_1 - n_2)^2 x^{-1}]/D$$

$$K_{c2} = [2(n_2^2 - n_1^2)x - (n_2 - n_1)^2 x^{-1}]/D$$

$$K_{m1} = K_{m2} = (n_1 - n_2)^2 x^{-1}/D$$

$$T_c = 4n_1 n_2/D \qquad\qquad (3.83)$$

$$T_m = 2(n_1^2 + n_2^2)x/D$$

$$K_{cm1} = -T_{c1m2} = 2n_1(n_1 - n_2)/D$$

$$K_{cm2} = -T_{c2m1} = 2n_2(n_2 - n_1)/D$$

where

$$x = (Z_m/Z_c)^{1/2}$$

$$D = 2(n_1^2 + n_2^2)x + (n_1 - n_2)^2 x^{-1} \qquad\qquad (3.84)$$

A small imbalance here allows small values of the reflection factor and of the insertion loss for both modes. It is thus easy to fit the characteristics of mode converters to the requirements of a specific application.

3.5 Leaky section as a directive mode converter

The mode converters described so far are non-directive in the sense that the mode conversion factors in reflection and transmission are equal. A problem may result from this property when several mode converters are inserted in a cable. Standing waves are observed in the tunnel space because of monofilar mode waves travelling in opposite directions. In the early stages of the use of annular-slot mode converters (De Keyser *et al.*, 1970; Delogne, 1970; De Keyser, 1972; Delogne, 1972), this problem was considered serious and it was solved by a classical technique used in waveguide directional couplers: the slots were used in pairs with a quarter-wavelength spacing between the elements of a pair. Later on, experience showed that standing waves frequently arise anyway because of inadvertent mode conversions in tunnels containing conductors other than the coaxial cable, and this method was abandoned.

We will now retrieve directional mode conversion as automatically provided in converters consisting of a short section of a leaky coaxial cable inserted in a non-leaky cable. The transition from a non-leaky cable to a leaky one was investigated in Section 3.2.3 and is described by the scattering matrix 3.63. Mode conversion occurs in the two directions at the transition but, of course, the coaxial and mono-filar modes are not identical on either side of the transition. A non-zero length mode converter can be built by inserting a leaky section of length L inside a non-leaky cable. In this case, the two transitions participate in the mode conversion process. The scattering matrix of the leaky section considered as a whole can easily be calculated, referring to Fig. 3.10 and eqn. 3.63. Neglecting the second-order terms as permitted by the weak-coupling assumption eqn. 3.52, we find[*]

$$|K_c| \simeq |K_m| \simeq 0 \tag{3.85}$$

$$|K_{cm}| \simeq 2C_2 \sin \left[(\beta_{c0} + \beta_{m0})L/2\right] \tag{3.86}$$

$$|T_{cm}| \simeq 2C_1 \sin \left[(\beta_{c0} - \beta_{m0})L/2\right] \tag{3.87}$$

$$|T_c^2| \simeq |T_m^2| \simeq 1 - |K_{cm}^2| - |T_{cm}^2| \tag{3.88}$$

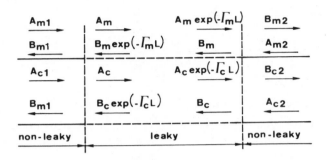

Fig 3.10   *Normalised waves in a leaky-section type mode converter*

The non-equality of the mode conversion factors in reflection K_{cm} and in transmission T_{cm} is the result of the non-zero electrical length of the device. As the phase constants β_{c0} and β_{m0} are proportional to the frequency, the bandwidth of T_{cm} around a maximum of the sine function is larger than that of K_{cm}. It is moreover maximum for

$$(\beta_{c0} - \beta_{m0})L = \pi \tag{3.89}$$

The 3 dB bandwidth of T_{cm} thus extends from $f_0/2$ to $3f_0/2$, where f_0 is the design frequency yielding eqn. 3.89. The choice of L is thus not at all critical. For the design of a mode converter, we may consider that

$$\beta_{m0} \simeq k_0 = \omega/(3 \times 10^8) \tag{3.90}$$

$$\beta_{c0} \simeq k_0 \kappa^{1/2} \tag{3.91}$$

where k_0 is the free-space wave number and κ is the dielectric constant of the cable insulation. It is seen from eqns. 3.50 and 3.51 that $C_1 > C_2$, and thus $T_{cm} > K_{cm}$. The mode conversion thereby exhibits an intrinsic directivity which can be enhanced by choosing a low value for κ; this yields an increase in T_{cm}.

---

[*] Here we take account of the fact that, in transmission line theory, the A and B waves are progressive and regressive waves, respectively. In network theory however they are the ingoing and outgoing waves, respectively, and the scattering matrix relates the latter to the former.

Several interesting properties of a coaxial cable containing periodic leaky sections are worth mentioning. As we have seen in Section 3.2.3 and in Fig. 3.2, for a continuous leaky cable the relative power of the leakage fields of the coaxial eigen-mode is C_1^2 (we assume that $C_1 \gg C_2$). A leaky section excites the monofilar mode with a power $4C_1^2$. We thus have a gain of 6 dB. This property as well as those of a continuous leaky cable are nicely illustrated by Fig. 3.11. Another useful character-istic is that the insertion loss of a leaky section for the coaxial mode, given by eqn. 3.88, remains small: for instance $C_1 = 0.15$ yields a conversion factor T_{cm} of -9 dB and an insertion loss T_c of 0.5 dB. As a matter of fact the global specific attenuation of a coaxial cable containing a leaky section every d metres will be

$$\alpha_c = \alpha_{c0} + \frac{10}{d} \log \frac{1}{1 - 4C_1^2} \tag{3.92}$$

This may be significantly smaller than the attenuation eqn. 3.53 of a continuous leaky cable. This property, together with the 6 dB gain mentioned above, should more than compensate for the fact that the radiated field suffers the attenuation α_{mo} of the monofilar mode between two successive leaky sections. The question however requires closer examination since the optimum value of C_1 for a con-tinuous leaky calbe and for a leaky section are not necessarily equal. Actually the optimum values depend on several factors as will be seen in Chapter 4.

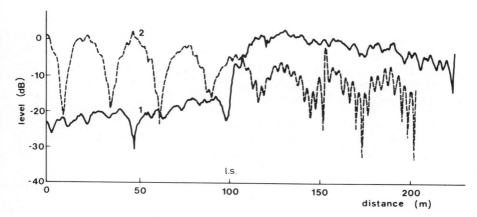

Fig. 3.11   *Field level measured along a continuous leaky feeder (curve 2) and along a non-leaky cable with a leaky section of the same type located at 100 m from the gener-ator (curve 1). The generator power has been adjusted to have the same power inside the cable at this point. The frequency is 35 MHz. The dielectric constant of the cable insulation is 1·69*

3.6 Mode converters for two-wire lines

It was pointed out in Chapter 1 that the classical ribbon feeder, which is very popular as an antenna feeder for television reception, was historically the first type of leaky feeder. There is however a fundamental difference from coaxial cables: the primary field of the balanced mode, i.e. that which would exist if the line were strung in free space, decreases with the square of the inverse distance to the line. This is much faster than for the leakage fields of an imperfectly shiedled coaxial cable. Actually we may say that a two-wire line is not fundamentally leaky when compared with a coaxial cable. When a transmission line of this type is strung at some distance from the wall in a tunnel, we may neglect the small asymmetry that the two wires may have with respect to the wall. Under this assumption, the eigenmodes of the structure are the classical balanced (or bifilar) and unbalanced (or monofilar) modes. In spite of the fundamental difference mentioned above, we will use the subscript c for the bifilar mode.

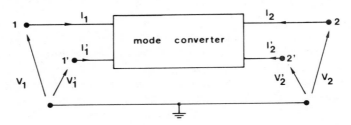

Fig. 3.12   *Wire currents and voltages in a mode converter for two-wire lines*

Groundless mode converters can be made by inserting lumped impedances or transformers into the transmission line in order to create asymmetry between the two wires. All the properties eqns. 3.72–3.80 of the scattering matrix are valid. The procedure for obtaining the scattering matrix is long and tedious, though straightforward. It involves the following steps:

Write the four circuit equations (Fig. 3.12);
Convert these into new equations relating the bifilar and monofilar currents and voltages defined by

$$I_m = I + I' \qquad V_m = (V + V')/2$$
$$I_c = (I - I')/2 \qquad V_c = V - V' \qquad (3.93)$$

Convert these new equations into ones relating the ingoing and outgoing waves, using

$$V_i = Z_i^{1/2}(A_i + B_i)$$
$$I_i = Z_i^{-1/2}(A_i - B_i) \qquad (3.94)$$

Solve these equations in the unknowns $B_{c1}, B_{m1}, B_{c2}, B_{m2}$ to obtain the scattering matrix of the mode converter.

Figures 3.13a and b show a selective and a wideband mode converter, respectively. They are very similar to those seen previously for a coaxial cable. The scattering matrix of these devices is given by Deryck (1975). We will recall here formulas valid at the centre frequency $\omega_0 = (LC)^{1/2}$ of the selective converter or for the wideband converter when $n_1 = n_4$ and $n_2 = n_3$:

$$K_c = K_m = d^2/(d^2 + 1)$$
$$T_c = T_m = 1/(d^2 + 1) \tag{3.95}$$
$$S = d/(d^2 + 1)$$

where d is an asymmetry factor given by

$$d = 2\omega_0 C(Z_c Z_m)^{1/2} \tag{3.96}$$

for the selective mode converter and by

$$d = \frac{n_1 - n_2}{2(n_1 + n_2)} (Z_c/Z_m)^{1/2} \tag{3.97}$$

for the wideband converter.

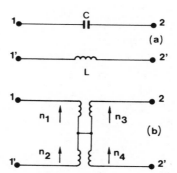

Fig. 3.13   *Selective (a) and wideband (b) mode converters for two-wire lines*

In principle, the ribbon feeder is equivalent to the coaxial cable for communication systems based on mode conversion techniques. It is however less popular because its specific attenuation is relativley sensitive to moisture, dust and so on.

3.7 Inadvertent mode conversions

We mentioned previously that any discontinuity along a tunnel containing axial conductors is likely to cause mode conversion. Numerous discontinuities may exist in any practical application. Though it is in general impossible to evaluate them quantitatively, it is nevertheless important and sometimes vital for predictions to have a clear picture of the possible effects.

Fig. 3.14 *Photograph of a tunnel under the river Schelde near Antwerp (Belgium). The coaxial cable with annular-slot mode converters is visible in the upper left part of the photograph*

A first type of discontinuity, which always exists, consists of the hanging devices used to support monofilar and bifilar lines or leaky coaxial cables. In general these devices act as small local shunt capacitors and they can in principle be studied by the method outlined in Section 3.2.5. If the spacing is much smaller than the wavelength, the effect is equivalent to an increase of the specific capacity and has negligible consequences, except that it disturbs symmetry. Deryck (1970, 1972, 1973) showed admirably that the apparently leaky character of the ribbon feeder was actually an inadvertent mode conversion process due to numerous slight asymmetries, among which are the hanging devices. Periodic spacing, even of small discontinuities, may however be a source of serious perturbation when the spacing is close to a multiple of the half wavelength, because of a filtering effect.

Objects with a small axial size are another type of discontinuity. They in general may be modelled as local capacities. The disturbance therefore increases with frequency. Experience has shown and theory confirms that the effect of an isolated object is mainly local. Indeed the shunt impedance due to the object is generally very small compared with the characteristic impedance of the modes. It has a negligible effect on propagation, although the local disturbance of the electromagnetic fields may be important. This does not remain true when numerous objects exist along the path, because of the resulting accumulation. For a large number of random objects with random spacing, the fields in the tunnel space rapidly become quite irregular.

Particular attention should again be paid to periodic objects. The photograph of Fig. 3.14 shows a tunnel located below the river Schelde near Antwerp, Belgium. This tunnel has a length of 1173 m and is used to transport oil products. The wall is made of concrete blocks and is supported by metallic structures with a spacing of 6 m. Radio communications at 36 MHz have been operational in this tunnel for many years. The system uses a coaxial cable with annular-slot mode converters. The total mass of metallic objects is impressive: if they were brought together, the transverse metallic structures would have a total length of about 40 m. In spite of this, they have a negligible effect on the specific attenuation of the monofilar mode since their spacing is 0·7 times the wavelength and this corresponds to the middle of a passband for a transmission line with periodic capacitive loading.

In the example of Fig. 3.14, the return path for the monofilar mode current is probably offered by the metallic foot-bridge and by the oil pipes rather than by the ground itself. This situation is comfortable since these conductors are continuous. In many cases axial conductors exist and play a role in the propagation mechanism. Their location may vary along the path. Frequently these inadvertent conductors are grounded by their hanging devices with a more or less regular spacing. These inhomogeneities, as well as tunnel non-uniformities (Seidel and Wait, 1979a) are sources of mode conversions. It would be a hopeless undertaking to calculate the effect of inadvertent mode conversions in actual situations, but it is nevertheless necessary to evaluate them coarsely or to design the system in such a way as to minimise their effects.

Chapter 4

Subsurface radio communication systems in the HF and lower frequency bands

4.1 Introduction

In this chapter we will investigate several aspects of subsurface radio communications at frequencies below a few tens of megahertz. Chapter 5 will be devoted to higher frequencies. This separation is mainly due to the propagation mechanisms involved. The frequencies considered here are below the cutoff frequencies of the tunnels in which communications need to be established. Consequently, one has necessarily to resort to some type of axial cable in order to guide electromagnetic waves along the tunnels. One exception to this general rule is waveguiding by low-conductivity coal seams at medium frequencies: Section 4.9 is devoted to some aspects of this type of communication. The cables considered include the monofilar wire, the induction loop, leaky coaxial cables and non-leaky cables with mode converters.

We will not devote much attention to the long induction loop. There are many reasons for this decision. Above a few hundreds of kilohertz, and even below if perfect symmetry of the two wires cannot be maintained, the specific attenuation becomes comparable to that of a monfilar wire. Use of an induction loop instead of a single wire cannot then be justified, since the installation cost is doubled. The useful frequency band itself is a source of limitation. Because it is relatively narrow, it does not lend itself to multichannel operations or to efficient modulation methods, which consume bandwidth. Furthermore manmade noise may be particularly intense at these low frequencies. In spite of these drawbacks the usefulness of the long induction loop should be recognised. Extrapolating the information contained in this chapter to the long induction loop is straightforward and is left to the reader.

Apart from retransmission of amplitude modulation radio broadcasting in road tunnels and from some special applications (private tunnels like the one shown in Fig. 3.14, subsurface military plants, underground collecting of waters, spelaeology, etc.), the transmission techniques described in this chapter mainly find application in mining. A short review of the main mining techniques has therefore been included in Section 4.2.

As a matter of fact, the frequency band considered here is crowded and there is little hope of obtaining frequency allocations for new services, unless this is justified by imperative reasons like safety. In some cases the problem to be solved is merely the extension of existing radio links into subsurface works and no new frequency allocation is required. In most applications to mining communications, it is possible to avoid any significant radiation above the earth's surface. As long as administrations responsible for frequency allocation do not feel concerned, and this is generally so, engineers are put in the exceptionally favourable position of freely choosing the best suitable frequency and bandwidth. The counterpart of this advantage is that special-purpose radio equipment has to be developed.

The use of waveguiding cables has of course certain implications for system design. According to the general rule, the specific attenuations of the guided modes increase with frequency. There are however reasons for avoiding low frequencies, namely the higher intensity of man-made noise and the reduced antenna efficiency of portable radio sets. As a result the largest communication range for repeaterless systems is obtained in the low HF band. This optimum is well exhibited by curves like those of Fig. 2.42 and is fortunately very broad. In continental Europe it was considered essential for safety that radio systems used in mines should continue to work at least for a few hours following a general failure of electrical power. In the selection to meet this requirement, repeated systems were rejected and radio communications were developed in the optimum frequency band 5—10 MHz. As a result it was necessary to design the systems in order to obtain maximum coverage. This kind of optimisation is far less imperative in repeated systems like those presently used in the United Kingdom and described in Chapter 5.

We will not review all aspects of underground mine communications and in particular we will ignore all systems which may not be considered as essentially based on the use of electromagnetic waves (the term wireless would be very restrictive here). The reader interested in a review of mine communication systems is referred to the excellent paper by Murphy and Parkinson (1978).

4.2 Mining techniques

Underground mines are extensive labyrinths that may extend over an area of many square kilometres. The communication requirements between people working in this area, and the relevant constraints, are strongly influenced by the structure of the mine. The aim of this section is restricted to a short description of some aspects of mining techniques that may influence the design of radio communication systems.

Though mining techniques may vary to some extent with many geological, technical and human factors, they may be classified into two types: room and pillar mining, and longwall mining. Nearly all coal mines in the US are of the first type whereas in Europe they are of the second.

In the room and pillar technique, several parallel entries are first driven as shown on Fig. 4.1. They start either from the surface or from the bottom of shafts. There may be two, three or four parallel main entries. Transverse or butt entries lead to areas where coal is mined. Coal mining of a panel is made in two steps. During first mining rooms are driven in the coal seam while pillars are left to support the roof. During second mining or robbing, pillars are drawn at least to some extent. Up to 95% of coal is recovered. Several sections may be worked out simultanesously. There are numerous variations of this general principle. The result is generally a quite complicated labyrinth, of which Figs. 4.2 and 4.3 are examples.

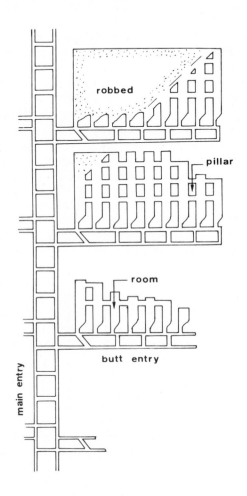

Fig. 4.1   *Principle of room and pillar mining technique*

The entries are mainly used as haulageways and for ventilation. Efficiency of the transportation system is of paramount importance since modern mines produce several thousands tons of coal daily. Communication with the transportation system is thus frequently required. The existence of several parallel entries with

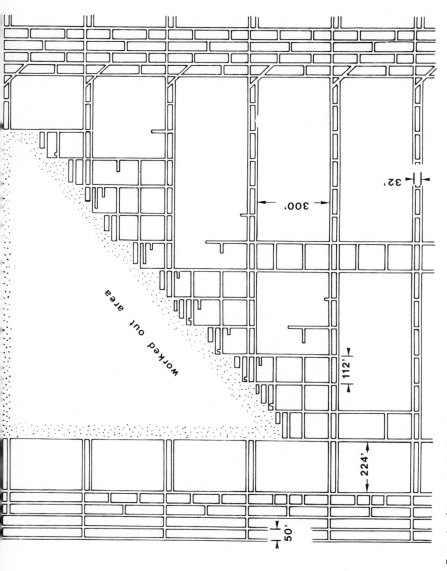

Fig. 4.2   Example of a room and pillar mine

distances of 10–20 m centres, leaving, for a 3 m entry, a 7–17 m pillar should be taken into account either by using distinct waveguiding cables for each entry or by choosing frequencies such that the wave has a sufficient lateral extent across the pillars. Communications may also be required between people working in a section or between them and a control centre. This type of communication is characterised by area rather than linear coverage.

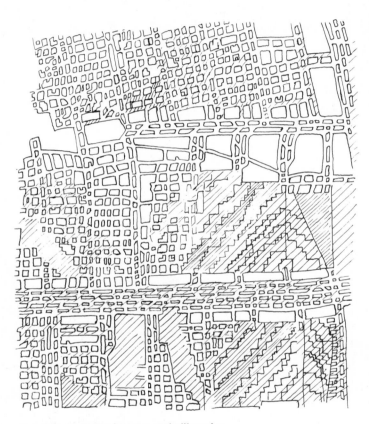

Fig. 4.3   *Another example of a room and pillar mine*

The fundamental principle in longwall methods is the complete removal of the entire seam in one operation, by carrying a continuous working face, leaving no pillars and allowing the roof to cave behind the face. Fig. 4.4 shows a general plan of the work.

Coal is left around the shafts to ensure mechanical stability. The main haulage-ways run away from the shaft toward the extraction perimeter across the caved area. The longwall advancing method has been shown. Alternatively one may use the retreating method in which the haulageways are driven through virgin coal and the extraction perimeter is shrinking instead of growing, but this does not change the communication requirements. The longwall face itself runs between two entries

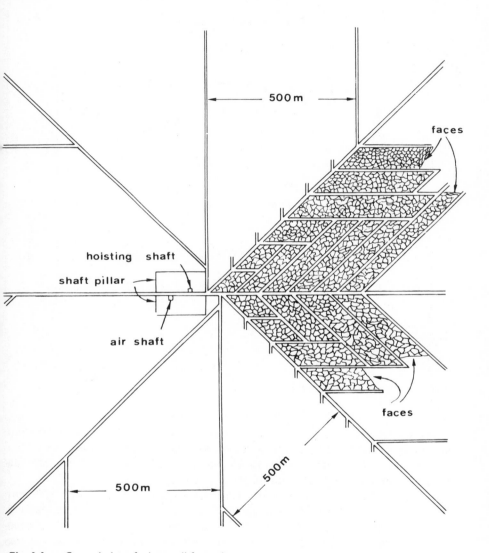

Fig. 4.4 *General plan of a longwall face mine*

which are called the top and bottom entries (Fig. 4.5), since the coal seam is frequently sloping.

Equipment used in the face is brought through the top entry and mined coal is evacuated through the bottom entry toward a loading point located at the end of the main haulageway. The length of the longwall face is typically 100 to 300 m. The top and bottom entries may be several hundreds of metres long.

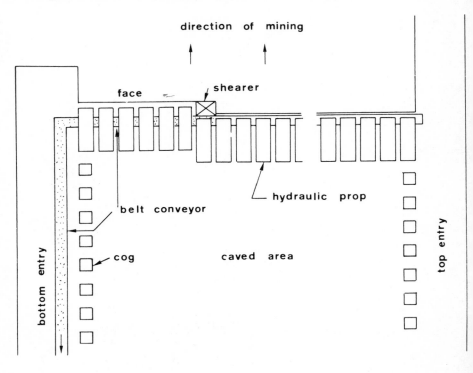

Fig. 4.5   *Plan of longwall face*

The photograph of Fig. 4.6 shows a longwall face. Coal is cut by a shearer running along the face and falls onto an armoured conveyor. The roof is supported in the work area by hydraulic props which follow the advancement of the face and provide a continuous metallic roof for the protection of the workers.

The communication requirements in a longwall face system are very different from those of room and pillar methods. The communication area is essentially linear, consisting of tunnels. Communications may be required for instance between people working in a face and some control centre, along haulageways, in the top entry, and so on. Note however that some lateral extent of the communication range may be required. In Europe in general single top and bottom entries are used, whereas US Federal mining laws impose triple entries.

The same general principles apply to non-coal mining as well. The mining environment is extremely complicated and constraining from the point of view of

Fig. 4.6 *Photograph of longwall face*

Fig. 4.7 *Mine tunnel distorted by ground pressure*

electromagnetic wave propagation. For instance, in longwall faces of the type shown in Fig. 4.6, the only place where a waveguiding cable may be located is together with hydraulic tubes and electric power cables in hooks located on the side of the armoured conveyor.

Similarly the photograph of Fig. 4.7 shows how actual tunnels may be distorted by mechanical effects and encumbered with various metal pieces playing the role of electrical conductors. That their findings, in combination with coarse engineering approximations, were nevertheless useful for solving the problem will undoubtedly be encouraging for theoreticians of electromagnetics.

4.3 The reference tunnel

Subsurface radio communications along waveguiding cables exhibit major differences from their free-space or above-the-surface equivalents. One of them is that the link attenuation is strongly dependent on the relative locations of the portable radio sets and of the waveguiding cables in the tunnel cross-section. Fundamentally, the operation of antennas in this environment is inductive or capacitive coupling rather than radiation. Consequently some parameters like transmitter power are meaningless unless they are defined under well-specified reference conditions.

These reference conditions may be highly idealised but need to be representative of the kind of structure in which radio communications are established. The following set of conditions has been used for many years by the author and his collaborators in private work:

The tunnel cross-section is circular, with a radius $a_0 = 3$ m, and has a perfectly conducting wall.

A perfectly conducting monofilar wire is located along the tunnel axis. For convenience its radius is taken equal to $5 \cdot 7$ mm in order to get a characteristic impedance Z_m for the monofilar mode of this structure equal to the intrinsic impedance of free space $\eta_0 = 377 \, \Omega$.

Portable transceivers are assumed to be located against the tunnel wall and oriented for maximum coupling to the monofilar mode.

In this tunnel, a monofilar mode carrying a power P_m with a voltage V_m and a current I_m has electromagnetic fields given by

$$E_{\text{ref}} = \frac{\eta_0 I_m}{2\pi a_0} = \frac{\eta_0 V_m}{2\pi a_0 Z_m} = \frac{\eta_0}{2\pi a_0} \left[\frac{P_m}{Z_m} \right]^{1/2} \tag{4.1}$$

$$H_{\text{ref}} = E_{\text{ref}}/\eta_0 \tag{4.2}$$

with

$$P_m = V_m I_m = Z_m I_m^2 = V_m^2/Z_m \tag{4.3}$$

Effective values are used throughout for voltages, currents and fields. Numerical expressions are

$$E_{\text{ref}} = 20 I_m = 5{\cdot}305 \times 10^{-2} V_m = 1{\cdot}030 P_m^{1/2} \qquad (4.4)$$

$$H_{\text{ref}} = 5{\cdot}305 \times 10^{-2} I_m = 1{\cdot}407 \times 10^{-4} V_m = 2{\cdot}732 \times 10^{-3} P_m^{1/2} \quad (4.5)$$

Quantities relevant to the reference tunnel will be referred to by the subscript 'ref'.

4.4 Electromagnetic fields in actual tunnels

4.4.1 Adequate location of waveguiding cables

Calculating the electromagnetic fields of modes propagating along waveguiding cables in actual tunnels is an extremely complicated task. It is thus interesting to develop some engineering models, however coarse they may be. The first task in designing a system consists of visiting the tunnels and choosing the location of the waveguiding cables. This location may of course vary along the path. The choice will be made in order to maximise either the electric or the magnetic field in the area where the portable radio sets are likely to be used. As a general rule, the cables should be located as far as possible from possible return paths for the monofilar mode current. These paths include other axial conductors which may exist in the tunnel and of course the tunnel walls. When the conductivity is not the same for all tunnel walls, the cable should preferably be located near to low-conductivity walls. Indeed a negative image is developed behind highly conducting walls and tends to cancel the primary fields of the waveguiding cable. For instance, in tunnels driven in coal seams, it is better to locate the cables near the side walls than close to the more conductive roof. This method aims not only to improve the coupling of portable transceivers to the electromagnetic field, but also to increase confidence in the engineering models which are used and which will now be described.

4.4.2 Engineering models for the monofilar mode

Let us first consider the case where a monofilar wire conductor is used. A general assumption is that the transmission line models developed in Chapter 3 are valid. Exact analytic expressions obtained in this context will be given. We will also give numerical results: in this case we will systematically assume that the characteristic impedance Z_m of the monofilar mode is equal to $\eta_0 = 377\,\Omega$. Actual values of Z_m may differ by up to 50% from this value, but this will not result in errors of more than two or three decibels.

A first model is the PCW (perfectly conducting wall) approximation. The transverse field distribution is modelled by that of the perfect TEM mode. The influence of the inadvertent return conductors present in the tunnel space is neglected as a consequence of an adequate choice of the monofilar wire location. Furthermore the cross-section of the tunnel is assimilated to a simple shape to allow calculations. The circular and rectangular shapes will be used below. The aim of the calculation is to obtain the *eccentricity ratio* (De Keyser *et al.*, 1978) defined as

$$C_e = \frac{H}{H_{ref}} = \frac{E}{E_{ref}} \tag{4.6}$$

Here H and E are the actual fields at the location considered for the antenna in the PCW model, whereas H_{ref} and E_{ref} are those obtained in the reference tunnel for the same power of the monofilar mode. Note that this is also for the same current and voltage, since we have assumed $Z_m = 377\ \Omega$ in both cases. It should be observed that $E/E_{ref} = H/H_{ref}$ is valid in the PCW model because the ratio E/H is equal to η_0 for the model as for the reference tunnel.

The *eccentricity loss* is defined by

$$L_e = -20 \log C_e \tag{4.7}$$

The term eccentricity has been used since C_e and L_e include the penalty which must be paid for an eccentric location of the waveguiding cable and of the antenna near to opposite and highly conducting walls. Note that these parameters are also influenced by the tunnel size.

The eccentricity ratio and loss can easily be calculated in the case of circular tunnels. Denoting the tunnel radius by a and the distance from the monofilar wire to the wall by d, a negative image arises in the external medium at a distance

$$d' = \frac{ad}{a-d} \tag{4.8}$$

from the wall. It is thus easy to calculate the eccentricity ratio for given locations of the cable and of the antenna.

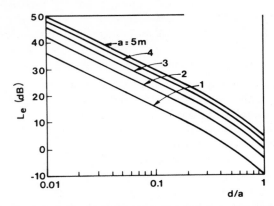

Fig. 4.8   *Eccentricity loss for a circular tunnel of radius a as a function of the ratio d/a, where d is the distance from the cable to the tunnel wall. The observation point is supposed to lie against the wall at the opposite side of the cable*

If we assume that the antenna is located in the same azimuth plane as the wire and at a distance D from the latter we obtain

$$C_e = \frac{3(d + d')}{D(D + d + d')} \tag{4.9}$$

The eccentricity loss is exhibited in Fig. 4.8 as a function of the d/a ratio for the case where the antenna is located against the wall opposite to the cable.

For a rectangular tunnel of width a and height b there is a double infinity of positive and negative images as shown in Fig. 4.9. Summing the contributions of all images can easily be done using a programmable pocket calculator.

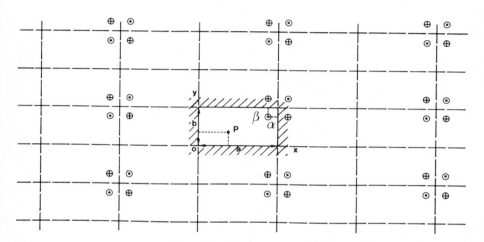

Fig. 4.9   *Double inifinity of images for a monofilar wire located in a rectangular tunnel*

The eccentricity loss may be cast into the form

$$L_e = 10 \log \frac{ab}{9} + f(\alpha/a, \beta/b) \tag{4.10}$$

where the function f depends on the b/a ratio and on the antenna location. As an example, Fig. 4.10 gives this function for $b/a = 0.5$ and for an antenna P located at the point of coordinates $(0.25\,a, 0.25\,b)$.

Figures 4.8 and 4.10 show that the choice of cable location is a critical step in the design of a communication system. Many decibels can be saved by an adequate choice, particularly in mobile-to-mobile communications where the eccentricity loss must be accounted for twice. Furthermore it should not be forgotten that the specific attenuation of the monofilar mode is strongly dependent on the cable-to-wall distance.

The PCW model is undoubtedly justified when electric-type antennas are used, since we have seen in Sections 2.12.2 and 3.1.1 that the electric field does not depart much from the TEM approximation. For magnetic-type antennas the model is adequate when the skin depth in the tunnel wall is very small, i.e. at the high end of the frequency range considered in this chapter. For lower frequencies, the PCW model generally provides a pessimistic estimation of the eccentricity loss. The error may be quite large at low frequencies for which the skin depth in the external medium is larger than the tunnel size. We may then use the LCW (low-conductivity wall) model in which the magnetic field is calculated by the Biot–Savart law. This yields for magnetic-type antennas

$$C_e = \frac{3}{D} \qquad (4.11)$$

where D is the distance from the antenna to the cable in metres. The eccentricity loss may be changed into a gain when D is small. Particular attention should be paid to inadvertent conductors at these low frequencies since they are likely to offer the main return path for the monofilar mode current. Their effect may be beneficial when the magnetic antenna is located between them and the cable.

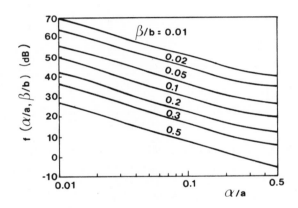

Fig. 4.10   *The function $f(\alpha/a, \beta/b)$ in eqn. 4.10 for a rectangular tunnel with $b/a = 0.5$. The geometrical parameters are defined in Fig. 4.9.*

4.4.3 Extension to continuous leaky feeders

If a continuous leaky coaxial cable is used instead of a monofilar wire, the main interest is in the leakage fields of the coaxial eigenmode. Since these leakage fields are the same as those of a monofilar wire conductor carrying a power C_1^2 times smaller than that of the coaxial eigenmode (Section 3.2.2), the foregoing theory applies to this case as well. We implicitly assumed that $C_1 \gg C_2$. We will refer to

$$L_1 = -20 \log C_1 \qquad (4.12)$$

as the *coupling loss* of the leaky coaxial cable. The same term will be used in Chapter 5 with another meaning.

4.5 Antennas and subsurface communications

4.5.1 Circuit modelling of antennas

Exact analysis of the transmitting and receiving properties of antennas in actual tunnels is an extremely complicated task. Some elements of the problem are contained in Chapter 2, Section 2.10.2. Though it is true that they radiate electromagnetic waves in the external medium, the primary function of antennas is to provide coupling to the monofilar mode or to the leakage fields of the coaxial eigenmode. In the frequency range considered in this chapter, antennas are necessarily very small compared with the wavelength and are confined to electric and magnetic dipoles. Consequently they are essentially reactive devices with however a small resistive part due to radiation and to coupling to the guided modes. The relative importance of this resistive component will be examined later. We will concentrate for the moment on the development of an engineering model for the coupling to monofilar wave fields.

(a) Electric-type antennas

An electric dipole is made essentially of two conducting bodies that will be referred to as poles 1 and 2. The effect of the dipole located inside a tunnel containing a monofilar wire conductor is completely determined by the capacity coefficients C_{12}, C_{1m}, C_{2w} shown in Fig. 4.11. This model is justified by the electrostatic nature of the field in the tunnel.

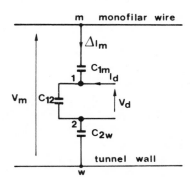

Fig. 4.11   *Modelling of electric dipole in tunnel containing a monofilar wire conductor*

The dipole thus appears as a circuit connected in parallel on the transmission line. It is not difficult to show that the admittance matrix of the two-part network $(m, w)–(1, 2)$ is given by

$$\Delta I_m = j\omega C_s V_m - j\omega C_s V_d$$
$$I_d = -j\omega C_s V_m + j\omega(C_{12} + C_s)V_d$$

$$(4.13)$$

where

$$C_s = \frac{C_{1m}C_{2w}}{C_{1m} + C_{2w}} \tag{4.14}$$

When the dipole is receiving, the open-circuit voltage is given by

$$e \triangleq (V_d)_{I_d=0} = \frac{C_s}{C_{12} + C_s} V_m \tag{4.15}$$

If the dipole is connected to a receiver with input impedance Z_r, it constitutes a shunt admittance

$$Y_s = j\omega C_s \frac{1 + j\omega C_{12}Z_r}{1 + j\omega(C_{12} + C_s)Z_r} \tag{4.16}$$

for the monofilar wire circuit.

If we assume that the monofilar wire is terminated on either side on its characteristic Y_m, the input admittance of the dipole is given by

$$Y_d = j\omega(C_{12} + C_s) + \frac{\omega^2 C_s^2}{j\omega C_s + 2Y_m} \tag{4.17}$$

$$\simeq j\omega(C_{12} + C_s) + \frac{\omega^2 C_s^2}{2Y_m} \tag{4.18}$$

since C_s is only a fraction of a picofarad, whereas Y_m is larger than 1 mS. When the dipole is used as a transmitting antenna, a current

$$-\Delta I_m \simeq j\omega C_s V_d \tag{4.19}$$

is delivered to the monofilar wire.

(b) Magnetic-type antennas

A magnetic dipole can be modelled as a loop of equivalent area A_e. For actual loop antennas A_e is equal to the loop area A times the number of turns; formulas for ferrite antennas will be given later. Coupling of the magnetic dipole to a monofilar wire can be modelled as shown in Fig. 4.12.

Indeed, if the magnetic dipole is open-circuited, we may admit that the monofilar mode is unperturbed and the effect of the loop thus reduces to a voltage generator in series with the wire. The electromotive force set up in the loop is given by Faraday's law and is thus proportional to ω and I_m. That the same coefficient of mutual induction M is used for the electromotive force induced by the dipole current in the monofilar wire is a consequence of reciprocity. The self-induction coefficient of the dipole in the actual tunnel condition is denoted by L_d.

When the magnetic dipole is receiving, the electromotive force developed in it by the monofilar wave is given by

$$e \triangleq (V_d)_{I_d=0} = j\omega M I_m \tag{4.20}$$

If the dipole is connected to a receiver with input impedance Z_r, the voltage generator appearing in the monofilar wire is equivalent to a series impedance

$$Z_s = \frac{\omega^2 M^2}{j\omega L_d + Z_r} \tag{4.21}$$

If we assume that the monofilar wire is terminated on either side with its characteristic impedance Z_m, the input impedance of the dipole is given by

$$Z_d = j\omega L_d + \frac{\omega^2 M^2}{2Z_m} \tag{4.22}$$

When the magnetic dipole is transmitting, an electromotive force

$$\Delta V_m = j\omega M I_d \tag{4.23}$$

is developed in the monofilar wire.

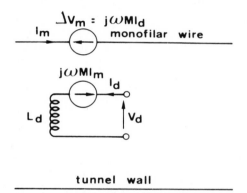

Fig. 4.12 *Modelling of magnetic dipole in tunnel containing a monofilar wire conductor*

4.5.2 Some general considerations

Eqns. 4.18 and 4.22 give the input admittance impedance of an electric magnetic dipole located in a tunnel containing a monofilar wire conductor. The second term of these expressions is real and is responsible for the active power delivered to the monofilar mode when the antenna is transmitting. Actually two other resistive parts should be added to take account of losses in the nearby tunnel wall and in the dipole itself. The exact equations are thus of the form

$$Y_d = jB_d + 2G_r + G_w + G_d \tag{4.24}$$

with

$$G_r = \frac{\omega^2 C_s^2}{4Y_m} \tag{4.25}$$

for the electric dipole and

$$Z_d = jX_d + 2R_r + R_w + R_d \tag{4.26}$$

with

$$R_r = \frac{\omega^2 M^2}{4Z_m} \tag{4.27}$$

for the magnetic dipole. It is of some importance to have a realistic assessment of the different terms and of their dependence on the tunnel environment, including the relative locations of the dipole and of the monofilar wire.

The quantities G_r, R_r correspond to the active power transferred to the monofilar mode in one direction when the antenna is transmitting and, since they are the useful terms, they will be referred to as the one-sided radiation conductance and resistance, respectively. The terms $(G_w + G_d)$ and $(R_w + R_d)$ are the loss conductance and resistance, respectively.

4.5.3 Assessment of the electric dipole

Let us first consider an electric dipole of length ds used as a receiving antenna. We will assume that the effective height of the dipole, defined as the ratio of the received electromotive voltage to the incident field, is unchanged with respect to that in free space and is thus equal to ds. In the reference tunnel, using eqns. 4.1 or 4.4 we obtain the electromotive force

$$e_{ref} = \frac{V_m ds}{6\pi} = 5\cdot305 \times 10^{-2} V_m ds \tag{4.28}$$

In the actual tunnel this reduces to

$$e = \frac{C_e V_m ds}{6\pi} = 5\cdot305 \times 10^{-2} C_e V_m ds \tag{4.29}$$

since the electric field is multiplied by the eccentricity ratio C_e. Referring for instance to Figs. 4.8 and 4.10, it appears that C_e is in practice smaller than unity. On the other hand, the length ds of electric dipoles usable in tunnels does not exceed one metre. Comparing eqn. 4.29 with eqn. 4.15 we see that the ratio C_s/C_{12} will not exceed a few per cent in practice. Actually we find the approximate formula

$$\frac{C_s}{C_{12}} = \frac{C_e ds}{6\pi} = 5.305 \times 10^{-2} C_e ds \tag{4.30}$$

The next step is to consider the dipole admittance eqn. 4.18, where the wall and dipole losses are neglected. We may define a 'radiation Q-factor' by

$$Q_r = \frac{|B_d|}{2G_r} \simeq \frac{2Y_m \omega C_{12}}{(\omega C_s)^2} \tag{4.31}$$

This parameter is obviously large compared with unity and we may write

$$Y_d \simeq j\omega C_{12}(1 - jQ_r^{-1}) \tag{4.32}$$

$$Z_d \triangleq R_d + jX_d \triangleq \frac{1}{Y_d} \simeq \frac{1}{\omega C_{12} Q_r} - \frac{j}{\omega C_{12}} \tag{4.33}$$

It is interesting to compare this expression with the input impedance

$$Z_0 = R_0 + jX_0 = \frac{|X_0|}{Q_0} + jX_0 \tag{4.34}$$

of the dipole in free space, which has a Q-factor

$$Q_0 = \frac{|X_0|}{R_0} \tag{4.35}$$

In order to evaluate Q_r we will assume that the input reactance of the dipole is not changed significantly by the tunnel environment, i.e. that $X_0 = -1/\omega C_{12}$. This assumption is based on physical arguments and also on some theoretical predictions reported in Section 2.10.2c. On the other hand, R_0 is given by

$$R_0 = \frac{2\pi}{3} \eta_0 \left(\frac{ds}{\lambda_0}\right)^2 \tag{4.36}$$

where λ_0 is the free-space wavelength. On combining eqns. 4.30, 4.32, 4.35 and 4.36 we obtain

$$\frac{Q_r}{Q_0} = \begin{cases} \dfrac{48\pi^3 Y_m \eta_0}{\lambda_0^2 C_e^2} = \left(\dfrac{38.58}{\lambda_0 C_e}\right)^2 \\[4mm] \dfrac{12\pi\omega^2 Y_m \eta_0}{c^2 C_e^2} = \left(\dfrac{2.0467 \times 10^{-8}\,\omega}{C_e}\right)^2 \end{cases} \tag{4.37}$$

A lower bound has been given for Q_0 (Chu, 1948):

$$Q_0 > \left(\frac{\lambda_0}{\pi ds}\right)^3 \tag{4.38}$$

from which we have

$$Q_r > 48 \frac{\lambda_0}{(ds)^3 C_e^2} \tag{4.39}$$

It is seen that Q_r will take on very high values below 30 MHz; for instance, if $ds < 0.2$ m, $C_e < 1$, we have $Q_r > 6 \times 10^4$.

On the other hand, a lower limit of X_0 and an upper limit of C_{12} can be obtained by combining eqns. 4.35, 4.36 and 4.38. This yields $C_{12} < 2 \times 10^{-12}$ ds. It is seen that the input capacitance of the electric dipole will not exceed a few picofarads for reasonable dipole lengths.

The next step in our evaluation will be to consider the conductance G_w due to losses in the nearby wall. Its effect is to increase the input resistance of the dipole to some value R_{rw}. As a result the Q-factor is reduced to

$$Q_{rw} = \frac{R_0}{R_{rw}} Q_0 \tag{4.40}$$

It is extremely difficult to evaluate the ratio R_{rw}/R_0 in a particular case. Values of this ratio are given in Figs. 2.32 and 2.34. Using these results and the lower bound 4.38, it may be seen that $Q_{rw} > 0.15 \lambda_0/ds^3$. The Q-factor thereby remains very high. For $ds = 0.2$ m it will be higher than 180 at frequencies below 30 MHz. As a matter of fact, actual electric dipoles have Q_0-factors at least ten times higher than the lower bound 4.38. We may thus conclude that coupling to the monofilar mode and losses in the tunnel wall do not reduce the Q-factor of an electric dipole below several thousand units. Finally it is probable that the Q-factor will be governed by the dipole loss rather than by the tunnel environment; it will nevertheless remain high since an electric dipole can be made with very low-loss elements.

The load admittance that the dipole constitutes for the monofilar mode is given by eqn. 4.16. As results from eqn. 4.30 the second factor of this expression is very close to unity for a realistic receiver impedance. The admittance load Y_s is thus completely negligible.

4.5.4 Assessment of the magnetic dipole

The same kind of evaluation as above will now be undertaken for the magnetic dipole. The latter behaves as a loop of area A_e. Let us assume that the dipole is receiving in the reference tunnel. The received electromotive force is given by $e = j\omega\mu_0 A_e H_{ref}$. On using eqn. 4.5 this yields

$$e_{ref} = \frac{\mu_0}{6\pi} j\omega A_e I_m = 6.67 \times 10^{-8} j\omega A_e I_m \tag{4.41}$$

In the actual tunnel we have

$$e = \frac{\mu_0}{6\pi} j\omega A_e C_e I_m = 6.67 \times 10^{-8} j\omega A_e C_e I_m \tag{4.42}$$

Comparing with eqn. 4.20, we obtain the mutual induction coefficient

$$M = \frac{\mu_0 A_e C_e}{6\pi} = 6.67 \times 10^{-8} A_e C_e \tag{4.43}$$

Hence we consider the dipole impedance, temporarily neglecting the wall and dipole losses

$$Z_d = j\omega L_d + 2R_r \qquad (4.44)$$

where R_r is given by eqn. 4.27. Again we will compare this to the dipole impedance in free space

$$Z_0 = j\omega L_0 + R_0 \qquad (4.45)$$

The loop self-induction L_d will not be changed significantly by the tunnel environment. This assumption is based on physical arguments and also on results reported in Section 2.10.2c. The only case where L_d could differ significantly from L_0 is when the loop is located parallel and very close to a highly conducting plane, for it is then tightly coupled to its antiphase image. For this effect to be significant the loop–plane spacing must be comparable to the loop size and the skin depth in the plane must be small compared with this spacing. These conditions will in practice never be fulfilled. Consequently we consider $L_d \simeq L_0$ as a very good assumption.

Combining eqns. 4.27 and 4.43 we have

$$2R_r = \begin{cases} \dfrac{2}{Z_m}\left[\dfrac{\mu_0 \omega A_e C_e}{12\pi}\right]^2 = (2{\cdot}428 \times 10^{-9} A_e C_e)^2 \\[2em] \dfrac{2}{Z_m}\left[\dfrac{\mu_0 c A_e C_e}{6\lambda_0}\right]^2 = \left(4{\cdot}756\dfrac{A_e C_e}{\lambda_0}\right)^2 \end{cases} \qquad (4.46)$$

The radiation resistance of the magnetic dipole in free space is given by

$$R_0 = \frac{8\pi^3 \eta_0}{3}\frac{A_e^2}{\lambda_0^4} = \frac{\eta_0}{6\pi c^4}\omega^4 A_e^2 \qquad (4.47)$$

where c is the speed of light. The radiation Q-factor is defined by

$$Q_r = \omega L_r/(2R_r) \qquad (4.48)$$

whereas

$$Q_0 = \omega L_0/R_0 \qquad (4.49)$$

When eqns. 4.46 to 4.49 are combined, we obtain eqn. 4.37 which was established for the electric dipole. That this result is universal is not surprising since it is due to our assumptions, namely that the antenna reactance and the effective height are unchanged with respect to their free-space values for both dipoles. Since Chu's lower bound 4.38 applies to the magnetic dipole as well, when ds is the largest geometrical size of the antenna, eqn. 4.39 and the relevant conclusions remain valid.

It is interesting to note that eqn. 4.49 combined with 4.38 yields a lower bound for the self-induction coefficient

$$L_0 > \frac{4\mu_0 A_e^2}{\pi^2 ds^3} \qquad (4.50)$$

Since the equivalent loop area A_e is likely to be proportional to ds^2, this lower bound increases as the antenna size.

The same reasons as for the electric dipole may be invoked to show that the losses in the tunnel wall do not reduce the Q-factor below several thousand units and that the actual Q-factor will be determined by the loss in the antenna material.

The series impedance eqn. 4.21 that a receiving magnetic dipole constitutes for the monofilar mode may be significant. To show this, let us suppose that the receiver input impedance and the antenna impedance realise a resonant circuit with a Q-factor of 70. Thus $Z_s \simeq 70\, \omega M^2 / L_0$. Eqns. 4.43 and 4.50 yield an upper bound of Z_s:

$$Z_s < 6\cdot 11 \times 10^{-7}\, \omega C_e^2\, ds^3$$

At 30 MHz this yields $Z_s < 115\, C_e^2\, ds^3$, which may be important if C_e is of the order of unity and the antenna size approaches one metre. In normal operating conditions however this effect is negligible.

4.5.5 Optimisation criterion for antennas

We have established some important results in the previous sections. Let us briefly recall them.

The input impedance of electric and magnetic dipoles in a tunnel containing a monofilar wire conductor may be written in the form

$$Z_d = jX_d + 2R_r + R_w + R_d \tag{4.51}$$

The reactance X_d does not differ much from that X_0 of the dipole in free space. When the dipole is transmitting, the active power apparently dissipated in the two resistances R_r is actually transferred to the two equal monofilar mode waves which the dipole excites in opposite directions. Since this is the useful power, the resistance R_r has been called the one-sided radiation resistance. The loss resistances R_w and R_d account for power dissipated in the tunnel wall and in the dipole itself. The 'radiation Q-factor'

$$Q_r = \frac{|X_d|}{2R_r} \simeq \frac{|X_0|}{2R_r} \tag{4.52}$$

is extremely high. It is related to that of the dipole in free space by the general formula

$$\frac{Q_r}{Q_0} = \frac{R_0}{2R_r} = \begin{cases} \dfrac{48\pi^3 \eta_0 Z_m}{(c\mu_0 C_e \lambda_0)^2} = \left[\dfrac{38\cdot 58}{\lambda_0 C_e}\right]^2 \\[3mm] \dfrac{12\pi \eta_0 Z_m \omega^2}{c^4 C_e^2} = \left(\dfrac{2\cdot 047 \times 10^{-8}\, \omega}{C_e}\right)^2 \end{cases} \tag{4.53}$$

This ratio does not depend on the dipole itself, but only on its location in the tunnel and on the frequency. Losses in the tunnel wall do normally not reduce the global Q-factor below several thousands of units.

Let us now suppose that the dipole has to be used for transmitting. The problem is to feed as much power as possible into the exceedingly small resistance $2R_r$. Obviously it will be necessary to eliminate the reactive effect by using the antenna as an element of a tuned circuit. For several reasons, namely the bandwidth required by the signal and by the temperature dependence of circuit elements, a high Q-factor for the tuned circuit is not acceptable. In practice, Q-factors above 100 are uncommon and it is not useful to design the antenna to have a very high Q-factor.

Hence we consider the global Q-factor of the tuned circuit as determined by signal and circuit considerations. It is easy to show that the radio of the active power $2P_r$ delivered to the two monofilar mode waves to the total active power P delivered to the tuned circuit is given by

$$\frac{2P_r}{P} = \frac{Q}{Q_r} \tag{4.54}$$

Obviously, apart from making P as large as possible by proper circuit design, *the optimisation objective is to make Q_r as low as possible.* Because of eqn. 4.53 this is equivalent to *optimising the antenna for free-space use by minimising Q_0.*

When the dipole has to be used for receiving, a reasonable objective is to maximise the signal/noise ratio. It will be shown in Section 4.6.2 that the received signal power is proportional to Q/Q_r (eqn. 4.69). The optimisation criterion thereby remains the same as for a transmitting dipole. The problem has thus been reduced to that of approaching the lower bound 4.38 as close as possible for a given maximum size ds of the antenna. It is quite remarkable that this limit is the same for both types of dipole. From this point of view the electric and magnetic dipoles appear to be strictly equivalent and will provide the same performance in subsurface radio communications. This is not completely true since the eccentricity ratio C_e may be significantly larger for the magnetic field than for the electric field. We thereby arrive at conclusions which are fully compatible with those of Section 2.11 (although the latter are unrealistic since R_w is considered as the only existing loss resistance). So far we conclude that magnetic dipoles have some superiority, particularly at low frequencies.

There are however more imperative practical reasons for discarding the electric dipole as an antenna for subsurface radio communications in the HF and lower frequency bands. As we showed in Section 4.5.3, the input capacitance of a short dipole antenna does not exceed a few picofarads. Increasing it artificially would be a nonsense, since it is just the contrary of the design objective. From a realistic circuitry viewpoint, it is not possible to tune efficiently such a small capacitance in the considered frequency band except perhaps at the high end of the HF band. But even if it were possible, small variable parasitic capacitances which unavoidably exist in operational use (e.g. operator's hand effect) would frequently detune the circuit. No such problems exist with the magnetic dipole. Indeed, since the self-inductance and the radiation resistance of loop and ferrite antennas are proportional to the square of the number of turns, Q_0 is independent of this parameter and the antenna reactance may take the most adequate value. Hence, apart from Section 4.10, we will restrict our attention to magnetic-type antennas.

4.5.6 Loop antennas

For a single-turn loop of thin wire, the equivalent area is equal to the actual area. The self-induction coefficient tends to be proportional to the length of the wire. Since the objective is to minimise the ratio $\omega L_0/R_0$, the circular loop is the best shape. For a single-turn loop of radius R, made of wire with radius r, one has

$$L_0 = \mu_0 R \left(\ln \frac{8R}{r} - 1{\cdot}75 \right) \tag{4.55}$$

Since, as far as $R \gg r$, $A_e = \pi R^2$, we have from eqn. 4.47

$$R_0 = \frac{8\pi^5 \eta_0}{3} \frac{R^4}{\lambda_0^4} \tag{4.56}$$

The free-space Q-factor is given by

$$Q_0 = \frac{3\lambda_0^3}{4\pi^4 R^3} \left(\ln \frac{8R}{r} - 1{\cdot}75 \right) \tag{4.57}$$

The lower bound $Q_{0\,min}$ is given by 4.38 with $ds = 2R$ and we have thus

$$\frac{Q_0}{Q_{0\,min}} = \frac{6}{\pi} \left(\ln \frac{8R}{r} - 1{\cdot}75 \right) \tag{4.58}$$

Numerical values of this ratio are 3·70, 5·03, 8·10 and 9·42 for $R/r = 5$, 10, 50 and 100, respectively. It is seen that a few decibels can be gained by using thick wire. It is of greater value to increase the loop size since Q_0 varies inversely with the cube of the radius.

For multiturn loops with n tightly coupled turns, L_0 and R_0 are both proportional to n^2 and Q_0 is independent of n. Adjusting L_0 to an adequate value may thus be done without affecting the antenna performance. However it may be interesting to space the turns somewhat. Indeed the effective area then remains proportional to n and R_0 to n^2, whereas L_0 increases more slowly than n^2. Actually there is not much difference between spaced turns made of thin wire and closer turns made of thicker wires.

Because of the factor R^3 in the denominator of eqn. 4.57, loop antennas are interesting compared with ferrite antennas only if the loop size can be made large enough. This is the case for antennas installed on machines, but not for individual portable radio sets, unless the users accept some form of bandolier antenna (Lagace and Emslie, 1978b).

4.5.7 Ferrite antennas

Ferrite antennas are very popular as receiving antennas for radio broadcasting. They have found use for transmitting as well as receiving in subsurface radio communications. The ferrite material is characterised by its relative permeability μ_i, also called initial permeability because it is assumed that no magnetic saturation occurs. This parameter may take values of several thousands for some materials.

A ferrite antenna is made of n turns of wire wound on a ferrite rod (Fig. 4.13). It behaves as a loop antenna of equivalent area A_e. The actual area of the turns is enhanced by the presence of magnetic material. One defines the effective permeability μ_e as the ratio of A_e to the actual area. The effective and initial permeabilities are related by

$$\mu_e = \frac{\mu_i}{1 + D(\mu_i - 1)} \tag{4.59}$$

where D is a parameter called the demagnetisation factor. It depends on the shape of the ferrite rod. For cylindrical rods D is a function of the length/diameter ratio l/d as shown in Fig. 4.14.

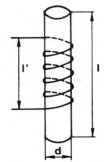

Fig. 4.13 *Ferrite antenna*

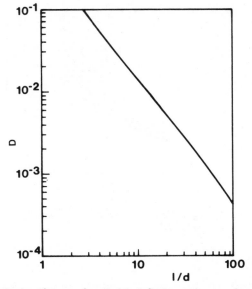

Fig. 4.14 *Demagnetisation factor of cylindrical ferrite rods as a function of the length/-diameter ratio*

The effective area is thus given by

$$A_e = n\mu_e \pi d^2 / 4 \tag{4.60}$$

The radiation resistance follows from eqn. 4.47.

The self-induction coefficient is given by

$$L_0 = \mu_0 \mu_e n^2 \frac{\pi d^2}{4l'} \tag{4.61}$$

where l' is the winding length. The radiation Q-factor in free space is thus given by

$$Q_0 = \frac{3\lambda_0^3}{\pi^3 d^2 l' \mu_e} \tag{4.62}$$

Since our purpose is to make Q_0 as small as possible, the winding length should be taken equal to the rod length l. For a given d, μ_e increases with l and it is obvious that d and l should both be taken as large as possible. For antennas which are incorporated in portable radio sets, l will be limited to the length of the apparatus. Hence we assume that l is fixed. The lower bound $Q_{0\,min}$ is then given by 4.38 with $ds = 1$ and we have

$$\frac{Q_0}{Q_{0\,min}} = \frac{3(l/d)^2 \, [1 + D(\mu_i - 1)]}{\mu_i} \tag{4.63}$$

This ratio is plotted against l/d for different values of μ_i in Fig. 4.15.

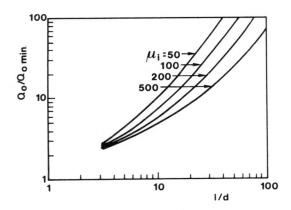

Fig. 4.15   *The ratio $Q_0/Q_{0\,min}$ for ferrite antennas as function of the length/diameter ratio and of the initial permeability*

With l/d ratios of the order of 15, which are acceptable, we obtain $Q_0/Q_{0\,min}$ close to 10. The ferrite antenna thus seems equivalent to a realistic loop antenna of diameter l.

The ratio Q/Q_{min} may be improved by connecting several ferrite antennas in series or in parallel. It has indeed been observed experimentally that the coupling between two parallel rods becomes very small when the spacing exceeds two or three times the rod diameter. When N rods are connected in series or in parallel, the effective area is multiplied by N and the radiation resistance by N^2. Since the self-induction coefficient is only multiplied by N, the ratio $Q_0/Q_{0\,\mathrm{min}}$ is reduced by a factor N. The theoretical limit can thus be approached very closely using four rods.

As can be seen from Fig. 4.15, the influence of initial permeability μ_i is rather limited for modest values of l/d. This is the result of demagnetisation. Another parameter which must be taken into account in the design of a ferrite antenna is the Q-factor. Assuming that losses are due to the ferrite material only, we will have a Q-factor for the antenna

$$Q = \frac{\mu_i}{\mu_e \tan \delta} \tag{4.64}$$

where δ is the loss angle of the material. Manufacturers give the ratio $\tan \delta/\mu_i$ — also called $(\mu_i Q)^{-1}$ — as a function of frequency. It is thus easy to select the right material.

4.6 Equipment performance

4.6.1 Transmitter power

There has been some confusion in the past about the definition of transmitter power in subsurface radio communications. It still happens that manufacturer's pamphlets do not provide any information other than some power P which is probably that supplied to the antenna circuit. As eqn. 4.54 shows, this does not provide any information on the power transferred to the monofilar mode, which is the useful information. Actually this power depends on the tunnel size and shape, on the relative locations of the antenna and of the monofilar wire, and on antenna performance, as follows from eqn. 4.53. Some normalisation is thus needed.

A useful form for this information is the power $P_{E\,\mathrm{ref}}$ which is supplied to one of the monofilar mode waves excited in opposite directions in the reference tunnel. Indeed the monofilar mode power excited in one direction in the actual tunnel can then be obtained by

$$P_{Em} = \frac{P_{E\,\mathrm{ref}}}{C_e^2} \tag{4.65}$$

For a continuous leaky cable the excited power of the coaxial eigenmode is given by

$$P_{Ec} = \frac{P_{E\,\mathrm{ref}}}{C_e^2 \, C_1^2} \tag{4.66}$$

The normalised transmitter power is given by $R_r I^2$ where I is the amplitude of the antenna current and R_r has to be calculated for the reference tunnel. For a magnetic dipole R_r is given by eqn. 4.46 with $C_e = 1$ and $Z_m = \eta_0$. This yields

$$P_{E\,ref} = 1 \cdot 16 \times 10^{-4} f^2_{MHz} (A_e I)^2 \qquad (4.67)$$

or, in decibels relative to one watt

$$(P_{E\,ref})_{dBW} = -39 \cdot 3 + 20 \log f_{MHz} + 20 \log (A_e I) \qquad (4.68)$$

The transmitter power may thus equivalently be specified by the magnetic moment $(A_e I)$ in A m^2. In practice $A_e I$ does not exceed a few mA m^2 and $P_{E\,ref}$ is of the order of one microwatt.

These considerations obviously do not apply to base stations. These have powers of a few watts and are directly connected to the monofilar wire or to the coaxial cable.

4.6.2 Receiver sensitivity

Fig. 4.16a shows the equivalent circuit of a tuned magnetic antenna connected to the input of a radio receiver. Three voltage generators appear in series with the magnetic dipole: e_s is the signal electromotive force and may be obtained from eqn. 4.42 for a tunnel containing a monofilar wire; e_e is the electromotive force due to external manmade noise. e_t is due to thermal noise in the resistance $R = 2R_r + R_1$, where R_r is the one-sided radiation resistance and R_1 is the loss resistance.

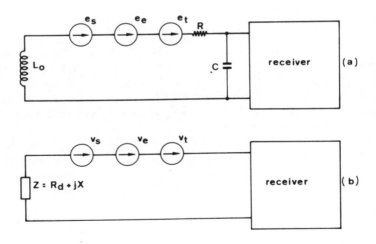

Fig. 4.16   *Equivalent circuits for a magnetic antenna connected to a receiver*

This circuit may be transformed by standard methods into the series representation of Fig. 4.16b. Assuming that the Q-factor of the resonant circuit is large, the series impedance Z resonates at $\omega_0 = (LC)^{-1/2}$ and reduces to a dynamic resistance

$R_d = Q^2 R$ at this frequency. We also suppose that the signal carrier is a sine wave at this frequency. We have consequently $v_s = Q e_s$, where effective values are used. The available carrier power is thus given by

$$C = \frac{v_s^2}{4R_d} = \frac{e_s^2}{4R} = \frac{e_s^2}{8R_r} \frac{Q}{Q_r} \tag{4.69}$$

since $Q_r = \omega L_0 / (2R_r)$ and $Q = \omega L_0 / R$. On using eqns. 4.42 and 4.46 we obtain

$$C = 0{\cdot}5 Z_m I_m^2 \frac{Q}{Q_r} = 0{\cdot}5 P_m \frac{Q}{Q_r} \tag{4.70}$$

This expression is remarkable. It shows that the received signal power should amount to 50% of the monofilar mode power if no other resistance than the radiation resistance $2R_r$ existed. Actually we know that loss resistances (R_w, R_d) are unavoidable and that furthermore Q is considered to be determined by bandwidth and temperature sensitivity considerations. This theoretical limit is drastically reduced by the very small factor Q/Q_r.

Unless the receiving antenna is very close to the noise source, external manmade noise reaches the receiver input at the same time as the useful signal, being propagated in the monofilar mode and received by the antenna. If we assume that this noise has a one-sided spectral power density in the monofilar mode $p_{me} \mathrm{W\,Hz^{-1}}$, the received external noise power in the receiver bandwith B will be given by

$$N_e = 0{\cdot}5 p_{me} B \frac{Q}{Q_r} \tag{4.71}$$

As usually the available thermal noise is given by

$$N_t = k T_0 B \tag{4.72}$$

where $k = 1{\cdot}38 \times 10^{-23} J K^{-1}$ is Boltzmann's constant and T_0 is the physical temperature in kelvins. We will use the standard value $T_0 = 290\,\mathrm{K}$. The last contribution to the total noise is that of the receiver; the latter is characterised by a noise factor F_0 or by an effective noise temperature

$$T_e = (F_0 - 1) T_0 \tag{4.73}$$

and the receiver noise is given by

$$N_r = k T_e B \tag{4.74}$$

The parameters F_0 and T_e are defined for the receiver fed by a generator of internal impedance equal to R_d.

We are now able to calculate the carrier/noise ratio

$$\frac{C}{N} = \frac{C}{N_e + N_t + N_r} = \frac{0{\cdot}5 P_m}{(0{\cdot}5 p_{me} + k T_0 F_0 Q_r / Q) B} \tag{4.75}$$

It is clear from this result that the best antenna is the one which has the lowest radiation Q-factor Q_r. Let $(C/N)_{min}$ be the minimum required carrier/noise ratio for acceptable receiving quality. Eqn. 4.75 allows us to determine the required monofilar mode power. The latter will depend on the spectral density of manmade noise p_{me} and also, through Q_r, on the eccentricity loss as shown by eqn. 4.53. It is thus necessary to define standard conditions if we intend to relate $(C/N)_{min}$ to a required minimum power of the monofilar mode and thereby define the receiver sensitivity. For this purpose we assume that the receiver is used in the reference tunnel and that manmade noise does not exist. The receiver sensitivity is defined as the required monofilar mode power $P_{R\ ref}$ in these conditions. It is given by

$$P_{R\ ref} = 2(C/N)_{min}kT_0F_0B\frac{Q_{r\ ref}}{Q} \qquad (4.76);$$

Here $Q_{r\ ref}$ may be evaluated by eqn. 4.53 with $C_e = 1$, provided the radiation Q-factor in free space Q_0 is known. An alternative definition of the receiver sensitivity is given by specifying the minimum required magnetic field H_{min}. The latter is related to $P_{R\ ref}$ by eqn. 4.5. In practical logarithmic units this yields

$$(P_{R\ ref})_{dBW} = 51\cdot3 + 20\log H_{min} \qquad (4.77)$$

These definitions of receiver sensitivity include not only the quality of the receiver input stages as a low-noise amplifier, but also the efficiency of the modulation method and of the antenna. It is a limiting sensitivity in the sense that manmade noise is supposed to be negligible. If this is not true the receiver sensitivity will have to be reduced accordingly. It should also be observed that, in eqn. 4.76, B is the noise bandwidth whereas Q is related to the 3 dB bandwidth B_3 of the antenna and to the carrier frequency f by the classical relation $B_3 = f/Q$.

Finally it should be clear that the required monofilar mode power in actual conditions is given by

$$P_{Rm} = \frac{P_{R\ ref}}{C_e^2} \qquad (4.78)$$

For a continuous leaky coaxial cable the required coaxial eigenmode power will be

$$P_{Rc} = \frac{P_{R\ ref}}{C_e^2\,C_1^2} \qquad (4.79)$$

Again it must be stressed that these considerations do not apply to base stations.

4.6.3 Example of equipment design

The example that will be described here is the design of radio equipment for mining applications recently undertaken by the Institut National des Industries Extractives, Liège, Belgium, for the frequency band 3–10 MHz.

(a) Choice of modulation
This is the first step in the design of a radio system. It is known that wideband frequency modulation provides excellent noise rejection.

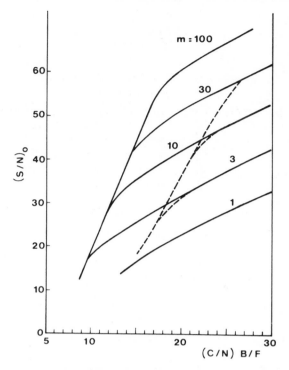

Fig. 4.17   *Performance curves of frequency modulation for a discriminator (continuous) and for a second-order phase-locked loop (broken)*

Fig. 4.17 shows the output signal/noise ratio S/N as a function of $(C/N)B/F$ for various values of the modulation index $m = \Delta f/F$, where Δf is the frequency excursion and $F = 3.5\,\text{kHz}$ is the highest frequency of the baseband. The dotted curves are valid when a discriminator is used for demodulation. They exhibit the well-known threshold effect. The latter is nearly absent in the continuous curves: these are valid for a properly designed phase-locked loop demodulator. A S/N value of 20 dB was considered as the minimum required quality for mine communications. The best choice for minimum bandwidth is thus a modulation index of 10, i.e. a frequency excursion of 35 kHz, provided a phase-locked loop demodulator is used. It results in

$$(C/N)_{\text{min}} = 2\,\text{dB} \tag{4.80}$$

which provides a 7 dB improvement compared with conventional narrowband FM ($m \simeq 1$). This choice is made possible by the non-existence of frequency allocations for mining applications, where no radiation above the earth's surface can be guaranteed. The required bandwidth may be estimated from Carson's rule

$$B = 2(\Delta f + F) = 77\,\text{kHz} \tag{4.81}$$

(*b*) *Antenna design*

In view of the results obtained in Section 4.5.7, a ferrite antenna was considered as adequate for portable radio sets. Available ferrite rods were carefully compared. It appeared that a good choice was a ferrite rod of 15 cm length, 1 cm diameter and initial permeability $\mu_i = 60$. Four ferrite rods are connected either in series or in parallel. We may then expect to have, according to Fig. 4.15, $Q_0/Q_{0\,min} = f_{MHz}/4$. On the other hand, 4.38 yields $Q_{0\,min} = 2{\cdot}58 \times 10^8\, f_{MHz}^{-3}$. Thus $Q_0 = 6{\cdot}45 \times 10^7\, f_{MHz}^{-2}$. Then, on using eqn. 4.37 with $C_e = 1$, we obtain $Q_{r\,ref} = 1{\cdot}07 \times 10^6$. This value is independent of frequency. On the other hand, the bandwidth requirement and the temperature sensitivity of ferrite rods allows the choice of a 3 dB bandwidth of 100 kHz for the antenna circuit throughout the whole frequency band 3–10 MHz. Thus the actual Q-factor of the antenna circuit is given by

$$Q = 10 f_{MHz} \tag{4.82}$$

and we have

$$\frac{Q_{r\,ref}}{Q} = 1{\cdot}07 \times 10^5\, f_{MHz}^{-1} \tag{4.83}$$

(*c*) *Receiver sensitivity*

On assuming a receiver noise factor of 3 dB and using eqn. 4.76 we obtain

$$(P_{R\,ref})_{dBW} = -96{\cdot}8 - 10 \log f_{MHz} \tag{4.84}$$

$$P_{r\,ref} = 2{\cdot}06 \times 10^{-10}\, f_{MHz}^{-1} \tag{4.85}$$

or, using eqn. 4.77

$$H_{min} = 3{\cdot}9 \times 10^{-8}\, f_{MHz}^{-1/2} \tag{4.86}$$

(*d*) *Transmitter power*

The transceivers are intended to be used in the frequency band 3–10 MHz. A desirable quality is that changing the frequency channel should be very simple. Consequently tuning the antenna circuit on another channel will be made by modifying the capacitor, the ferrite antenna itself remaining unchanged. Since the Q-factor is given by $Q = R_d/(\omega L)$, eqn. 4.82 implies that the dynamic resistance R_d needs to be changed too. It will thus vary in a ratio of 9 to 100, i.e. about one decade. Suitable load impedances for the collector circuit of a transistor are in the range 1–10 kΩ. Let us assume that $R_d = 100\, f_{MHz}^2$ and that the current delivered to the antenna circuit is 10 mA at all frequencies. The active power delivered to the antenna circuit is then given by $P = 0{\cdot}01\, f_{MHz}^2$. The power $P_{E\,ref}$ of one of the two monofilar mode waves excited in the reference tunnel may then be obtained from eqns. 4.54 and 4.83. This yields

$$P_{E\,ref} = 4{\cdot}67 \times 10^{-8}\, f_{MHz}^3 \tag{4.87}$$

or

$$(P_{E\,ref})_{dBW} = -73{\cdot}3 + 30 \log f_{MHz} \tag{4.88}$$

Having fixed the Q-factor and the dynamic resistance R_d, we automatically know the total self-inductance $L = 1.59\,\mu\text{H}$ of the antenna circuit and thus that of the four ferrite rods used either in series or in parallel. The number of turns may then be obtained from eqn. 4.61. The total magnetic moment of the antenna follows from eqns. 4.67 and 4.87:

$$A_e I = 0.02\,f_{\text{MHz}}^{-1/2} \tag{4.89}$$

(e) Final remarks

It may be observed that the receiver sensitivity and transmitter power increase with frequency. In the configuration of the reference tunnel ($C_e = 1$) the maximum allowable attenuation for the monofilar mode in a mobile-to-mobile transmission is obtained from eqns. 4.84 and 4.88. This yields

$$A_{\text{m ref}} = 22.5 + 40\log f_{\text{MHz}} \tag{4.90}$$

This frequency dependence is beneficial since it tends to compensate for the increase of the eccentricity loss (for magnetic antennas) and of specific attenuation with frequency. To be complete we also need to consider base-station-to-mobile and reverse communications, but this is the subject of a further study.

4.6.4 Manmade noise in mining environments

With the advent of highly mechanised and automated mining techniques, the intensity of manmade noise has tended to increase. This is particularly true since, as radio communications are not yet intensively used in most mines, little attention has been devoted to this problem in the design of mining machines.

Information on the level of manmade noise in mines is rather scarce. The only systematic investigation in this field was undertaken by the US National Bureau of Standards under the sponsorship of the US Bureau of Mines. The results were published in a series of reports (Adams *et al.*, 1971; Kanda, 1974; Kanda *et al.*, 1974; Bensema *et al.*, 1974; Scott *et al.*, 1974). Manmade noise is frequently non-stationary with respect to time. The noise level may be very high in the proximity of mining machines. At other places the noise level will be strongly influenced by the propagation properties of the tunnels separating this location from the noise sources. Most measurements reported in the cited literature were made in relatively busy areas (working area in room and pillar mines, end of longwall face, bottom of shaft, etc.)

The continuous curves on Fig. 4.18 were obtained by averaging published results and may be considered as typical for this kind of location. They show the magnetic field measured in a 1 kHz bandwidth. For a bandwidth B it may be assumed that the square of the magnetic field is proportional to B. Also shown is the limit sensitivity of a high-performance portable receiver designed to withstand thermal noise only. It was obtained by reducing eqn. 4.86 to a 1 kHz bandwidth instead of 77 kHz. That manmade noise will be dominant in mines seems an obvious conclusion. This is however not completely true. Measurements show a large spread about the typical curves of Fig. 4.18. One has further to take into consideration that these are representative of busy areas, i.e. for locations which are relatively close to the noise sources. Experience has shown that working close to the thermal noise limit was frequently possible. A conservative rule might be that manmade noise requires the reduction of the sensitivity of portable receivers by a few decibels.

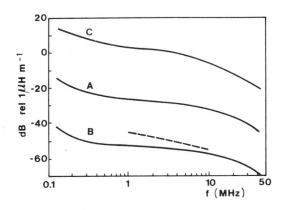

Fig. 4.18   *Typical manmade noise in busy areas of mines. The curves exhibit the magnetic field measured in a 1 kHz bandwidth as a function of frequency. Curve A shows the r.m.s. average of measurements at a given location, B and C the values exceeded during 99 and 0·001 % of time. The broken curve is the limit sensitivity of a high-performance receiver*

The same rule is valid for base stations connected to a monofilar wire conductor. Receivers connected to a coaxial cable may be operated at the sensitivity limit since the screening properties of the cable are active for manmade noise as for useful signals. The receiver noise will then be the limiting factor.

4.7 Mine communication system design

4.7.1 General principles

When a mine communication system has to be installed the communication requirements are defined by the mine operator. The locations specified for the base stations are in general those of the control points, i.e. of the audio frequency part of the base stations. Remote control may be used to locate the radio frequency part as close as possible to the centre of the system of tunnels. The first step in designing the system consists of visiting the tunnels and defining the location of waveguiding cables in order to minimise the eccentricity loss and the monofilar mode attenuation (Section 4.4.1). After having chosen the operating frequency, these parameters are known for all tunnels. A particular problem occurs for longwall faces, because the only acceptable cable location is together with other electric cables and water pipes in the hooks on the side of the armoured conveyor. The monofilar and coaxial modes are obviously not useable in these conditions. The only solution available today, at least to the author's knowledge, is to use a coaxial cable with annular-slot mode converters spaced by about 20 m (Delogne, 1973). The near-field radiation of the slots is then the useful mechanism.

The next step is to select a waveguiding cable together with the system configuration. Simple monofilar wires are useful only for short ranges. They may however be used to give some lateral extension. For instance, in room and pillar work, it is possible to use a fish-bone configuration with a coaxial cable and slot mode converters for the axis and monofilar wires connected to the cable external conductor for the crosscuts. For long tunnels the choice will in general be between a coaxial calbe with annular-slot mode converters and a continuous leaky feeder. The former has the advantages of providing a great design flexibility and of being usable with different types of cable, even in the same system. For instance large-diameter cables with plain outer conductor provide low specific attenuations and are suitable for long haulageways, while cheaper small-diameter braided cables may be used in relatively short tunnels, including longwall faces. In some cases it may be useful to insert annular-slot mode converters in a leaky cable, for instance to obtain a high level of the monofilar mode in a short section where the eccentricity loss is very high. Leaky section cables are not frequently used at the low end of the HF band since the sections are too long.

The network configuration may be a rather complicated star network as will be seen from examples described below. Loops may provide redundancy in the event of cable break. Network branching and base station connection is made through power splitters and hybrids (directional couplers). Power division is chosen as a function of the length of the various branches. It is necessary to calculate the balance of losses for each branch of the network. Designing the system this way may require several trials. Mine operators should greatly benefit by planning their future needs since multichannel operation is possible on a properly designed network of cables.

4.7.2 *Example 1*

The example that will be described now is that of a very large network, probably the largest in the world. It covers about 19 km of tunnels in a potash mine near Mulhouse, France. The conductivity of potash is very low, maybe 10^{-5} S m^{-1}. This results in a rather high specific attenuation for the monofilar mode, but also in a low coupling loss. The only radio equipment available at the time of design was working at 7 MHz. A coarse theoretical estimate of the specific attenuation of the monofilar mode at this frequency was 6 dB/100 m. In order to know the effect of existing conductors in the mine tunnel, a preliminary test was carried out. It consisted in drawing about 1 km of coaxial cable with a single mode converter in the middle. The cable location was varied along the path. The cable was fed by a generator and the relative level of the monofilar mode was measured along the path.

This test showed that the attenuation would not exceed 4 dB/100 m for the worst location. Consequently a value of 5 dB/100 m was used in the design.

Fig. 4.19 shows the configuration of the network. The base station denoted by Y is located at the centre of the network. The latter is made from coaxial cables with annular-slot mode converters. The power distribution is made by 3 dB power splitters and by hybrids providing different degrees of power division. It was carefully studied to provide a balanced repartition of power in all branches.

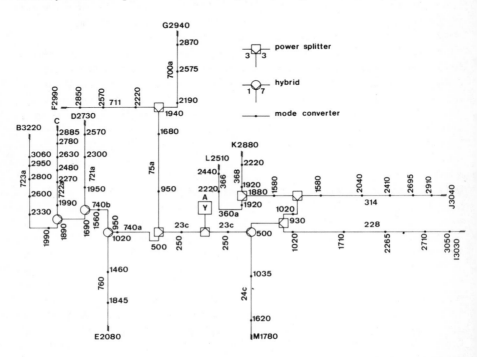

Fig. 4.19   *Network in a potash mine near Mulhouse, France. Letters are used to indicate locations. The figures give the total distance from a point to the base station Y located at point A. The power division (in decibels) realised by power splitters and hybrids is indicated*

As an example of the methodology which was followed, we will discuss the calculation for the path AJ. The characteristics of the radio equipments are:

Base station: transmitter power 3 dBW
                 receiver sensitivity -130 dBW
Portable sets: transmitter power $P_{E \, ref} = -58$ dBW
                 receiver sensitivity $P_{R \, ref} = -73.5$ dBW

The maximum allowable attenuations from the coaxial cable input at the base

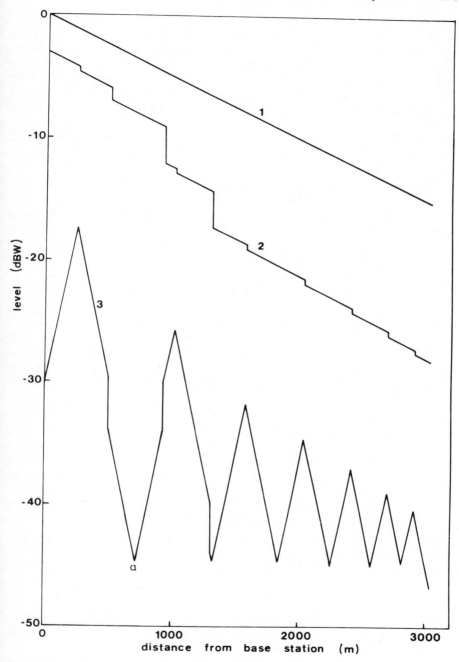

Fig. 4.20 *Balance of losses along the path AJ of Fig. 4.16. Line 1 shows the power inside a coaxial cable without any device, line 2 takes account of power splitters, hybrids and mode converters. Line 3 exhibits the power of the monofilar mode.*

station to the monofilar mode in the reference tunnel near the portable set are thus of 76·5 dB for base-station-to-portable-receiver communications and 72 dB for the reverse link. Because of the low conductivity of potash and of the suitable cable location, the eccentricity loss is expected to be slightly negative. Standing waves will arise due to monofilar mode waves travelling in opposite directions. Consequently it was estimated that the maximum allowable attenuation from the coaxial input at the base station to the monofilar mode anywhere in the tunnels should not exceed 45 dB.

The coaxial cable has a specific attenuation of 0·5 dB/100 m. The mode converters were designed to have a transmission loss $T = -0·45$ dB and a mode conversion factor $S = -13·2$ dB.

Fig. 4.20 shows the balance of losses along the path AJ. Straight line 1 shows the level inside a coaxial cable without any device connected to it. Each time a power splitter, a hybrid or a mode converter is crossed, the level of the coaxial mode drops by the attenuation of this device, as shown by line 2.

The first mode converter is located at 250 m from the base station. At this point two monofilar mode waves are excited in opposite directions at a level of 13·2 dB below that of the incident coaxial mode. They propagate with an attenuation $\alpha_m = 5$ dB/100 m as shown by line 3. A hybrid circuit is encountered at 500 m by the wave travelling away from the generator; since this device acts as a T-junction for the monofilar mode, the latter suffers a loss of about 4 dB at this point. The threshold level of -45 dBW is reached at 730 m (point a). A mode converter will be needed ahead: to know where, we draw a straight line from the point a with a slope of 5 dB/100 m, making a 4 dB step when we cross the power divider located at 930 m. The next mode converter will be located at the point 1020 m, where this line is 13·2 dB below line 2. This procedure is continued up to the end of the path.

It is seen that the spacing of mode converters becomes small with distance. The mean spacing for the whole system is about 370 m, dropping to 100 m at the end of the branches. Actually the limit of range was reached in this application. Increasing the range further would require either the use of higher-performance equipment, like that described in Section 4.6.3, or the use of coaxial cables with a lower specific attenuation.

It should be observed that the coaxial cable is useless for portable-to-portable communications with the equipment used here. Indeed $P_{E\ ref} - P_{R\ ref} = 15.5$ dB is less than twice the mode conversion factor. This type of communication thus uses the coaxial cable only as a single monofilar wire. This would not be true for the high-performance equipment described in Section 4.6.3.

4.7.3 Example 2

This is an example of superposition of several frequency channels in coaxial cables. It was installed in the mine of Gardanne, Houillères du Bassin de Provence, France. This mine was among the most advanced in the world, being highly automated. The coal production is about 9000 tonnes d^{-1}. Efficiency of the transportation system is vital in such cases. In Gardanne, coal mined in a longwall face (and also in a room

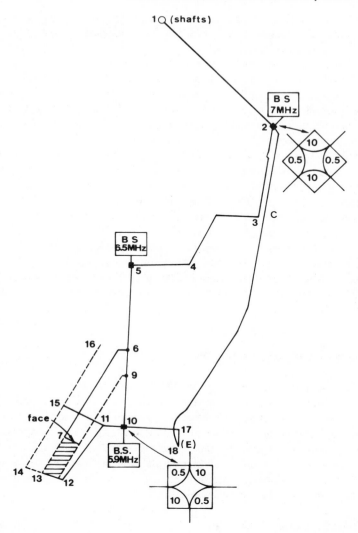

Fig. 4.21   *Three radio networks realised at the mine of Gardanne. Coaxial cables are shown as continuous lines, haulageways without cables as broken lines. The power division of four-part hybrids is indicated in decibels*

and pillar section not considered here) is brought by a belt conveyor to the loading point E (Fig. 4.21), and from there to the shafts — path E–C–2–1 — by a system of belt conveyors and of moving buckets pulled by linear motors. Staff and equipment are transported from the shafts to the working tunnels on tyre motor cars along the tunnels 1–2–3–4–5–6–9–10.

Three radio networks work in this mine. Network I which was first installed is operated at 7 MHz and provides radio communications from a base station located at point 2 and remote controlled from the surface, with patrols supervising the coal transportation along the path 1–2–C–E. The length is 4800 m. A Cellflex coaxial cable with annular-slot mode converters is used along this path.

Network II works at 6·5 MHz. It provides communications from a base station located at point 5 and remote controlled from point 9 with the motor cars running along the paths 1–2–3–4–5–6–9–10–11–12–13 and 6–7. The total length of the path is 8·5 km. This network reuses the coaxial cable of network I on a length of 1·5 km between points 1 and 2. A low-attenuation cable with mode converters was used from 2 to 13. A cheap braided cable type RG-217/U was used for section 6–7: this is indeed a haulageway that will be replaced by 14–15–16 when the next longwall face is started. A simple T-junction is used at point 6.

Network III works at 5·9 MHz. It provides communications from a base station located at point 10 with workers along the path 15–11–10–17–18. This path is only 1200 m long and uses RG-217/U cable, except between 10 and 11 where the cable of network II is reused. Connection at point 11 is made with a T-junction.

The base stations at points 2 and 10 are connected to the cables through four-port hybrids. The power distribution realised by these devices was selected in order to obtain adequate signal transmission at the relevant frequencies as functions of the ranges to be covered.

4.8 Optimisation of leaky coaxial cables

The use of leaky coaxial cables to establish subsurface radio communications has been amply discussed in previous sections. Hence we are able to answer an important question: how leaky should the cable be made? This question does not arise for mode converters, since they are easily tailored to fit a specific application, but does for continuous leaky feeders and cables with leaky sections. The answer to this question depends on a number of factors and will be discussed below. A related question is: what will be the performance of a leaky feeder when these factors differ from those used for the optimisation?

4.8.1 Continuous leaky feeders

Let us assume the following characteristics for the radio equipment: P_E, P_R, transmitter power and receiver sensitivity for the base station; $P_{E\ ref}, P_{R\ ref}$, normalised transmitter power and receiver sensitivity. Let A be the smallest of the ratios $P_E/P_{R\ ref}$ and $P_{E\ ref}/P_R$. This is the maximum allowable power attenuation between the coaxial mode at the base station and the monofilar mode at the portable set in the reference tunnel. To calculate the range R_c of communications between the base station and portable sets, we must take into account the coupling factor C_1 of the leaky feeder, the eccentricity factor C_e for the cable in the actual tunnel and the specific attenuation eqn. 3.53 of the coaxial eigenmode in the actual tunnel. Assuming $C_2 \ll C_1$, the range is thus given by

$$AC_1^2 C_e^2 \exp\left[-2(\alpha_{co} + C_1^2 \alpha_{mo})R_c\right] = 1 \tag{4.91}$$

where α_{co} and α_{mo} are expressed in nepers per unit length. Solving for R_c and setting $dR_c/dC_1 = 0$ we obtain the transcendental equation in C_1

$$C_1^2 = \frac{\alpha_{co}}{\alpha_{mo}\left[\ln(AC_1^2 C_e^2) - 1\right]} \tag{4.92}$$

It thus appears that the optimum coupling coefficient C_1 depends on the equipment performance, on the specific attenuation of the non-leaky version of the cable and, through α_{mo} and C_e, on the use that will be made of the leaky cable. Some of these parameters depend on frequency. It is obviously not possible to manufacture leaky feeders optimised for each set of parameters. The cable will be optimised for average conditions.

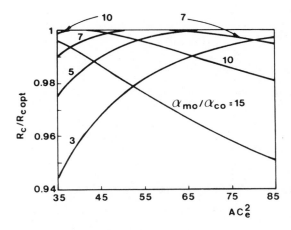

Fig. 4.22   *Range reduction $R_c/R_{c\,opt}$ of continuous leaky feeders as a function of AC_e^2 and of α_{mo}/α_{co} Here, R_c is the range obtained using a cable optimised for $AC_e^2 = 55\,dB$ and $\alpha_{mo}/\alpha_{co} = 7$, whereas $R_{c\,opt}$ assumes that the cable is optimised for each value of these parameters*

As an example we took the following set of parameters: $AC_e^2 = 55$ db, $\alpha_{mo}/\alpha_{co} = 7$. The optimum is then for $C_1 = 0.11$. This would correspond to a specific transfer inductance of about $20\,\mathrm{nH\,m^{-1}}$. Subsequently we calculated the range reduction that results from using this cable instead of one that would be optimum for the actual conditions. As Fig. 4.22 shows, the reduction does not exceed a few per cent over a relatively wide range of parameters. Optimisation of continuous leaky feeders does not appear to be critical. It may be interesting to note that the optimum value of C_1 increases with α_{co}/α_{mo} and with AC_e^2.

These conclusions are valid for communications from the base station to portable sets. For communications between the latter, we must use the parameter $A_p =$

$P_{E\,ref}/P_{R\,ref}$ instead of A. Furthermore the coupling coefficient C_1 and the eccentricity coefficient C_e must be counted twice. The range is thus given by

$$A_p C_1^4 C_e^4 \exp\left[-2(\alpha_{c0} + C_1^2 \alpha_{m0})R_c\right] = 1 \tag{4.93}$$

The optimum is for the C_1^2 solution of equation

$$C_1 = \frac{2\alpha_{c0}}{\alpha_{m0}\left[\ln\left(A_p C_1^4 C_e^4\right) - 2\right]} \tag{4.94}$$

This equation has no solution unless $A_p C_e^4$ is large enough. Furthermore, if a solution exists, it is useful only if the range provided by the coaxial mode is larger than that of the monofilar mode $R_m = \ln(A_p C_e^2)/(2\alpha_m)$. It appears that this is the case only for $A_p C_e^2$ larger than about 30 dB and α_{c0} small. The first condition is practically never met for portable radio equipment in the HF band. Consequently continuous leaky feeders are in general not better than a simple monofilar wire for portable-to-portable communications. Long range communications of this type require the use of the base station as a relay.

4.8.2 Cable with leaky sections

Hence we consider communications with a base station connected at the input of a cable with leaky sections. The latter are characterised by a coupling factor C_1 and a spacing d. If the portable set is located just before the nth leaky section, the useful signal crosses $(n-2)$ leaky sections in the coaxial mode and is converted to the monofilar mode at the $(n-1)$th section. The range thus corresponds to a number n of spacings d such that

$$AC_e^2 (1 - 4C_1^2)^{n-2} (2C_1)^2 \exp\{-2\left[(n-1)\alpha_{c0} + \alpha_{m0}\right]d\} = 1 \tag{4.95}$$

Solving this equation for n we obtain the communication range

$$R = \frac{\ln(AC_e^2) + \ln x - 2\ln(1-x) - 2(\alpha_{m0} - \alpha_{c0})d}{2\alpha_{c0} - d^{-1}\ln(1-x)} \tag{4.96}$$

where $x = 4C_1^2$.

It is possible to calculate the values of x and d that maximise R for given values of AC_e^2, α_{c0} and α_{m0}. A systematic analysis of the problem shows that a cable with leaky sections and a continuous leaky cable are roughly equivalent. For the cable with leaky sections the optimum value of C_1 increases with α_{c0} and decreases with α_{m0} and AC_e^2. The optimum distance d depends mainly on α_{m0}, to which it is roughly inversely proportional. For α_{m0} larger than 5 dB/100 m the spacing becomes too small for practical use.

Here again it is not possible to optimise the parameters for each application. A spacing of about 150 m is considered to be adequate. The coupling factor C_1 should ideally increse with α_{c0}: values of -26 and -19 dB are adequate for $\alpha_{c0} = 0.5$ and 3 dB/100 m, respectively. In these conditions the cable with leaky sections remains equivalent to a properly designed continuous leaky cable. Actually the former becomes advantageous when α_{c0} and α_{m0} are both high, as will be the case at higher frequencies.

4.9 Medium-frequency communications in room and pillar mines

In room and pillar mines the communication requirements are for an area rather than for a linear tunnel path as in longwall face mines. Drawing cables in all rooms is considered impractical. A search was therefore made for truly wireless communication systems. Two types of solution are possible. Frequencies well above the cutoff of subsurface cavities (rooms, cross-cuts, haulageways, etc.) may propagate with low attenuation in these cavities: this type of solution will be considered in Chapter 5. The alternative is to use frequencies low enough to benefit by a small attenuation in through-the-ground propagation.

In the particular case of coal mines, advantage is gained from the low conductivity of coal compared with the surrounding rock. Such a low-conductivity seam is able to guide electromagnetic waves. This subject was studied extensively in Section 2.6 and the specific attenuation of the dominant mode has been calculated for a wide range of parameters (Fig. 2.12). By making somewhat crude approximations in eqn. 2.245, it may be shown that the magnetic field radiated by a magnetic dipole is given by

$$H = \frac{\pi \kappa^{3/4} IA_e}{a \lambda_0^{3/2} d^{1/2}} \exp(-\alpha d) \tag{4.97}$$

where I is the dipole current, A_e the equivalent loop area, κ the dielectric constant of coal, a the seam thickness, λ_0 the free-space wavelength, α the specific attenuation in $Np\,m^{-1}$ and d the distance in metres. Low frequencies are unfavourably affected by the denominator of eqn. 4.97 whereas higher frequencies suffer from a large specific attenuation. Furthermore manmade noise may be intense at low frequencies.

The medium-frequency band ($300\,kHz - 3\,MHz$) was consequently found adequate for this type of communication in the United States, where intense research efforts have been devoted to this problem. The reader is referred to the relevant literature (Lagace *et al.*, 1975; 1977; Lagace and Emslie, 1978a, b; Emslie and Lagace, 1978; Emslie *et al.*, 1978; Cory, 1978). Because of shortage of space we will limit ourselves to some comments.

This type of communication is successful only if the specific attenuation of the seam wave is low enough. Some comments on the dependence of attenuation on the electrical parameters of coal and rock were made in Section 2.6.1. The theoretical and experimental investigations made in the United States mainly concern communication between portable radio sets. Optimum frequencies are between 500 and 1000 kHz. The communication range is 100 to 200 m in conductor-free areas. It increases up to 500 m in the low-conductivity Pittsburgh seam. It should however be noted that calculations were made for a transmitter with a magnetic moment IA_e of $2 \cdot 5\,A\,m^2$. This is about 40 dB above the values considered as realistic for light portable sets in the previous sections. Base stations could use a large loop as the antenna. Alternatively they can be connected to bolts driven into the floor and roof; this provides about 20 dB improvement with respect to magnetic antennas.

The existence of conductors in some entries may considerably increase the communication range for radio sets which are not located too far from the conductors. This is not surprising in view of the methods developed in the other sections of this chapter. Skin depth in the rock is indeed large at medium frequencies and the eccentricity loss for magnetic antennas may remain acceptable when the portable radio set is relatively distant from the conductor. Whether or not a synthesis of the MF and guided wave communication techniques can be made remains at present an open question. Undertaking this study would undoubtedly lead to considerable progress.

4.10 Retransmission of AM broadcasting in road tunnels

Retransmitting radio broadcast programmes in road tunnels becomes particularly useful when these signals convey traffic information. An example of a rather complex system will be described in Chapter 5. We will here discuss only some technical aspects of the retransmission of amplitude modulation broadcasting in the long and medium-wave frequency bands, i.e. between 150 and 1500 kHz.

Although the methods described in the present chapter apply, the problem shows some peculiarities. Of course we cannot act on the on-board receiving installations. Car radio antennas and receivers are designed for surface use and they have electric-type antennas. Consequently the specification will be to obtain the same minimum electric field, e.g. 1 mV m^{-1}, as above the surface.

Since the electric field is the quantity of interest, the PCW (perfectly conducting wall) model provides a good approximation for the calculation of eccentricity loss. Road tunnels in which radio retransmission is required have in general very large cross-sections. In spite of this, gauge considerations frequently impose a very eccentric location of the cable. As Fig. 4.10 and eqn. 4.10 show, the eccentricity loss will frequently amount to 50 to 60 dB. According to eqn. 4.4, the required monofilar mode power is 1 μW in the reference tunnel and consequently 0·1 to 1 W in the actual tunnel. This is obviously a per channel power. Fortunately road tunnels are most frequently relatively short and a simple monofilar wire may be used, propagation attenuation being negligible.

In order to avoid a beating note due to interference, it is necessary to retransmit the stations exactly on their original frequency. Signals are received by an external antenna with a level of a few millivolts and thus need to be amplified by 60 to 70 dB. Wideband or channel per channel amplification may be used. Intermodulation is critical in the former case and amplifiers with output powers of more than 10 W at saturation may be required.

A further problem results from the use of high gains. The monofilar mode excited in the tunnel flows back toward the external antenna along the outer surface of the antenna feeder, causing instability. This problem may be solved by means of monofilar mode blocking circuits but this requires some experience.

Chapter 5

Subsurface radio communications in the VHF and UHF frequency bands

5.1 Introduction

The propagation properties of electromagnetic waves in subsurface works at VHF (30–300 MHz) and UHF (300 MHz–3 GHz), compared with lower frequency bands, are governed by the fact that these frequencies are above the cutoff of the waveguide modes.

The monofilar and similar modes, which can be guided by conductors parallel to the tunnel axis, still exist but their attenuations become so high that they are not directly useful. In principle it is possible to reduce the specific attenuation of the monofilar mode by coating the conductor with a dielectric layer thereby transforming it into a Goubau waveguide. The attenuation will be drastically reduced if the effective radius of the Goubau wave becomes smaller than the distance from the wire to the tunnel wall and as long as this effective radius remains unobstructed. It is clear that this technique may not be efficient if the mobile radio sets are used outside the effective radius of the surface wave, but it may be interesting in cases where the mobile antennas move along well-defined paths parallel to the tunnel axis, as occurs in transportation systems. Though potentially attractive, the Goubau waveguide does not seem to have been used for this type of application. High installation costs and also the strong influence of moisture and dust on attenuation are probable reasons for this lack of success.

Communication systems at VHF and UHF will thus most frequently be based on the propagation properties of waveguide modes, i.e. on natural propagation in tunnels. It will appear that leaky or non-leaky coaxial cables, if any, are used essentially to convey and radiate electromagnetic energy, the main difference with lower frequencies being that direct coupling of the portable radio sets to the leakage fields of the coaxial eigenmode is no longer the essential mechanism.

The techniques and the system structures used at VHF and UHF are various and more dependent on the application context than at lower frequencies. Reasons for this can be found in the wider spectrum of applications as well as in propagation characteristics: higher specific attenuation of coaxial cable, influence of tunnel bends and crossings, of obstacles in the tunnel and so on.

5.2 Natural propagation in tunnels

By natural propagation we mean that no special-purpose cables are strung in the tunnel to guide electromagnetic waves. Propagation is then that of waveguide modes. Unintentional axial conductors existing in the tunnel may influence these modes but this effect is considered as of second order at the frequencies considered here.

The classical theory of hollow waveguides was used as an introduction to the problem in Section 1.2.1 but the weakness of this approach was also stressed. Comparison with some experimental results in Section 1.2.2 required lenience in the interpretation of measurements. The situation was greatly clarified by the theoretical investigations reported in Section 2.7. Let us summarise some important conclusions for those readers who skipped Chapter 2.

Classical theory is relatively accurate for the calculation of cutoff frequencies but it fails in the prediction of specific attenuations. It appears that the specific attenuation of waveguide modes above their critical frequency is a steadily decreasing function of frequency. The attenuation is essentially governed by the tunnel size and shape and by the dielectric constant of the tunnel walls rather than by their conductivity. All six components of the electromagnetic fields are in general non-zero.

For a rectangular tunnel, the modes are called $E_{mn}^{(v)}$ and $E_{mn}^{(h)}$, where the superscript indicates that the main component of the electric field is either vertical or horizontal. The mode indices m, n run from zero to infinity. Under conditions 2.301 the characteristics of the $E_{mn}^{(v)}$ modes are given by eqns. 2.296–2.300, where a is the width of the tunnel, b the height, and κ_v' and κ_h' the complex dielectric constant of the vertical and horizontal walls, respectively. The coordinates (x, y) are defined in Fig. 2.16. The results for the $E_{mn}^{(h)}$ modes may be obtained by interchanging the x and y axes. The lowest attenuation is obtained for the $(0, 0)$ mode with the electric field parallel to the largest side of the rectangle.

For a circular tunnel, there are TE_{0n}, TM_{0n} and hybrid $EH_{mn}^{\pm}$ modes. The specific attenuations are given by eqns. 2.333 to 2.336 and the electromagnetic fields by 2.315–2.317, 2.320–2.322 and 2.327–2.329. All formulas mentioned here are in SI units.

Some properties demonstrated for these shapes appear to be general. Well above the critical frequency, the specific attenuation of the modes is inversely proportional to the square of frequency and to the cube of tunnel size. Furthermore the dominant field components of a mode behave as those of a plane wave in free space, apart from a transverse amplitude distribution imposed by the boundary condition on the tunnel walls. Losses are mainly due to dielectric refraction on the tunnel walls.

The fact that all six components of the electromagnetic fields are non-zero should always be kept in mind since it can explain effects which might otherwise seem surprising. As an example let us consider a large road tunnel with width $a = 17$ m, height $b = 4 \cdot 9$ m, and assume that the walls have a dielectric constant $\kappa =$

10. The $E_{0n}^{(v)}$ modes show attenuations between 5 and 6 dB/100 m for several values of n. Many $E_{mn}^{(h)}$ modes have much lower attenuations, starting from 0·677 dB/100 m for $m = 0$, $n = 0$. A vertical whip antenna is mainly coupled to the $E_{mn}^{(v)}$ modes but also slightly to the $E_{mn}^{(h)}$ modes, since these have a small vertical electric field component. The attenuation of a radio link between two vertical antennas as a function of distance will therefore show two slopes. For small distances the $E_{0n}^{(v)}$ modes are dominant and the slope of the attenuation curve will be 5 to 6 dB/100 m. For large distances however the horizontally polarised waves become dominant because of their small attenuation, and the slope will be 0·677 dB/100 m. Similar considerations apply to a radio link between a vertical and a horizontal antenna. They are well confirmed by experimental observations (Degauque *et al.*, 1979; Chiba *et al.*, 1978). It is therefore sometimes said that polarisation effects are not marked in tunnel propagation. Actually the constructive conclusion which can be drawn from this discussion is that circularly polarised antennas are attractive since they provide an efficient coupling to the lowest-attenuation mode.

Knowing the specific attenuation of waveguide modes does not allow the calculation of the range of a communication link. We also need to know the coupling of the transmitting and receiving antennas to these modes. The problem is relatively simple if we consider coupling to the main field components of a mode. For instance we may use eqns. 2.299 and 2.300 for a vertically polarised mode in a rectangular tunnel. Poynting's vector at a point of coordinates $(x, y, 0)$ is given by

$$S_{mn}(x, y) = \tfrac{1}{2} \eta_0 E_y^2 \tag{5.1}$$

and the mode power, calculated as the flux of this vector through the tunnel cross-section, is given by

$$P_{mn} = \frac{\eta_0 ab}{4} \tag{5.2}$$

Let us now consider a vertical receiving antenna located at the point $(x, y, 0)$. Since the wave is locally plane, we may assume that coupling occurs as in free space. This is deemed to be true if the effective area A_e of the antenna is small compared with the tunnel cross-sectional area. The received power is then given by $P_r = A_e S_{mn}(x, y)$. On using the well-known relation between the effective area and the antenna gain G

$$G = \frac{4\pi}{\lambda_0^2} A_e \tag{5.3}$$

we obtain the coupling loss

$$\frac{P_{mn}}{P_r} = \frac{2\pi ab}{\lambda_0^2 G} \sin^{-2} \frac{(m+1)\pi x}{a} \sin^{-2} \frac{(n+1)\pi y}{b} \tag{5.4}$$

By reciprocity the coupling loss also gives the ratio of the power supplied to a transmitting antenna to that of the modes excited in one direction. Eqns. 2.299–2.300 and consequently 5.4 do not remain accurate when the antenna is located close to the tunnel wall.

It is thus in principle easy to calculate the range of radio links in unobstructed straight-line tunnels, as long as we consider coupling of the antennas to the main field components. A safety margin should be included to account for standing waves. These cannot be avoided in practice: they are due to the existance of several modes travelling with different phase velocities.

It is far more difficult to predict the effect of tunnel bends and corners, and of obstacles present in the tunnel. Apart from general rules given in Section 1.2.3, data and theoretical models available at present (Emslie *et al.*, 1975; Degauque *et al.*, 1979) do not allow the making of accurate predictions.

5.3 Leaky feeders at VHF and UHF

5.3.1 General

The principle of radiation of leaky feeders when used well above the tunnel cutoff frequency, i.e. scattering of leakage fields by numerous inhomogeneities randomly distributed along the cable, was explained in Section 1.5.3. Here we would like to refine somewhat this analysis.

Because of the random mechanism involved, one has necessarily to resort to experimental data. Information on this question is scattered about in a large number of articles, starting with the early and rather accidental discovery by Monk and Winbigler (1956), dug out by Martin (1970a, b, 1973) sometime after the use of slotted coaxial cables by Hilscher and Plischke (1969). The first systematic experimental investigation of the properties of various types of leaky feeder was undertaken by Cree and Giles (1975).

It is not easy to extract quantitative data from these early measurements, although many useful observations were reported. One reason for this is that the actual mechanism of radiation was not yet explictly elucidated (Delogne, 1976). Consequently investigators did not always pay the required attention to some important operating conditions, nor did they report the relevant parameters. Further confusion may stem from the fact that most results reported by the important British team centred on the National Coal Board (Martin) and the University of Surrey (Davis, Critchley, Haining) are relative to the low VHF band and in conditions where the monofilar and waveguide modes may play comparable roles. Although this choice may be justified for some practical applications, conclusions drawn in these conditions have sometimes been extrapolated inappropriately to other frequencies and operating conditions.

By far the most systematic and comprehensive experimental investigation was carried out by Jouan (1977) in an alas nearly inaccessible student's thesis. It includes a rational discussion of specific attenuations and coupling losses measured for fifteen cables located at six different places in a tunnel of the Paris Underground, at the frequencies of 36, 79, 160 and 450 MHz. The cables are shown in the photograph of Fig. 5.1. Most of the information supplied here is based on this remarkable work. The interesting quantities are the coupling loss and the increase of the specific propagation attenuation of the cable as functions of operating conditions. The cable locations considered are realistic for most applications, i.e. at a maximum of 15 cm from the wall.

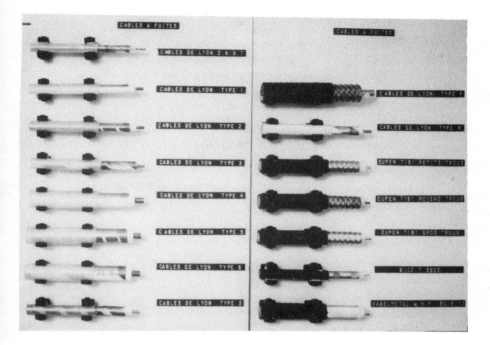

Fig. 5.1   *Leaky coaxial cables tested by Jouan in the Underground of Paris*

5.3.2 Statistical distribution of radiation

For a given location and mounting of the leaky feeder, the field amplitude along a path parallel to the tunnel axis shows random variations about a general exponential decay due to the specific attenuation of the cable. This attenuation depends on the cable location and mounting. For a given power level of the signals propagating inside the cable, the probability density function of the field amplitude along a path parallel to the tunnel axis is close to a Rayleigh distribution.

This behaviour can be justified as follows. Let us assume that an arbitrary complex field component E_c at an abscissa z along the tunnel axis is due to scattering of the near leakage field of the coaxial eigenmode by inhomogeneities located at abscissae z_i. We have then

$$E_c(z) = \sum_i a_i \exp\left(-\Gamma_c z_i\right) F(z - z_i, d)$$

where a_i is the scattering coefficient of the ith inhomogeneity, Γ_c is the propagation constant of the coaxial eigenmode, d is a parameter describing the location of the observation point in the tunnel cross-section and F is a function describing the propagation of the scattered wave. Two assumptions are made. We suppose that the abscissae z_i are randomly distributed, i.e. follow a Poisson process, and that the scattering coefficients a_i are independent random complex numbers all having the same probability distribution and being independent of the z_i. It can be shown that the real and imaginary parts of $E(z)$ are then two centered independent Gaussian variables with the same variance. Consequently (Korn and Korn, 1968) the amplitude $E(z) = |E_c(z)|$ is Rayleigh distributed.

As a result the probability that some amplitude E_0 will not be exceeded for a given power of the coaxial eigenmode is given by

$$p(E < E_0) = 1 - \exp\left[-E_0^2/(2\sigma^2)\right] \tag{5.5}$$

where σ^2 is the variance of E. This yields

$$20 \log E_0 = 10 \log(2\sigma^2) + 10 \log\left[-\ln(1-p)\right] \tag{5.6}$$

The coupling loss A_c being defined as the ratio expressed in decibels of the power of the coaxial mode to that received by a mobile antenna, the coupling loss $A_c(P)$ which is not exceeded at a percentage P of places is given by a formula of the type

$$A_c(P) = K - 10 \log\left[-\ln(P/100)\right] \tag{5.7}$$

where K is some constant related to σ^2. This may be eliminated by referring to the median $A_c(50)$: this yields

$$A_c(P) = A_c(50) - 1 \cdot 6 - 10 \log\left[-\ln(P/100)\right] \tag{5.8}$$

The vertical scale on Fig. 5.2 has been chosen so as to represent this equation by a straight line.

Jouan's measurements, confirmed by others (Grüssi and König, 1978; Suzuki *et al.*, 1980), have shown that the model of Rayleigh distribution is in general correct in the limits of measurement accuracy. The median coupling loss is thus a suitable descriptor of leaky feeder radiation. Of course this parameter depends on the operating conditions, namely on the frequency, on the tunnel shape and size, on the leaky feeder type and mounting, and on the relative locations of the leaky feeder and of the mobile antenna w.r.t. the tunnel cross-section.

5.3.3 Parameters influencing coupling loss

Some dependence of the coupling loss on frequency is observed. At 36 MHz the waveguide modes still suffer a high attenuation. The monofilar mode, though itself seriously attenuated, is the main mechanism by which scattered fields propagate. Consequently the median coupling loss remains high, e.g. 110 dB. Not surprisingly it depends on the cable and observation point locations in a manner very similar to that of the eccentricity loss studied in the previous chapter. For instance it increases when the cable-to-wall distance is diminished or when the

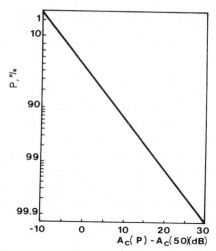

Fig. 5.2   *Rayleigh distribution: $A_c(P) - A_c(50)$ is the required margin to cover P% of locations instead of 50%*

antenna-to-cable distance increases. Variations due to the cable location amount to 20 dB. Departure from Rayleigh distribution is also maximum for this frequency, since beatings of the monofilar eigenmode with the coaxial eigenmode of the type shown in Fig. 3.11 are observed.

The median coupling loss drops sharply by about 40 dB when the frequency is raised from 36 to 79 MHz. It then increases again from about 10 dB between 79 and 160 MHz and from another 10 dB between 160 and 450 MHz. The reason for this behaviour is as follows. The decrease of the attenuation of waveguide modes with frequency should normally result in a decrease of the coupling loss. However this effect is inverted by the evolution of the effective radius of the leakage fields with frequency. At 79 MHz the effective radius is comparable with the tunnel size. All homogeneities present in the tunnel cross-section contribute to scattering and thus minimum coupling loss. It is also not surprising that no definite relation could be found between cable location and coupling loss at this frequency. At 160 and 450 MHz however it was clearly observed that the coupling loss was governed by the importance of scattering inhomogeneities located in the effective radius. The variation of the median coupling loss as a function of the cable mounting and location was about 15 dB. Previous tests had shown that radiation was virtually suppressed when the mounting and location were such that the effective radius was free of inhomogeneities.

Although this was not tested explicitly by Jouan, the median coupling loss does not appear to depend notably on the distance from the mobile antenna to the leaky feeder, at least at frequencies where the waveguide modes are the main vehicle for scattered fields. This had already been observed by other investigators and is of course due to the field distribution of waveguide modes in the tunnel cross-section. This expected result is specific to tunnels and contrasts with the coupling loss increase of 6 dB per doubling distance observed by Cree and Giles (1975) above the earth's surface.

It was of course interesting to relate the median coupling loss to some global parameter describing the imperfect cable shield. The best correlation was found with the coupling coefficient C_1 defined in Chapter 3, rather than with parameters like optical coverage or even specific transfer inductance. The parameter C_1^2 is a measure of the relative power of the coaxial eigenmode leakage fields in the transmission line model and, although the latter is no longer valid at VHF and UHF, this result was expected. Note that some crude approximations were made in order to calculate C_1 from eqn. 3.50: the characteristic impedance Z_m and phase velocity v of the monofilar mode were taken equal to 300 Ω and 3×10^8 m s^{-1}, which yields $l_m = 1000$ nH m^{-1}. Furthermore the value of m_t measured by the method described in Section 2.5.5 was used, even in the controversial case of slotted cables. Jouan's results and $-45°$ tilted regression lines are shown in Fig. 5.3. The dispersion of experimental points is of the order of measurement accuracy.

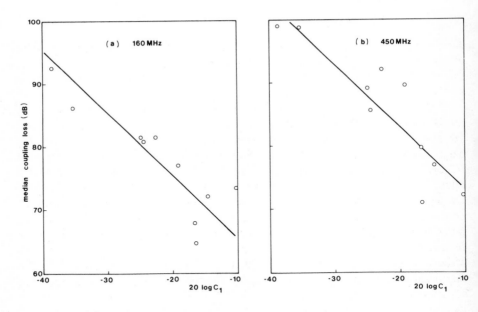

Fig. 5.3   *Relation between coupling loss and coefficient C_1*

The equation of the regression line is for Jouan's data

$$A_c(50) = \begin{cases} 55 - 20 \log C_1 & \text{at 160 MHz} \\ 63 - 20 \log C_1 & \text{at 450 MHz} \end{cases}$$

$$(5.9)$$

This exercise was repeated using the data obtained by Martin (1973) in coal mines. This yields

$$A_c = \begin{cases} 65 \cdot 6 - 20 \log C_1 & \text{at } 85 \text{ MHz} \\ 71 \cdot 7 - 20 \log C_1 & \text{at } 160 \text{ MHz} \end{cases} \qquad (5.10)$$

for an unspecified percentage P of places. Martin however stated that this coupling loss is valid for reliable communications: although the tunnels and cable mounting were different, the agreement with Jouan's data at 160 MHz is excellent if this means $P = 98\%$, as results from Fig. 5.2.

5.3.4 Specific attenuation

The specific attenuation of the coaxial eigenmode is influenced by the mounting in a tunnel. The increase in attenuation w.r.t. an equivalent non-leaky cable was expressed by eqn. 3.53 for low frequencies at which the waveguide modes are cut-off. The effect is by far more severe at higher frequencies. It appears that the attenuation increase $(\alpha_c - \alpha_{co})$ is quite sensitive to cable mounting. Even at 160 MHz, running the leaky feeder together with other cables in a common cable path may cause a threefold increase in attenuation. The increase remains moderate for careful mounting. Mounting precautions are however not critical at this frequency since the effective radius of the leakage fields is still appreciable. The effect is more serious at 450 MHz, where attenuation increases of 50% are currently met for very careful mounting. Proximity of the tunnel wall is critical.

There should obviously be a relation between $(\alpha_c - \alpha_{co})$ and $A_c(50)$ since attenuation increase and radiation are both due to scattering by objects present inside the effective radius of the leakage fields. Fig. 5.4 shows scattergrams of Jouan's data for several leaky feeders with the same location in the tunnel. For a location closer to the tunnel wall, at 450 MHz, the coupling losses should be reduced by perhaps 10 dB, but this might be paid for by a further 2 or 3 dB/100 m increase in attenuation. Actually the dependence of α_c on $A_c(50)$ is nonlinear, as the scattergram suggests.

A grading technique is sometimes proposed to limit the effect of coupling loss on attenuation. In this method the cable is made progressively more leaky with increasing distance away from the base station. This technique of course requires that grading should be useful for the mobile-to-base station direction too. Actually this is not possible for some system structures (Section 5.5).

The increase in attenuation for a given coupling loss is not the same for all types of cables. Cree and Giles (1975) observed that axial slot cables were the most sensitive but this was not fully confirmed by Jouan's tests. The truth may be that they are indeed more sensitive when objects with poor scattering properties, like a smooth wall, are close to the cable and on the side of the slot. Indeed, it is known that the leakage fields rapidly approach rotational symmetry when the distance from the cable increases (Fernandes, 1976). It should also be stressed that Jouan's tests showed that no correlation exits between $(\alpha_c - \alpha_{co})$ and the shield optical coverage.

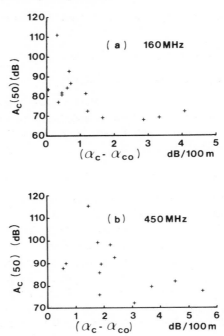

Fig. 5.4   *Scattergrams of the median coupling loss against the increase in specific attenuation of leaky cables*

5.3.5 Further information

Gale and Beal (1980) developed an interesting and novel method for testing leaky feeders. The direct relation with radiation properties has however not yet been explicitly illustrated.

It is worth while making one final remark. Until now we have considered that the radiation of leaky feeders is due only to inadvertent scattering by unavoidable inhomogeneities. From a theoretical point of view this is undoubtedly the only possible explanation of the observed phenomena, as long as the leaky feeder is itself homogeneous, like the axially slotted cable, or when the spatial period of apertures is small compared with the wavelength. Japanese investigators have however developed leaky feeders in which the spatial period of the apertures is directly related to the wavelength (Nakahara *et al.*, 1971). It is true that this kind of structure possesses intrinsic radiation properties, but no quantitative assessment of the radiation level is available. As a matter of fact, the efficient radiation of waves with polarisation perpendicular to the tunnel axis, as these authors intended, does not seem possible if the cable diameter remains small w.r.t. the wavelength.

Subsequent experiments showed that these cables actually did not behave differently from other types with regard to the sensitivity of attenuation and coupling loss to mounting, or Rayleigh distribution (Okada *et al.*, 1975; Suzuki *et al.*, 1980). It is therefore thought that random scattering by inhomogeneities located nearby remain the predominant mechanism of radiation.

5.4 Coaxial cable with leaky sections

5.4.1 General

It was stressed that the radiation of continuous leaky feeders is due to scattering of the coaxial eigenmode leakage fields by inhomogeneities located in the immediate vicinity of the cable. Actually this is an effect of mode conversion similar to those described in Section 3.7, but with the monofilar mode replaced by the spectrum of the tunnel waveguide modes, or by the continuous spectrum of free-space radiation modes according to the case. This mode conversion process is obviously not optimum and one has to pay for it by an excessive increase in the cable attenuation. There is little doubt that a well-controlled radiation process would be more efficient. Furthermore this would avoid excessive safety margins in system design required to face unexpected values of median coupling loss and cable attenuation.

This leads the author (1976a) to extend the concept of mode conversion developed in Chapter 3 to higher frequencies. That there is no fundamental difference between mode conversion and radiation should be stressed again. Consequently any discontinuity along an open waveguiding structure will act as an antenna fed by this structure. The mode converters described previously may thus also be considered as radiating devices.

Particular attention was paid to leaky sections inserted in a non-leaky cable. The reason for this was mainly one of cost and reliability. In the VHF and UHF bands the spacing of mode converters should drop to 70 to 100 m. Inserting these devices into the cable by means of connectors is expensive, particularly at these frequencies where thicker cables are used to reduce the attenuation rate. Furthermore using a large number of connectors in cascade is not favourable to reliability. These problems are completely solved by building leaky sections into the cable directly in manufacture.

However the main advantage of a non-leaky cable with widely spaced short leaky sections, compared with a continuous leaky feeder, is a drastic reduction in mounting costs. Mounting instructions of continuous leaky feeders must be followed strictly along the whole cable length. A cable with leaky sections may on the contrary be unrolled without any precaution together with other cables into existing cable paths, except for the leaky sections themselves and a few metres on either side of them. This is an important advantage since the installation costs are sometimes significantly higher than the cable price itself, for instance in existing underground railways where only a few night hours are available for work.

5.4.2 Radiation properties of leaky section

The simplest approach to studying the radiation of a leaky section is to first suppose that the cable is strung in free space. The leaky section of length L extends on the interval $|z| < L/2$ of the z-axis and is described by a specific transfer inductance m_t. Using a weak-coupling assumption, we suppose that the current of the coaxial mode is unperturbed by the leaky section and is given by $I_{co} \exp(-j\beta_{co} z)$. We also neglect the attenuation of this mode on the small length L. An axial electric field

$$E_z = j\omega m_t I_{co} \exp(-j\beta_{co} z)$$

thus arises on the external surface of the leaky section. The latter may be seen as a continuous array of annular slots of length dz excited by a voltage $E_z dz$.

The radiation of an annular slot was studied extensively in Section 2.10.3. In particular eqn. 2.457 yields the magnetic field radiated by a slot fed by a voltage V. Here $x = 2\pi b/\lambda_0$, where b is the cable radius, R is the distance from the observation point to the slot and θ is the angle w.r.t. the z-axis. Summing the contributions of all elementary slots and calculating Poynting's vector, it is easy to show that the leaky section is equivalent to an antenna that would be fed by the coaxial mode power, with an antenna gain given by

$$G(\theta) = \frac{\omega^2 m_t^2 G_r}{Z_{co} \sqrt{\kappa}} f_a^2(\theta) D(\theta) \tag{5.11}$$

where G_r and $D(\theta)$ are the radiation conductance (Tables 2.3 and 2.4) and directive gain (Figs. 2.40 and 2.41) of an elementary slot, Z_{co} is the characteristic impedance of the cable, κ is the dielectric constant of the cable insulation, and $f_a(\theta)$ is the array factor given by

$$f_a(\theta) = \frac{\sin\left[(\beta_{co} - k_0 \cos\theta) L/2\right]}{(\beta_{co} - k_0 \cos\theta)/2} \tag{5.12}$$

$$\simeq \frac{\sin\left[\pi L/\lambda_0 \left(\sqrt{\kappa} - \cos\theta\right)\right]}{\pi/\lambda_0 \left(\sqrt{\kappa} - \cos\theta\right)} \tag{5.13}$$

Note that $G(\theta)$ is defined as if the whole power carried by the coaxial mode was fed into the antenna.

The length L may be chosen so as to obtain interesting characteristics of the radiation pattern over some frequency band. First it should be observed that the ω^2 factor in eqn. 5.11 is exactly compensated by the frequency dependence of the denominator of $f_a^2(\theta)$. The frequency and length dependence of $G(\theta)$ are thus only governed by the function $\sin^2\left[\pi L/\lambda_0 \left(\sqrt{\kappa} - \cos\theta\right)\right]$.

The array factor is maximised for the endfire direction $\theta = 0$ if L is taken equal to

$$L = \frac{\lambda_0}{2(\sqrt{\kappa} - 1)} \tag{5.14}$$

which corresponds exactly to eqn. 3.89. If this condition is used, the 3 dB band-width of the gain will extend from $0 \cdot 5 \, f_0$ to $1 \cdot 5 \, f_0$, where f_0 is the design frequency. However additional frequency bands centred at $(2n + 1) \, f_0$, with the same peak gain and absolute bandwidth f_0, are available and may be useful. For instance a cable designed to provide maximum endfire radiation at 140 MHz will provide 3 dB bandwidths of 70–210 MHz and 350–490 MHz.

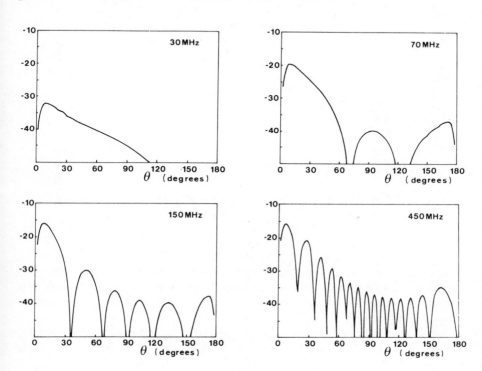

Fig. 5.5   *Radiation patterns of a leaky section at different frequencies for $m_t = 30 \, nH \, m^{-1}$, $Z_{c0} = 75 \, \Omega$, $\kappa = 1 \cdot 5$ and $L = 4 \cdot 77 \, m$*

Figure 5.5 shows radiation patterns at different frequencies of interest, for a leaky section designed in this way with $m_t = 30$ nH m^{-1}, $Z_{c0} = 75 \, \Omega$, $\kappa = 1 \cdot 5$. The endfire effect is quite remarkable. Results at 30 MHz, which is outside the bandwidth, have been shown to illustrate the decrease of radiation at low frequencies.

5.4.3 Applications

A leaky section is equivalent to a directive antenna connected to the coaxial cable through a power divider. The method developed in Section 5.2 may thus be used for range predictions. Only a small part of the power propagated inside the cable is extracted and converted into radiation. Actually some power is also converted into the monofilar mode. The total loss suffered by the signals carried by the coaxial cable however does not exceed a few tenths of a decibel, according to the design.

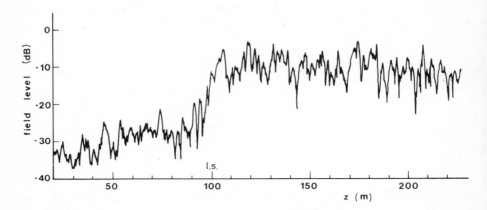

Fig. 5.6   *Field level recorded at 450 MHz along a tunnel containing a cable with a leaky section located at z = 100 m*

Figure 5.6 shows the field level registered at 450 MHz along a tunnel containing an axial cable with a leaky section located at $z = 100$ m. Standing waves arise from reflections on the tunnel walls or, equivalently, from the excitation of numerous waveguide modes. It is now recognised that these effects cannot be avoided at these frequencies (Haining *et al.*, 1978).

Cables with built-in leaky sections are now available. The spacing between leaky sections had to be chosen in order to provide acceptable results at various frequencies and for tunnels which may be curved. It appeared that a spacing of 70 to 100 m was adequate to meet all conditions. Cables with leaky sections designed in this way may be used under the same conditions as continuous leaky feeders. They currently provide a median coupling loss as low as 55 dB for an increase of attenuation below $0 \cdot 5$ dB/100 m. The result is a range increase of about 30%.

The cables available have mainly been designed for tunnel use. This is why end-fire radiation was maximised. It is however possible to design cables to favour transverse radiation. This could be useful to provide radio communications in the cross-cuts of room and pillar mines. In this case the spacing of leaky sections should ideally be equal to that of cross-cuts in order to have a leaky section in front of each crosscut.

5.5 Systems structure

5.5.1 General

There is a wide variety of needs for subsurface radio communications at VHF and UHF, for instance:

Communications in mine tunnels and in room and pillar areas.
Retransmission of frequency modulation (88–108 MHz) in road tunnels.
Communications in underground railways and in railway tunnels.
Relaying radio communications of services in road tunnels.
Relaying of radiotelephony in road tunnels. This differs from the previous case in having a large number of duplex channels, with the transmit and receive frequencies grouped in two distinct frequency bands.
Communication in subsurface areas like car parks and in large buildings.

These applications differ in certain operational conditions and requirements, namely:

Coverage area consisting either of a single tunnel or of several with crossings and branchings.
One-way or two-way communications.
Single or multichannel operation.
Simplex or duplex channels.
Speech and control signals available on telephone lines or relaying of open-space radio communications.
Location of frequency channels in open-space relaying: grouped in an exclusive frequency band or spread and mixed with other signals in a wide band.

Several system structures have been proposed to comply with these different situations. Each of them has its own advantages and drawbacks regarding cost, reliability, complexity and technical efficiency. We will briefly review them below. We will however not go into detailed calculations of important effects like intermodulation, since these are relevant to standard telecommunications practice.

Range predictions themselves are not complicated. For discrete antenna systems they involve estimations of the coupling of an antenna to the waveguide modes (e.g. eqn. 5.4 for the rectangular tunnel) and of the specific attenuation of the latter. A safety margin should be taken into account for standing waves. For continuous leaky feeders and cables with leaky sections, it is necessary to estimate the specific attenuation of the cable and the coupling loss for the required percentage of locations. The remaining calculations are simple and do not need to be described here.

It should be clear that the coupling loss from a mobile to either a leaky feeder or to a discrete antenna connected to a cable is far too high to allow direct communication between mobiles via the cable. Some form of relaying will thus eventually be required.

5.5.2 Antennas against leaky feeders

It has been pointed out that the specific attenuation of the waveguide modes decreases with increasing frequency, whereas that of coaxial cables increases. It may thus be thought that truly wireless radio links based on natural propagation are preferable to leaky feeders, particularly at the highest frequencies. This is only true for straight-line unobstructed tunnels. Wireless systems have indeed been found competitive for railway and underground communications at 450 MHz, but only for relatively short (maximum 500 m) straight-line paths.

Natural propagation has also been considered for communications in room and pillar mines. In such a complex labyrinth, a link between two antennas involves a certain number of corners and some length of straight-line tunnels. The link attenuation may be calculated by the method described in Section 1.2.3, namely eqn. 1.17 and Table 1.1. That the optimum frequencies were found between 70 and 150 MHz is mainly due to the effect of corners, which may not exceed two or three in number. It is however possible to extend the range by means of passive repeaters located at some intersections of cross-cuts with the entries which are directly illuminated by the base station. Corner reflectors were found efficient by Chufo and Isberg (1978). The optimum frequency may in this way be shifted to 450 MHz and perhaps higher.

For some geometrical configurations of tunnels it may be interesting to use discrete antennas connected to non-leaky coaxial cables by means of directional couplers or power dividers. Obviously the cables are here used only to convey signals from the base station to the antennas. This structure may be preferable to leaky feeders when the latter do not provide enough lateral coverage. In some early and unreported tests we found that the best coverage was obtained in room and pillar areas at 150 MHz, with a coaxial cable running along a centrally located entry and simple monopole antennas connected to it at the intersection with the cross-cuts. The leaky section principle was discovered later and would probably provide similar properties.

5.5.3 Multiple base station system

The principle of a multiple base station system is illustrated in Fig. 5.7. Several base stations BS are remotely controlled from the same control point C through a telephone line. They are connected to leaky feeders through power dividers P. This system was used very early with axial-slot cables in undergrounds (Hilscher and Plischke, 1968) and with other types of leaky feeders (Martin 1970). For current equipment and leaky feeders the ranges covered on either side of the base stations do not exceed 1300 to 2500 m at 150 MHz and 800 to 1500 m and 450 MHz.

The base stations may be operated either in different frequency channels or at the same nominal frequency. In the latter case the actual carrier frequencies will unavoidably differ slightly and heterodyne beating effects may result in the areas covered by two base stations. It is however possible to reduce these areas by interrupting the leaky feeders between the stations, as shown on the figure. The

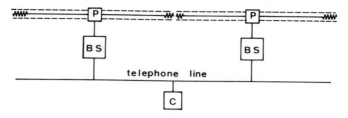

Fig. 5.7   *Multiple base station system*

same system structure may be used with discrete antennas connected to non-leaky cables. Reducing the beating areas is then difficult.

An interesting but somewhat different multiple base station system has been described by Chufo and Vancura (1976). It involves base stations operated as demodulating–remodulating duplex radio relays at 150 and 450 MHz. Control of the base stations is thus via the radio signals carried by the leaky cables.

The cost of multiple base station system is seriously affected by that of the base stations themselves. The system allows branching of the leaky feeders in a star network.

5.5.4 Wideband in-line repeaters

The term repeater is borrowed from telephony to name radio frequency amplifiers which are inserted at regular intervals to compensate the propagation attenuation of a leaky feeder. The term wideband is used here to indicate that the amplifier bandwidth may be significantly larger than that of a radio frequency channel. Several system architectures based on in-line repeaters have been studied in the United Kingdom for mining applications (Martin, 1975, 1977, 1978a, b; Davis *et al.*, 1977; Isberg, 1978).

In the daisy chain principle (Fig. 5.8), the transmitting and receiving parts of the base station are located at opposite ends of the leaky feeder. The operation mode is full or half duplex: the base station transmits on frequency f_1 and receives on frequency f_2, whereas the mobile transceivers do the converse. At least one of the two base station halves is necessarily remote controlled through a telephone line. Communication between mobiles is made possible if the signals received from the mobiles by the base receiver remodulate the base transmitter.

The daisy-chain principle has been used in the UK mainly for single-channel operation. The amplifiers may then be operated close to or even at saturation, which provides a self-regulation of amplifier gain. The economic optimum was found for low power (a few milliwatts) and medium gain (15 dB) amplifiers. Since these devices are not very expensive, they may be associated with a leaky feeder of modest performance and fairly high attenuation rate (3·2 dB/100 m at 80 MHz). This would however not remain true for very long systems, since the accumulated internal noise of the amplifiers is the limiting factor for mobile-to-base-station communications. This leads to minimising the number of repeaters by using amplifiers with higher gains and output powers, and leaky feeders with a smaller attenuation.

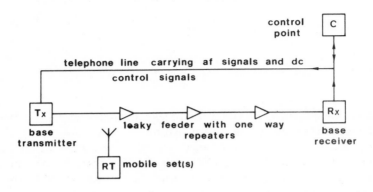

Fig. 5.8   *The daisy-chain principle*

The daisy-chain principle is potentially capable of multichannel operation. Several base stations operated at different frequencies may indeed be connected to the same repeatered leaky feeder. The dominant problem here is the intermodulation due to the non-linearity of amplifiers. Since the level of the signals transmitted by the base station is much higher than that of those transmitted by the mobiles, intermodulation due to the former onto the latter is the most critical. It will therefore be necessary to operate the amplifiers well below saturation, and thus to use devices with a high saturated output power. This undoubtedly raises the cost, but also the difficulty of delivering the required d.c. power through the leaky feeder itself. Furthermore, automatic gain control may be required in long chains. Actually the intermodulation problems are very similar to those encountered in cable television.

Reliability of long daisy chains is obviously conditioned by that of amplifiers. It may be improved by redundant amplifiers but this again raises seriously the global cost of the system.

The telephone lines used to link the two halves of the base station may also be an important element in the global clost. They may be avoided by retransmitting the received signal upstream along the leaky feeder at an intermediate frequency, provided the amplifiers are made transparent in the reverse direction for this frequency. A further element to appreciate is that the daisy-chain principle is well suited for long single runs but not for star networks.

A more attractive structure for star networks is the double daisy chain (Fig. 5.9a) which however requires two separate leaky feeders and doubles the number of amplifiers. The tailback arrangement proposed by Martin (Fig. 5.9b) has the advantage of keeping the same number of amplifiers, since (apart from the 4 dB loss of a T-junction) the path between a mobile transceiver and the input or output of an amplifier is always smaller than half the amplifier spacing. Of course this costs a third leaky feeder.

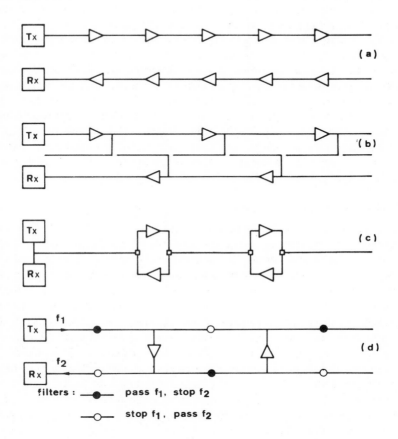

Fig. 5.9   *Two-way systems*
a Double daisy chain
b With tailback arrangement
c Two-way repeaters
d BDR system

The classical structure with two-way repeaters (Fig. 5.9c) does not seem to have been evaluated seriously hitherto. It is undoubtedly attractive when the transmit and receive bands are separated, since repeater stability can easily be ensured by fairly simple filtering. It may also work when these bands overlap, provided some general decrease of the levels along the path can be accepted, but this would certainly require close impedance matching. There is little doubt that work remains to be done on this subject.

Another structure proposed by Martin is the BDR (bi-directional routing) system shown in Fig. 5.9d. The amplifiers are common to the transmit and receive signals. Filters are inserted in the two leaky feeders to ensure stability. An advantage claimed for the structure is its intrinsic tailback arrangement.

Each of the structures shown in Fig. 5.9 has its own advantages and drawbacks. They are all suitable for star networks and, to some extent, for multiple channel operation. Intermodulation problems are more serious with structure (d) than with the others, since these use separate amplifiers for the transmit and receive signals. Overlapping of the transmit and receive bands is not possible for structure (d) and raises some problems for structure (c). It is fully possible for structures (a) and (b) provided cross-talk between neighbouring leaky cables is low enough. This is undoubtedly garanteed if these cables are distinct, but not necessarily if they are located under the same jacket, as considered by Martin (1976). Similar problems can arise in structure (d).

5.5.5 Open air to subsurface relaying

It was implicitly assumed in the previous sections that the subsurface network was completely isolated, starting from audio and control signals. In many applications however one has to relay signals available as electromagnetic waves in open air into tunnels and, for two-way communications, in the reverse sense. In most cases, the mobile users may travel on the surface as well as into the tunnels. It is therefore necessary to keep the carrier frequencies rigorously unchanged. Otherwise heterodyne beating effects will result in an important area in the vicinity of tunnel ends. The signal processing that must take place between the external antenna and the tunnel system, and in the reverse sense, must consequently be equivalent to an amplification. The required gain is quite high since it must compensate the coupling loss and propagation attenuation of the leaky feeders, at least partly.

If the signals to be relayed are located in an exclusive frequency band, i.e. if no other users are allowed to perturb the system, the whole band may be amplified in a common amplifier. This solution is undoubtedly the simplest but not without certain problems. The gain should be fixed as a function of the weakest signal, taking into account the existence of standing waves above the surface as well as in the tunnels. However the intermodulation to be considered is from the strongest signals onto the weakest ones, in the relay itself as well as in the subsurface repeaters. Particular attention should be paid in this respect to other users of the frequency spectrum, and namely to mobiles which may transmit in the immediate vicinity of the external antenna and in the tunnels. Strict filtering of the useful frequency band is in general necessary.

Selective channel per channel relaying is therefore attractive and is not systematically more expensive. It greatly simplifies intermodulation problems not only in the relay station itself, but also in the repeaters. Indeed automatic gain control is then possible for each channel separately. Selective relaying imposes itself when the channels to be relayed are dispersed in a frequency band containing many other signals, particularly when the latter are strong. It becomes strictly imperative for simplex channels where switching of the direction of amplification is required. A sort of daisy chain with two distinct external antennas for transmitting and for receiving would in fact constitute an unstable loop because of coupling between the two antennas.

5.6 Some examples of existing and projected systems

Several examples of applications to mining can be found in the papers by Martin and colleagues previously mentioned. Systems used in the London underground were described by Isberg (1978). Some tunnels are covered by a multiple base station system similar to Fig. 5.7, while a daisy-chain technique (Fig. 5.8) was used elsewhere. These single-channel systems are relatively simple and will not be described here.

Grüssi and König (1978) described a system which is proposed for retransmission of FM broadcast and of two-way service and radio telephone communications at 80 and 160 MHz in Swiss highway tunnels. The proposal is basically a multichannel daisy-chain system. The paper contains a good analysis of the required signal levels and of the intermodulation problems of in-line amplifiers. It is however not very explicit on the relaying of signals from the open air into the tunnels and vice versa. Apparently it was assumed that two-part base stations remotely controlled by tele-phone lines as in Fig. 5.8 were used. We assume that wideband amplification was considered for FM broadcast signals.

More recently Suzuki *et al.* (1980) proposed the extension of radio telephony in the 800 MHz band into road tunnels. The channels are of course all duplex. They are grouped in two 15 MHz wide frequency bands, one for the down-link (from base station to mobile) and the other for the up-link (reverse direction), separated by a 15 MHz interval. The system uses two-way repeaters according to the principle of Fig. 5.9c, stability being ensured by selective duplexers. A very interesting amplifier structure providing intrinsic redundancy properties is used. The drawback of this structure is however that the amplifiers are common to the up and down-links, which is not desirable from the point of view of intermodulation. Relaying from the open air into the tunnels and vice versa is made through wideband amplifiers. The reader is referred to the paper for a very complete analysis of the intermodulation problem (we however disagree with the authors on the dependence of the output intermodulation levels on the number of in-line repeaters).

We will devote somewhat more attention to a rather complex system which has been installed in a junction of tunnels at the centre of Brussels. Compared with the road tunnel systems mentioned hereabove, which are only planned, it has the merit of existing. It is therefore a good example for the description of the sometimes unexpected problems encountered in this kind of application. A somewhat more detailed description of the system has been given by Sarteel *et al.* (1980). It is planned to install progressively similar systems in the most inportant tunnels of Belgian cities.

The tunnels to be covered are not very long, maximum 395 m. Two tunnels cross at right angles, but at different levels. They are joined by two other tunnels having the form of a bend and of a loop, respectively. The cross-section is large: 8·0 m by 4·8 m and 13·5 m by 4·7 m for the one-way and two-way tunnels, respectively.

The main specifications were:

To retransmit AM broadcasting in the band 150 KHz–1·5 MHz.
To retransmit FM broadcasting in the band 88–108 MHz.
To offer two-way communications to seven users (police, fire services, etc.) using simplex equipment at the frequencies 70·25 MHz (AM), 152·80, 152·90, 153·05, 153·25, 166·33 and 169·46 MHz (FM).

For the AM and FM broadcasting a field level higher than 1 mV m^{-1} was required for the four strongest channels of each band, with severe intermodulation specifications. It was also specified that all signals should be relayed exactly at their original frequency in order to avoid heterodyne beating in or near the tunnels. Fig. 5.10 shows a general block diagram of the system. As aesthetics is important in the centre of the city, it was decided to design a single antenna for the whole range 150 KHz–170 MHz. As the photograph of Fig. 5.11 shows, it is a discone antenna for the band 70–170 MHz; the disc and the cone are reduced to a few rods; a whip is

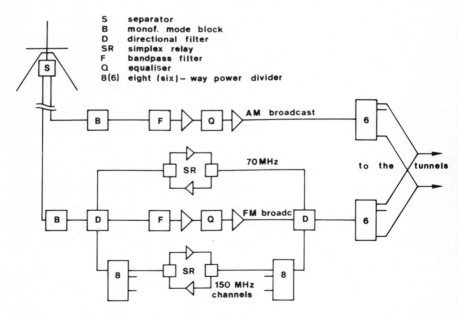

Fig. 5.10   *General diagram of road tunnel communication system in Brussels*

Fig. 5.11 *Photograph of external antenna used in road tunnel communication system*

mounted above the disc for the 150 KHz-1·5 MHz band. The AM broadcasting and VHF signals are brought down to a small underground room on two separated coaxial cables; a FET amplifier is used at the antenna base to match the impedance in the AM broadcasting band. In principle the broadcasting bands simply need to be amplified for the retransmission into the tunnels. As the latter are short, a simple monofilar wire would solve the problem for the AM band. Two major difficulties were encountered here. The first regards stability. The output signal had to be applied between the monofilar wire and a ground connection. No good ground connection could be found in the control room with the result that the amplified signals were retransmitted towards the antenna in the monofilar mode on the feeder, yielding instability. This problem was solved by suitable monofilar mode blocking circuits in the antenna feeders and by transmitting the AM band towards the tunnels inside the coaxial cables used for the VHF band. A specially designed mode converter used at the entry of each tunnel effects a complete transformation of the AM band from the coaxial mode into the monofilar mode between the cable outer conductor and the tunnel wall. This mode converter is transparent for the VHF band in the coaxial mode, while the subsequent mode converters used for the VHF band are transparent for the AM band in the monofilar mode.

The second difficulty with the AM band originates from the tunnel gauge. It was required to strung the coaxial cables at 15 cm from the roof. With this very eccentric position the field lines of the monofilar mode tend to concentrate towards the roof and it was necessary to use an amplifier gain of 74 dB to obtain a field of 1 mV m^{-1} above the tunnel floor. Taking into account intermodulation effects in the wideband amplifier, a saturated output of 15 W was necessary for the latter. Similar problems were encountered with the FM broadcasting signals. Many problems had to be solved for the two-way simplex channels. Some carrier frequencies in the 150–170 MHz band are only 100 KHz apart and it is impossible to separate them efficiently with filters. A total loss of 30 dB for each channel in the power dividers and directional filters could not be avoided, with the result that a maximum gain of 109 dB and an output power of 22 dBm were required for each simplex relay.

Figure 5.12 shows the block diagram of the simplex relays that were specially constructed for this application. The problems encountered in the design included:

Existence of high-gain loops at the local oscillator frequency: this was solved by a suitable choice of the local oscillator configuration and by the use of crystal filters in the IF section providing 120 dB attenuation at this frequency.

Intermodulation and capture effects when all channels apart from one are active in the same direction: in spite of the close channel separation, this was solved by a careful choice of the input mixers, by the use of crystal filters at IF and of isolators at the output stages.

A further stability problem appears on the general block diagram of Fig. 5.10. The simplex relays in the tunnel-to-antenna direction are backed by the FM broadcasting amplifier and it was necessary to insert filters providing more than 140 dB attenuation at the repeater frequencies in this amplifier.

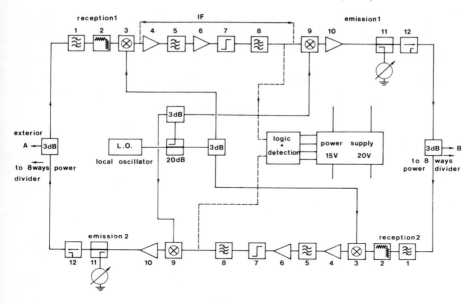

Fig. 5.12 *Functional diagram of simplex radio relay*

This enumeration of problems shows that the major problems were relative to system rather than wave propagation aspects. Much experience was gained from the realisation and subsequent use of this complex system. Wideband amplification should be questioned for the retransmission of AM and FM broadcasting. Indeed open-air propagation is particularly subject to standing waves in an urban environment. The external antenna is not necessarily at the best location for all stations and may even be located accidentally at a deep node for some of them. It was therefore necessary to equalise the levels of the most important stations in the wideband amplifier. Important modifications of the urban environment, if they occur subsequently, may perturb the standing wave patterns and make the equaliser's tuning obsolete. Selective retransmission of a limited number of stations instead of wideband amplification would avoid these problems and also the backing of simplex channels through the wideband amplifier.

Furthermore the structure would need major adjustments if the tunnel were longer, to the point of requiring in-line repeaters.

5.7 Further studies

It is felt that much remains to be done on the system aspects of subsurface radio communications in the VHF and UHF bands. The variety of requirements (Section

5.5.1) is so wide that they can probably not all be met by a single system structure. A systematic analysis of the question has hitherto not been made but would be a valuable contribution to the subject.

Finally we must mention that some interesting questions, such as data transmission on leaky feeders (Davis *et al.*, 1979), diversity systems (Treen, 1978) and perimeter intrusion radar based on leaky feeders (Mackay and Beal, 1978b), have been beyond the scope of this book.

References

Preliminary notes

(a) NTIS is the National Technical Information Service, 5285 Port Royal Road, Springfield VA 22161, USA.
(b) The abbreviation WEGWME is used for the Proceedings of the Workshop on Electromagnetic Guided Waves in Mine Environments, held in Boulder, Colo., USA, 28–30 March 1978, available from NTIS.

ADAMS, J. W., TAGGART, H. R., and SPAULDING, A. D. (1971): 'Survey report of the US Bureau of Mines electromagnetic noise measurement program', NBS, available from NTIS under acc. no. PB-226773

ANDERSEN, J. B., BERNTSEN, S., and DALSGAARD, P. (1975): 'Propagation in rectangular waveguides with arbitrary internal and external media', *IEEE Trans.,* MTT-23, pp. 555–560

BALANIS, C. A., JEFFREY, J. L., and YOON, Y. K. (1978): 'Electrical properties of eastern bituminous coal as a function of frequency, polarisation and direction of the electromagnetic wave, and temperature of the sample', WEGWME, pp. 277–287

BANNISTER, P. R., and DUBE, R. L. (1978): 'Approximate expressions for the quasi-static fields within and outside a conducting slab', WEGWME, pp. 251–259

BAÑOS, A (1966): 'Dipole radiation in the presence of a conducting half-space' (Pergamon)

BEAL, J. C., and DEWAR, W. J. (1968): 'Coaxial-slot surface-wave launcher', *Electron. Lett.,* 4, pp. 557–559

BECKMANN, P., and SPIZZICHINO, A. (1964): 'The scattering of electromagnetic waves from rough surfaces' (Pergamon)

BENSEMA, W. D., KANDA, M., and ADAMS, J. W. (1974): 'Electromagnetic noise in Itmann mine', NBS, available from NTIS under acc. no. COM-74-11718

BETHE, H. A. (1944): 'Theory of diffraction by small holes', *Phys. Rev.,* 66, pp. 163–182

CASEY, K. F. (1976): 'On the effective transfer impedance of thin coaxial cable shields', *IEEE Trans.,* EMC-18, pp. 110–117

CASEY, K. F., and VANCE, E. F. (1978): 'EMP coupling through cable shields', *IEE Trans.,* AP-26, pp. 100–106

CHIBA, J., INADA, T., KUWAMOTO, Y., BANNO, O., and SATO, R. (1978): 'Wireless communication in the tunnels and underground streets', WEGWME, pp. 218–226

CHU, L. J. (1948): 'Physical limitations of omni-directional antennas', *J. Appl. Phys.,* 19, pp. 1163–1175

CHUFO, R. L. and ISBERG, R. A. (1978): 'Passive reflectors as a means for extending UHF signals down intersecting cross cuts in mines', WEGWME, pp. 204–217

CHUFO, R. L., and VANCURA, P. D. (1976): 'The installation, operation, and maintenance of an underground mine wide ratio system', IAS–76 Conf. Rec., Printcraft, Inc., Conneaut, Ohio 44030, USA, pp. 1196–1202

COGGLESHALL, E. J., FELEGY, E. W., and HARRISON, L. H. (1948): 'Some studies on emergency mine communications', Rep. Inv. US Bur. Mines 4135

COHN, S. B. (1951): 'Determination of aperture parameters by electrolytic-tank measurements', *Proc. IRE*, **39**, pp. 1416–1421

COLBURN, C. L., BOUTON, C. M., and FREEMAN, H. B. (1922): 'Underground signalling with radio sets', Rep. Inv. US Bur. Mines 2407

COLLIN, R. E. (1960): 'Field theory of guided waves' (McGraw–Hill)

COLLIN, R. E., and ZUCKER, F. J. (1969): 'Antenna theory' (McGraw–Hill)

COMSTOCK, W. D. (1971): 'Electromagnetic wave propagation in an underground duct', *Proc. IREE*, Australia, pp. 159–162

COOK, J. C. (1975): 'Radar transparencies of mine and tunnel rocks', *Geophysics*, **40**, pp. 865–885

CORY, T. S. (1978): 'Wireless radio transmission at medium frequencies in underground coal mines. Summary of measurements and expected system propagational effects', WEGWME, pp. 184–202

CREE, D. J., and GILES, L. J. (1975): 'Practical performance of radiating cables', *Radio & Electron. Eng.*, **45**, pp. 215–223

DAVIS, Q. V., HAINING, R. W., CRITCHLEY, B. L., and MARTIN, D. J. R. (1977): 'Developments in leaky feeder radio communication', Proc. Conf. Eurocon 1977, Venezia, Italy, pp. 2.9.1.1–5

DAVIS, Q. V., HAINING, R. W., and MOTLEY, A. J. (1979): 'Reliable high-speed data transmission from a moving vehicle via a leaky feeder', IERE Conf. on Civil Land Mobile Radio, Lancaster, UK, September 1979, pp. 59–71

DEGAUQUE, P., DEMOULIN, B., CAUTERMAN, M., and GABILLARD, R. (1978): 'Experimental results obtained in a mine on a network of leaky cables', WEGWME, pp. 2–11

DEGAUQUE, P., DEMOULIN, B., FONTAINE, J., and GABILLARD, R. (1976): 'Theory and experiment of a mobile radiocommunication in tunnels by means of a leaky braided coaxial cable', *Radio Sci.*, **11**, pp. 305–314

DEGAUQUE, P., MANGEZ, P., DEMOULIN, B., and GABILLARD, R. (1979): 'Résultats expérimentaux concernant la propagation libre des ondes UHF dans les tunnels', Proc. AGARD Conf. No. 263 on 'Special topics in HF propagation', Lisbon, Portugal, 28 May–1 June 1979, Paper 40

DE KEYSER, R. (1972): 'La radio sous le sol. Mesures et réalisations pratiques avec le système INIEX/Delogne', Bull. Tech. Mines et Carrières, INIEX, Liége, Belgium, no. 135

DE KEYSER, R. (1973): 'La radio sous le sol. Installations avec le système INIEX/Delogne', Bull. Tech. Mines et Carrières, INIEX, Liége, Belgium, no. 138

DE KEYSER, R. (1974): 'Résultats obtenus avec le système INIEX/Delogne en dehors de la Campine', *Ann. Mines de Belg.*, (1), pp. 95–103

DE KEYSER, R., DELOGNE, P., DERYCK, L., and LIEGEOIS, R. (1978): 'Comparative analysis of leaky feeder techniques for mine communications', WEGWME, pp. 12–23

DE KEYSER, R., DELOGNE, P., JANSSENS, J., and LIEGEOIS, R. (1970): 'Radiocommunication and control in mines and tunnels', *Electron. Lett.*, **6**, pp. 767–768

DELOGNE, P. (1969a): 'Problèmes de diffraction sur guide unifilaire', *Ann. Télécommun.*, **24**, pp. 405–426

DELOGNE, P. (1969b): 'Diaphonie et orthogonalité des modes', *Rev. HF*, Belgium, **VIII**, pp. 255–258

DELOGNE, P. (1970): 'Les liaisons radioélectriques par câble coaxial dans la mine', *Ann. Mines de Belg.*, (7–8), pp. 967–975

DELOGNE (1971): 'Système breveté INIEX/Delogne', *Ann. Mines de Belg.*, (5), pp. 531–533

DELOGNE, P. (1972): 'Système INIEX/Delogne de télécommunications et de télécommande par radio', *Ann. Mines de Belg.*, (11), pp. 1069–1081

DELOGNE, P. (1973): 'Les télécommunications par radio en milieu souterrain', *Rev. HF*, Belgium, IX, pp. 18–26

DELOGNE, P. (1974a): 'Le système de transmission radio recommandé par l' INIEX', *Ann. Mines de Belg.*, (1), pp. 85–93

DELOGNE, P. (1974b): 'Les systèmes INIEX de communication par radio', *Ann. Mines de Belg.*, (10), pp. 951–962

DELOGNE, P. (1976a): 'Basic mechanisms of tunnel propagation', *Radio Sci.*, 11, pp. 295–303

DELOGNE, P. (1976b): 'Comments on 'Calculated channel characteristics of a braided coaxial cable in a mine tunnel', *IEEE Trans.*, COM-24, pp. 1212–1213

DELOGNE, P. (1979): 'Mode converters for HF tunnels transmission', Proc. AGARD Conf. No. 263 on 'Special topics in HF propagation', Lisbon, Portugal, 28 May–1 June, paper 37

DELOGNE, P., DERYCK, L., and LIEGEOIS, R. (1973): 'Guided propagation of radio waves', Proc. Thru-the-Earth EM Workshop, Colorado School of Mines, Golden, Colo., USA, August 1973, available from NTIS under acc. no. 231154, pp. 49–53

DELOGNE, P., and LALOUX, A. (1980): 'Theory of the slotted coaxial cable', *IEEE Trans.*, MTT-28, pp. 1102–1107

DELOGNE, P., and LIÉGEOIS, R. (1971): 'Le rayonnement d'une interruption du conducteur extérieur d'un câble coaxial', *Ann. Télécommun.*, 26, pp. 85–100

DELOGNE, P., and SAFAK, M. (1975): 'Electromagnetic theory of the leaky coaxial cable', *Radio & Electron. Eng.*, 45, pp. 233–240

DERYCK, L. (1970): 'Etude des modes de propagation d'ondes electromagnétiques susceptibles d'exister sur une ligne bifilaire en milieu souterrain', *Ann. Mines de Belg.*, (7–8), pp. 939–949

DERYCK, L. (1972): 'Exchanges d'énergie entre modes de propagation sur la ligne bifilaire', *Ann. Mines de Belg.*, (2), pp. 109–123

DERYCK, L. (1973): 'Etude de la propagation des ondes électromagnétiques guidées dans les galeries souterraines', Thèse de doctorat en sciences, Univ. Liége, Belgium, 1973. Also available from Commission of European Communities as 'Recueils de recherche charbon, Technique minière, Recueil no. 64, 1974'

DERYCK, L. (1975): 'Control of mode conversions on bifilar line in tunnels', *Radio & Electron. Eng.*, 45, pp. 241–247

DERYCK, L. (1979): 'Effective use of natural modes in VHF and UHF tunnel propagation', Proc. AGARD Conf. No. 263 on 'Special topics in HF propagation', Lisbon, Portugal, 28 May–1 June 1979, paper 43

DE SMEDT, R., and VAN BLADEL, J. (1980): 'Magnetic polarisability of some small apertures', *IEEE Trans.*, AP-28, pp. 703–707

DEWAR. W. J., and BEAL, J. C. (1970): 'Coaxial-slot surface wave launchers', *IEEE Trans.*, MTT-18, pp. 449–455

DE WATTEVILLE, M. (1964), *Mines*, 111, p. 127

EMSLIE, A. G., and LAGACE, R. L. (1978): 'Medium frequency radio propagation and coupling in coal mines', WEGWME, pp. 142–152

EMSLIE, A. G., LAGACE, R. L., SPENCER, R. H., and STRONG, P. F. (1978): 'Use of auxiliary dedicated wire as a means of aiding carrier current propagation on a trolley wire/rail transmission line', WEGWME, pp. 26–41

EMSLIE, A. G., LAGACE, R. L., and STRONG, P. F. (1975): 'Theory of the propagation of UHF radio waves in coal mine tunnels', *IEEE Trans.*, AP-23, pp. 192–205

EMSLIE, A. G., LAGACE, R. L., and STRONG, P. F. (1973): 'Theory of propagation of

UHF radio waves in coal mine tunnels', Proc. 'Thru-the-Earth Electromagnetics Workshop', Colorado School of Mines, August 15–17, 1973, available through NTIS, pp. 38–48

ERDÉLEYI, A., MAGNUS, W., OBERHETTINGER, F., and TRICOMI, F. G. (1953): 'Higher transcendental functions' (McGraw–Hill)

EVE, A. S., KEYS, D. A., and LEE, F. W. (1929), *Nature*, **124**, p. 178

FARMER, R. A., and SHEPHERD, N. H. (1965): 'Guided radiation . . . the key to tunnel talking', *IEEE Trans.*, **VC-14**, pp. 93–102

FELEGY, E. W. (1953), *Min. Eng.*, **5**, p. 518

FERNANDES, A. S. (1976): 'Studies of the propagation and radiation characteristics of leaky cables', Ph. D. thesis, Univ. of Surrey, Guildford, UK, October 1976

FONTAINE, J., DEMOULIN, B., DEGAUQUE, P., and GABILLARD, R. (1974): 'Feasibility of a two-way mobile radiocommunication in tunnels by means of a leaky braided coaxial cable', Proc. Coll. on 'Leaky feeders radiocommunication systems', Univ. of Surrey, Guildford, UK, April 1974, pp. 38–48

FONTAINE, J., SOIRON, M., DEGAUQUE, P., and GABILLARD, R. (1974): 'Propagation d'une onde électromagnétique le long d'une ligne à un ou plusieurs conducteurs placés dans une galerie cylindrique', C.R. Acad. Sci., France, série B, 17 juin 1974

FOOT, J. B. L., (1950): 'Transmission through tunnels', *Wireless World*, **56**, pp. 456–458

FRANKEL, S. (1977): 'Multiconductor transmission line analysis' (Artech House)

GABILLARD, R. (1969): 'Propagation des ondes radio dans les ouvrages souterrains', CERCHAR, France, AEL 69-74-02

GALE, D. J., and BEAL, J. C. (1980): 'Comparative testing of leaky coaxial cables for communications and guided radar', *IEEE Trans.*, **MTT-28**, pp. 1006–1013

GRÜSSI, O., and KÖNIG, P. (1978): 'Radio links for highway tunnels', *Bull. Tech. PTT*, Bern, Switzerland, (7), pp. 285–293

HAINING, R. W., CRITCHLEY, B. L., and MOTLEY, A. J. (1978): 'Field measurements of leaky feeder signals', WEGWME, pp. 76–86

HILL, D. A., and WAIT, J. R. (1974a): 'Excitation of monofilar and bifilar modes on a transmission line in a circular tunnel', *J. Appl. Phys.*, **45**, pp. 3402–3406

HILL, D. A., and WAIT, J. R. (1974b): 'Gap excitation of an axial conductor in a circular tunnel', *J. Appl. Phys.*, **45**, pp. 4774–4777

HILL, D. A., and WAIT, J. R. (1975a): 'Electromagnetic fields of a coaxial cable with an interrupted shield located in a circular tunnel', *J. Appl. Phys.*, **46**, pp. 4352–4356

HILL, D. A., and WAIT, J. R. (1975b): 'Coupling between a radiating coaxial cable and a dipole antenna', *IEEE Trans.*, **COM-23**, pp. 1354–1357

HILL, D. A., and WAIT, J. R. (1976a): 'Scattering from a break in the shield of a braided coaxial cable. Numerical results', *Arch. Elektron. & Uebertragungstech.*, **30**, pp. 117–121

HILL, D. A., and WAIT, J. R. (1976b): 'Calculated transmission loss for a leaky feeder communication system in a circular tunnel', *Radio Sci.*, **11**, pp. 315–321

HILL, D. A., and WAIT, J. R. (1976c): 'Propagation along a braided coaxial cable located close to a tunnel wall', *IEEE Trans.*, **MTT-24**, pp. 476–480

HILL, D. A., and WAIT, J. R. (1976d): 'Analysis of radio frequency transmission along a trolley wire in a mine tunnel', *IEEE Trans.*, **EMC-18**, pp. 170–174

HILL, D. A., and WAIT, J. R. (1977a): 'Effect of a lossy jacket on the external fields of a coaxial cable with an interrupted shield', *IEEE Trans.*, **AP-25**, p. 726

HILL, D. A., and WAIT, J. R. (1977b): 'Analysis of radio frequency transmission in a semi-circular mine tunnel containing two axial conductors', *IEEE Trans.*, **COM-25**, pp. 1046–1050

HILL, D. A., and WAIT, J. R. (1978a): 'The impedance of dipoles in a circular tunnel with an axial conductor', *IEEE Trans.*, **GE-16**, pp. 118–126

HILL, D. A., and WAIT, J. R. (1978b): 'Bandwidth of a leaky coaxial cable in a circular tunnel', *IEEE Trans.*, **COM-26**, pp. 1765–1771

HILL, D. A., and WAIT, J. R. (1979): 'Comparison of loop and dipole antennas in leaky feeder communication systems', Proc. AGARD Conf. No. 263 on 'Special topics in HF propagation', Lisbon, Portugal, 28 May–1 June 1979, paper 44. See also *Int. J. Electron.*, **47**, pp. 155–166

HILL, D. A., and WAIT, J. R. (1980a): 'Propagation along a coaxial cable with a helical shield', *IEEE Trans.*, **MTT-28**, pp. 84–89

HILL, D. A., and WAIT, J. R. (1980b): 'Electromagnetic theory of the loosely braided coaxial cable: numerical results', *IEEE Trans.*, **MTT-28**, pp. 326–331

HILSCHER, G., and PLISCHKE, A. (1968): 'Liaisons radiotéléphoniques VHF dans les tunnels du métropolitain de Munich', *Nahverkehrpraxis* 16, cahier 5, AEG-Telefunken, Ulm, Germany, pp. 169–179

ISBERG, R. A. (1978): 'Radio communication in subways and mines through repeater amplifiers and leaky transmission lines', 28th IEEE Conf. on Vehicular Technology, Denver, Colo., March 1978

ISLEY, L. C., FREEMAN, H. B., and ZELLERS, D. H. (1928): 'Experiments in underground communication through earth strata', Tech. Paper US Bur. of Mines 433

JONES, D. S. (1964): 'The theory of electromagnetism' (Pergamon)

JOUAN, G. (1977): 'Etude de la couverture radioélectrique des tunnels', Conservatoire National des Arts et Métiers, Paris

KADEN, H. (1957): 'Wirbelströme und Schirmung in der Nachrichtentechnik' (Springer-Verlag)

KANDA, M. (1974): 'Time and amplitude statistics for electromagnetic noise in mines', NBS, available through NTIS under acc. no. COM-74-11450

KANDA, M., ADAMS, J. W., and BENSEMA, W. D. (1974): 'Electromagnetic noise in McElroy mine', NBS, available through NTIS under acc. no. COM-74-11717

KEITH-MURRAY, P. I. (1933): 'Radio communications applied to mines', *Trans. Mining and Geological Inst. India*, **28**, p. 67

KORN, G. A., and KORN, T. M. (1961, 1968): 'Mathematical handbook for scientists and engineers' (McGraw–Hill)

KRÜGEL, L. (1956): 'Abschirmwirkung von Aussenleitern flexibler Koaxialkabel', *Telefunken Z.*, Germany, **29**, p. 114

KUESTER, E. F., and SEIDEL, D. B. (1979): 'Low-frequency behavior of the propagation along a thin wire in an arbitrarily shaped mine tunnel', *IEEE Trans.*, **MTT-27**, pp. 736–741

LAAKMAN, K. D., and STEIER, W. H. (1976): 'Waveguides: characteristic modes of hollow rectangular dielectric waveguides', *Appl. Opt.*, **15**, pp. 1334–1340

LAGACE, R. L., COHEN, M. L., EMSLIE, A. G., and SPENCER, R. H. (1975): 'Propagation of radio waves in coal mines', Report to US Bureau of Mines Contract No. HO-346045, task F, A. D. Little Inc., Cambridge, Mass. 02140, USA

LAGACE, R. L., CURTIS, D. A., FOULKES, J. D., and ROTHERY, J. L. (1977): 'Transmit antennas for portable VLF to MF wireless mine communications', Report to US Bureau of Mines Contract no. HO-346045, Task C, A. D. Little Inc., Cambridge, Mass. 02140, USA

LAGACE, R. L., and EMSLIE, A. G. (1978a): 'Coupling of the coal-seam mode to a cable in a tunnel at medium radio frequencies', WEGWME, pp. 170–183

LAGACE, R. L., and EMSLIE, A. G. (1978b): 'Antenna technology for medium-frequency portable radiocommunication in coal mines', WEGWME, pp. 153–169

LANDAU, L., and LIFSHITZ, E. M. (1960): 'Electrodynamics of continuous media' (Pergamon)

LATHAM, R. W. (1972): 'An approach to certain cable shielding calculations', Interaction Notes, Note 90, AFWL, Kirtland AFB, NM, USA

LEE, K. S. H., and BAUM, C. E. (1975): 'Application of modal analysis to braided-shield cables', *IEEE Trans.*, **EMC-17**, pp. 159–169

LIÉGEOIS, R. (1968): 'Télécommunications souterraines et télécommande par radio dans la mine', Bull. Tech. Mines et Carrières, INIEX, Liége, Belgium, no. 117

MACKAY, N. A. M., and BEAL. J. C. (1978a): 'The testing of leaky coaxial cables and their application to guided radar', WEGWME, pp. 227–241

MACKAY, N. A. M., and BEAL, J. C. (1978b): 'Guided radar for obstacle detection in ground transportation systems', IEEE AP-S Intl. Symp. Digest, Washington DC, 15–19 May 1978, pp. 302–305

MAHMOUD, S. F. (1974a): 'On the attenuation of monofilar and bifilar modes in mine tunnels', *IEEE Trans.*, **MTT-22**, pp. 845–847

MAHMOUD, S. F. (1974b): 'Characteristics of electromagnetic guided waves for communication in coal mine tunnels', *IEEE Trans.*, **COM-22**, pp. 1547–1554

MAHMOUD, S. F., and WAIT, J. R. (1974a): 'Theory of wave propagation along a thin wire inside a rectangular waveguide', *Radio Sci.*, 9, pp. 417–420

MAHMOUD, S. F., and WAIT, J. R. (1974b): 'Guided electromagnetic waves in a curved rectangular mine tunnel', *Radio Sci.,* 9, pp. 567–572

MAHMOUD, S. F., and WAIT, J. R. (1974c): 'Geometrical optical approach for electromagnetic wave propagation in rectangular mine tunnels', *Radio Sci.,* 9, pp. 1147–1158

MAHMOUD, S. F., and WAIT, J. R. (1976): 'Calculated channel characteristics of a braided coaxial cable in a mine tunnel', *IEEE Trans.,* **COM-24**, pp. 82–87

MARCATILI, E. A. J., and SCHMELTZER, R. A. (1964): 'Hollow metallic and dielectric waveguides for long distance optical transmission and lasers', *Bell Syst. Tech. J.,* 43, pp. 1783–1809

MARCUVITZ, N. (1951): 'Waveguide handbook' (McGraw-Hill)

MARTIN, D. J. R. (1970a): 'Radiocommunication in mines and tunnels', *Electron. Lett.,* 6, pp. 563–564

MARTIN, D. J. R. (1970b): 'Full coverage of the mine by line-propagated VHF radio', *Mine. Technol.,* 52, pp. 7–15

MARTIN, D. J. R. (1973): 'Very-high-frequency radio communication in mines and tunnels', Ph. D. thesis, Univ. of Surrey, Guildford, UK

MARTIN, D. J. R. (1975): 'Systems aspects of leakage-field radio communication', IERE Conf. on Civil Land Mobile Radio, Teddington, UK, 18–20 November 1975

MARTIN, D. J. R. (1976): 'The bicoaxial leaky line and its application to underground radio communication', *Radio Sci.,* pp. 779–885

MARTIN, D. J. R. (1977): 'Radio communication in mines', *The Mining Eng.*, December 1977/January 1978, pp. 275–282

MARTIN, D. J. R. (1978a): 'Some aspects of leaky feeder radio communication being studied by the UK National Coal Board', WEGWME, pp. 66–75

MARTIN, D. J. R. (1978b): 'The use of in-line repeaters in leaky feeder radio systems for coal mines', 28th IEEE Conf. on Vehicular Technology, Denver, Colo., USA, March 1978

MIKOSHIBA, K., and NURITA, Y. (1969): 'Guided radiation by coaxial cable for train wireless systems in tunnels', *IEEE Trans.,* **VT-18**, pp. 66–69

MONK, N., and WINBIGLER, H. S. (1956): 'Communication with moving trains in tunnels', *IRE Trans.,* **PGVC-7**, pp. 21–28

MORSE, P. M., and FESHBACH, H. (1953): 'Methods of theoretical physics' (McGraw–Hill)

MURPHY, J. N., and PARKINSON, H. E. (1978): 'Underground mine communications', *Proc. IEEE,* 66, pp. 26–50

NAKAHARA, T., KURAUCHI, N., YOSHIDA, K., and MIYAMOTO, Y. (1971): 'Extensive applications of leaky cables', *Sumitomo Electric Tech. Rev.,* Japan, 15, pp. 27–31

NOBLE, B. (1958): 'Methods based on the Wiener–Hopf technique for the solution of partial differential equations' (Pergamon)

OKADA, S., KISHIMOTO, T., AKAGAWA, K., MIKOSHIBA, K., and OKAMOTO, K. (1975): 'Leaky coaxial cable for communication in high-speed railway transportation', *Radio & Electron Eng.,* 45, pp. 224–228

POGORZELSKI, R. J. (1979): 'Electromagnetic propagation along a wire in a tunnel. Approximate analysis', *IEEE Trans.,* **AP-27**, pp. 814–819

SARTEEL, F., DELOGNE, P., and DERYCK, L. (1980): 'Les communications par radio dans les tunnels routiers', *Rev. HF,* Belgium, **XI**, pp. 221–239

SAVKINE, M. M. (1964), Rec. Acad. Sci. URSS, Filiale de Sibérie, Inst. Mines, 7

SCHELKUNOFF, S. A. (1934): 'The electromagnetic theory of coaxial transmission lines and cylindrical shields', *Bell Syst. Tech. J.,* **13**, 532–579

SCOTT, W. W., ADAMS, J. W., BENSEMA, W. D., and DOBROSKI, H. (1974): 'Electromagnetic noise in Lucky Friday mine', NBS, available from NTIS under acc. no. COM-75-10258

SEIDEL, D. B., and WAIT, J. R. (1978a): 'Modes in braided coaxial cables in circular and elliptical tunnels', WEGWME, pp. 24–25

SEIDEL, D. B., and WAIT, J. R. (1978b): 'Transmission modes in a braided coaxial cable and coupling to a tunnel environment', *IEEE Trans.,* **MTT-26**, pp. 494–499

SEIDEL, D. B., and WAIT, J. R. (1978c): 'Role of controlled mode conversion in leaky feeder mine-communication systems', *IEEE Trans.,* **AP-26**, pp. 690–694

SEIDEL, D. B., and WAIT, J. R. (1979a): 'Mode conversions by tunnel non-uniformities in leaky feeder communications ststems', *IEEE Trans.,* **AP-27**, pp. 560–563

SEIDEL, D. B., and WAIT, J. R. (1979b): 'Radio transmission in an elliptic tunnel with a contained axial conductor', *J. Appl. Phys.,* **50**, pp. 602–605

SHANKLIN, J. P. (1947): 'VHF railroad communications', *Communications,* **27**, pp. 16–19

STRATTON, J. A. (1941): 'Electromagnetic theory' (McGraw–Hill)

SUZUKI, T., HANAZAWA, T., and KOZONO, S. (1980): 'Design of a tunnel relay system with a leaky coaxial cable in an 800-MHz band land mobile telephone system', *IEEE Trans.,* **VT-29**, pp. 305–316

TREEN (1978): 'Radiating cable system, proposal for multiple-diversity voice or data communication', *Wireless World,* August, pp. 42–44

TRETIAKOV, M. (1959), *Revue de l'Industrie Minérale,* **41**, p. 551

VAN BLADEL, J. (1970): 'Small-hole coupling of resonant cavities and waveguides', *Proc. IEE,* **117**, pp. 1098–1104

VANCE, E. F. (1975): 'Shielding effectiveness of braided-wire shields', *IEEE Trans.,* **EMC-17**, pp. 71–77

WAIT, J. R. (1962): 'Electromagnetic waves in stratified media' (Pergamon)

WAIT, J. R. (1969): 'Image theory of a quasistatic magnetic dipole over a dissipative half-space', *Electron. Lett.,* **5**, pp. 281–282

WAIT, J. R. (1975a): 'Theory of transmission of electromagnetic waves along multiconductor lines in the proximity of walls of mine tunnels', *Radio & Electron. Eng.,* **45**, pp. 229–232

WAIT, J. R. (1975b): 'Theory of E.M. wave propagation through tunnels', *Radio Sci.,* **10**, pp. 753–759

WAIT, J. R. (1975c): 'The scattering from a break in the shield of a braided coaxial cable – Theory', *Arch. Elektron. & Uebertragungstech.,* **29**, pp. 467–473

WAIT, J. R. (1976a): 'Note on the theory of transmission of electromagnetic waves in a coal seam', *Radio Sci.,* **11**, pp. 262–265

WAIT, J. R. (1976b): 'Electromagnetic theory of the loosely braided coaxial cable: part I', *IEEE Trans.,* **MTT-24**, pp. 547–593

WAIT, J. R. (1977): 'Quasi-static limit for the propagating mode along a thin wire in a circular tunnel', *IEEE Trans.,* **AP-25**, pp. 441–443

WAIT, J. R. (1978): 'Guided electromagnetic waves in a periodically nonuniform tunnel', *IEEE Trans.,* **AP-26**, pp. 623–625

WAIT, J. R. (1980): 'Propagation in a rectangular tunnel with imperfectly conducting walls', *Electron. Lett.,* **16**, pp. 521–522

WAIT, J. R., and HILL, D. A. (1974a): 'Guided electromagnetic waves along an axial conductor in a circular tunnel', *IEEE Trans.,* **AP-22**, pp. 627–630

WAIT, J. R., and HILL, D. A. (1974b): 'Coaxial and bifilar modes on a transmission line in a circular tunnel', *Appl. Phys.,* **4**, pp. 307–312

WAIT, J. R., and HILL, D. A. (1975a): 'Electromagnetic fields of a dielectric coated coaxial cable with an interrupted shield', OT Technical Memorandum 75-192, Office of Telecommunications, Boulder, Colo., USA

WAIT, J. R., and HILL, D. A. (1975b): 'Propagation along a braided coaxial cable in a circular tunnel', *IEEE Trans.*, **MTT-23**, pp. 401–405

WAIT, J. R., and HILL, D. A. (1975c): 'Theory of the transmission of electromagnetic waves down a mine hoist', *Radio Sci.*, **10**, pp. 625–632

WAIT, J. R., and HILL, D. A. (1975d): 'On the electromagnetic field of a dielectric coated coaxial cable with an interrupted shield', *IEEE Trans.*, **AP-23**, pp. 470–479

WAIT, J. R., and HILL, D. A. (1975e): 'Electromagnetic fields of a dielectric coated coaxial cable with an interrupted shield – Quasi-static approach', *IEEE Trans.*, **AP-23**, pp. 679–682

WAIT, J. R., and HILL, D. A. (1976a): 'Low-frequency radio transmission in a circular tunnel containing a wire conductor near the wall', *Electron. Lett.*, **12**, pp. 346–347

WAIT, J. R., and HILL, D. A. (1976b): 'Attenuation on a surface wave G-line suspended within a circular tunnel', *J. Appl. Phys.*, **47**, pp. 5472–5473

WAIT, J. R., and HILL, D. A. (1976c): 'Impedance of an electric dipole located in a cylindrical cavity in a dissipative medium', *Appl. Phys.*, **11**, pp. 351–356

WAIT, J. R., and HILL, D. A. (1977a): 'Radio frequency transmission via a trolley wire in a tunnel with a rail return', *IEEE Trans.*, **AP-25**, pp. 248–253

WAIT, J. R., and HILL, D. A. (1977b): 'Electromagnetic fields of a dipole source in a circular tunnel containing a surface wave line', *Int. J. Electron.*, **42**, pp. 377–391

WAIT, J. R., and HILL, D. A. (1978): 'Analysis of the dedicated communication line in a mine tunnel for a shunt-loaded trolley wire', *IEEE Trans.*, **COM-26**, pp. 355–361

WAIT, J. R., and SPIES, K. P. (1969): 'On the image representation of the quasi-static fields of line current source above the ground', *Can. J. Phys.*, **47**, pp. 2731–2733

WALDRON, R. A. (1967): 'The theory of waveguides and cavities' (Maclaren and Sons)

WYKE, P. N., and GILL, R. (1955): 'Applications of radio type communication equipment at collieries', *Proc. IMEME*, **36**, pp. 128–137

'Improvements in underground radiocommunication systems', Final Report on ECSC Research Project 6220-AE/8/802, MRDE, Burton-on-Trent, UK, 1977

'Underground mine communications:
  Part I, no. IC 8742: Mine telephone systems.
  Part II, no. IC 8743: Paging systems.
  Part III, no. IC 8744: Haulage systems.
  Part IV, no. IC 8745: Section-to-place communications';
  US Bureau of Mines, 4800 Forbes Av., Pittsburgh, Pa. 15213, USA, 1977

Appendixes

Appendix A   Calculation of fields from potentials in the Cartesian coordinate system

The sourceless $(J_a = 0, J_{ma} = 0)$ part of eqns. 2.24 and 2.25 yields, for each cartesian component of the potentials,

$$\pi'_x: \quad E = \left(-\frac{\partial^2}{\partial y^2} - \frac{\partial^2}{\partial z^2} \quad , \quad \frac{\partial^2}{\partial x \partial y} \quad , \quad \frac{\partial^2}{\partial x \partial z} \right) \pi'_x \tag{A.1}$$

$$H = j\omega\epsilon \left(0 \quad , \quad \frac{\partial}{\partial z} \quad , \quad -\frac{\partial}{\partial y} \right) \pi'_x \tag{A.2}$$

$$\pi'_y: \quad E = \left(\frac{\partial^2}{\partial x \partial y} \quad , \quad -\frac{\partial^2}{\partial x^2} - \frac{\partial^2}{\partial z^2} \quad , \quad \frac{\partial^2}{\partial y \partial z} \right) \pi'_y \tag{A.3}$$

$$H = j\omega\epsilon \left(-\frac{\partial}{\partial z} \quad , \quad 0 \quad , \quad \frac{\partial}{\partial x} \right) \pi'_y \tag{A.4}$$

$$\pi'_z: \quad E = \left(\frac{\partial^2}{\partial x \partial z} \quad , \quad \frac{\partial^2}{\partial y \partial z} \quad , \quad -\frac{\partial^2}{\partial x^2} - \frac{\partial^2}{\partial y^2} \right) \pi'_z \tag{A.5}$$

$$H = j\omega\epsilon \left(\frac{\partial}{\partial y} \quad , \quad -\frac{\partial}{\partial x} \quad , \quad 0 \right) \pi'_z \tag{A.6}$$

$$\pi''_x: \quad E = -j\omega\mu_0 \left(0 \quad , \quad \frac{\partial}{\partial z} \quad , \quad -\frac{\partial}{\partial y} \right) \pi''_x \tag{A.7}$$

$$H = \left(-\frac{\partial^2}{\partial y^2} - \frac{\partial^2}{\partial z^2} \quad , \quad \frac{\partial^2}{\partial x \partial y} \quad , \quad \frac{\partial^2}{\partial x \partial z} \right) \pi''_x \tag{A.8}$$

$$\pi_y'': \quad E = -j\omega\mu_0 \quad \left(-\frac{\partial}{\partial z} \quad , \quad 0 \quad , \quad \frac{\partial}{\partial x} \right) \quad \pi_y'' \qquad (A.9)$$

$$H = \left(\frac{\partial^2}{\partial x \partial y} \quad , \quad -\frac{\partial^2}{\partial x^2} - \frac{\partial^2}{\partial z^2} \quad , \quad \frac{\partial^2}{\partial y \partial z} \right) \quad \pi_y'' \qquad (A.10)$$

$$\pi_z'': \quad E = -j\omega\mu_0 \quad \left(\frac{\partial}{\partial y} \quad , \quad -\frac{\partial}{\partial x} \quad , \quad 0 \right) \quad \pi_z'' \qquad (A.11)$$

$$H = \left(\frac{\partial^2}{\partial x \partial z} \quad , \quad \frac{\partial^2}{\partial y \partial z} \quad , \quad -\frac{\partial^2}{\partial x^2} - \frac{\partial^2}{\partial y^2} \right) \quad \pi_z'' \qquad (A.12)$$

For the calculation of a mode with a propagation factor $\exp(-\Gamma z)$, and taking into account that

$$\left(\frac{\partial^2}{\partial x^2} + \frac{\partial^2}{\partial y^2} + \Gamma^2 + k^2 \right) \pi = 0$$

one has:

$$\pi_x': \quad E = \left(-\frac{\partial^2}{\partial y^2} - \Gamma^2 \quad , \quad \frac{\partial^2}{\partial x \partial y} \quad , \quad -\Gamma \frac{\partial}{\partial x} \right) \quad \pi_x' \qquad (A.13)$$

$$= \left(\frac{\partial^2}{\partial x^2} + k^2 \quad , \quad \frac{\partial^2}{\partial x \partial y} \quad , \quad -\Gamma \frac{\partial}{\partial x} \right) \quad \pi_x' \qquad (A.14)$$

$$H = j\omega\epsilon \quad \left(0 \quad , \quad -\Gamma \quad , \quad -\frac{\partial}{\partial y} \right) \quad \pi_x' \qquad (A.15)$$

$$\pi_y': \quad E = \left(\frac{\partial^2}{\partial x \partial y} \quad , \quad -\frac{\partial^2}{\partial x^2} - \Gamma^2 \quad , \quad -\Gamma \frac{\partial}{\partial y} \right) \quad \pi_y' \qquad (A.16)$$

$$= \left(\frac{\partial^2}{\partial x \partial y} \quad , \quad \frac{\partial^2}{\partial y^2} + k^2 \quad , \quad -\Gamma \frac{\partial}{\partial y} \right) \quad \pi_y' \qquad (A.17)$$

$$H = j\omega\epsilon \quad \left(\Gamma \quad , \quad 0 \quad , \quad \frac{\partial}{\partial x} \right) \quad \pi_y' \qquad (A.18)$$

$$\pi_z': \quad E = \left(-\Gamma \frac{\partial}{\partial x} \quad , \quad -\Gamma \frac{\partial}{\partial y} \quad , \quad -\frac{\partial^2}{\partial x^2} - \frac{\partial^2}{\partial y^2} \right) \quad \pi_z' \qquad (A.19)$$

$$= \left(-\Gamma \frac{\partial}{\partial x} \quad , \quad -\Gamma \frac{\partial}{\partial y} \quad , \quad k^2 + \Gamma^2 \right) \quad \pi_z' \qquad (A.20)$$

$$H = j\omega\epsilon \left(\frac{\partial}{\partial y} \quad , \quad -\frac{\partial}{\partial x} \quad , \quad 0 \right) \pi'_z \quad (A.21)$$

$$\pi''_x: \quad E = -j\omega\mu_0 \left(0 \quad , \quad -\Gamma \quad , \quad -\frac{\partial}{\partial y} \right) \pi''_x \quad (A.22)$$

$$H = \left(-\frac{\partial^2}{\partial y^2} - \Gamma^2 \quad , \quad \frac{\partial}{\partial x \partial y} \quad , \quad -\Gamma\frac{\partial}{\partial x} \right) \pi''_x \quad (A.23)$$

$$= \left(\frac{\partial^2}{\partial x^2} + k^2 \quad , \quad \frac{\partial}{\partial x \partial y} \quad , \quad -\Gamma\frac{\partial}{\partial x} \right) \pi''_x \quad (A.24)$$

$$\pi''_y: \quad E = -j\omega\mu_0 \left(\Gamma \quad , \quad 0 \quad , \quad \frac{\partial}{\partial x} \right) \pi''_y \quad (A.25)$$

$$H = \left(\frac{\partial^2}{\partial x \partial y} \quad , \quad -\frac{\partial^2}{\partial x^2} - \Gamma^2 \quad , \quad -\Gamma\frac{\partial}{\partial y} \right) \pi''_y \quad (A.26)$$

$$= \left(\frac{\partial^2}{\partial x \partial y} \quad , \quad \frac{\partial^2}{\partial y^2} + k^2 \quad , \quad -\Gamma\frac{\partial}{\partial y} \right) \pi''_y \quad (A.27)$$

$$\pi''_z: \quad E = -j\omega\mu_0 \left(\frac{\partial}{\partial y} \quad , \quad -\frac{\partial}{\partial x} \quad , \quad 0 \right) \pi''_z \quad (A.28)$$

$$H = \left(-\Gamma\frac{\partial}{\partial x} \quad , \quad -\Gamma\frac{\partial}{\partial y} \quad , \quad -\frac{\partial^2}{\partial x^2} - \frac{\partial^2}{\partial y^2} \right) \pi''_z \quad (A.29)$$

$$= \left(-\Gamma\frac{\partial}{\partial x} \quad , \quad -\Gamma\frac{\partial}{\partial y} \quad , \quad k^2 + \Gamma^2 \right) \pi''_z \quad (A.30)$$

Appendix B   Calculation of fields from potentials $U = \pi'_z$ and $V = \pi''_z$ in the cylindrical coordinate system

We recall that U and V satisfy the equations

$$\left(\frac{\partial^2}{\partial \rho^2} + \frac{1}{\rho}\frac{\partial}{\partial \rho} + \frac{1}{\rho^2}\frac{\partial^2}{\partial \phi^2} + \frac{\partial^2}{\partial z^2} + k^2 \right) U = (j\omega\epsilon)^{-1} J_{az} \quad (B.1)$$

$$\left(\frac{\partial^2}{\partial \rho^2} + \frac{1}{\rho}\frac{\partial}{\partial \rho} + \frac{1}{\rho^2}\frac{\partial^2}{\partial \phi^2} + \frac{\partial^2}{\partial z^2} + k^2 \right) V = (j\omega\mu)^{-1} J_{maz} \quad (B.2)$$

The sourceless ($J_a = 0, J_{ma} = 0$) part of eqns. 2.24 and 2.25 yields

$$E_\rho = \frac{\partial^2 U}{\partial \rho \partial z} - \frac{j\omega\mu_0}{\rho} \frac{\partial V}{\partial \phi} \tag{B.3}$$

$$E_\phi = \frac{1}{\rho} \frac{\partial^2 U}{\partial \phi \partial z} + j\omega\mu_0 \frac{\partial V}{\partial \rho} \tag{B.4}$$

$$E_z = \left(\frac{\partial^2}{\partial z^2} + k^2 \right) U \tag{B.5}$$

$$H_\rho = \frac{j\omega\epsilon}{\rho} \frac{\partial U}{\partial \phi} + \frac{\partial^2 V}{\partial \rho \partial z} \tag{B.6}$$

$$H_\phi = -j\omega\epsilon \frac{\partial U}{\partial \rho} + \frac{1}{\rho} \frac{\partial^2 V}{\partial \phi \partial z} \tag{B.7}$$

$$H_z = \left(\frac{\partial^2}{\partial z^2} + k^2 \right) V \tag{B.8}$$

For a mode with a propagation factor $\exp(-\Gamma z)$, one has:

$$E_\rho = -\Gamma \frac{\partial U}{\partial \rho} - \frac{j\omega\mu_0}{\rho} \frac{\partial V}{\partial \phi} \tag{B.9}$$

$$E_\phi = \frac{-\Gamma}{\rho} \frac{\partial U}{\partial \phi} + j\omega\mu_0 \frac{\partial V}{\partial \rho} \tag{B.10}$$

$$E_z = (\Gamma^2 + k^2) U \tag{B.11}$$

$$H_\rho = \frac{j\omega\epsilon}{\rho} \frac{\partial U}{\partial \phi} - \Gamma \frac{\partial V}{\partial \rho} \tag{B.12}$$

$$H_\phi = -j\omega\epsilon \frac{\partial U}{\partial \rho} - \frac{\Gamma}{\rho} \frac{\partial V}{\partial \phi} \tag{B.13}$$

$$H_z = (\Gamma^2 + k^2) V \tag{B.14}$$

Next we examine the development of modes in cylindrical harmonics. All quantities are expanded in a Fourier series of the type

$$F(\rho, \phi, z) = \sum_{m=-\infty}^{\infty} F_m(\rho) e^{jm\phi} e^{-\Gamma z} \tag{B.15}$$

In a sourceless region the solution of eqns. B.1 and B.2 yields

$$U_m = Q_m I_m(u\rho) + B_m K_m(u\rho) \tag{B.16}$$

$$V_m = P_m I_m(u\rho) + A_m K_m(u\rho) \tag{B.17}$$

where A_m, B_m, P_m, Q_m are arbitrary constants and I_m, K_m are modified Bessel functions. We defined

$$u = \sqrt{-k^2 - \Gamma^2} = \sqrt{\gamma^2 - \Gamma^2}; \qquad -\frac{\pi}{2} < \arg u \leqslant \frac{\pi}{2} \tag{B.18}$$

On using these expressions in eqns. B. 9 to B.14 we obtain:

$$E_{\rho m} = -\Gamma u \left[Q_m I'_m(u\rho) + B_m K'_m(u\rho) \right] + \frac{\omega\mu_0 m}{\rho} \left[P_m I_m(u\rho) + A_m K_m(u\rho) \right] \tag{B.19}$$

$$E_{\phi m} = -\frac{j\Gamma m}{\rho} \left[Q_m I_m(u\rho) + B_m K_m(u\rho) \right] + j\omega\mu_0 u \left[P_m I'_m(u\rho) + A_m K'_m(u\rho) \right] \tag{B.20}$$

$$E_{zm} = -u^2 \left[Q_m I_m(u\rho) + B_m K_m(u\rho) \right] \tag{B.21}$$

$$H_{\rho m} = -\frac{\omega\epsilon m}{\rho} \left[Q_m I_m(u\rho) + B_m K_m(u\rho) \right] - \Gamma u \left[P_m I'_m(u\rho) + A_m K'_m(u\rho) \right] \tag{B.22}$$

$$H_{\phi m} = -j\omega\epsilon u \left[Q_m I'_m(u\rho) + B_m K'_m(u\rho) \right] - \frac{j\Gamma m}{\rho} \left[P_m I_m(u\rho) + A_m K_m(u\rho) \right] \tag{B.23}$$

$$H_{zm} = \qquad\qquad\qquad -u^2 \left[P_m I_m(u\rho) + A_m K_m(u\rho) \right] \tag{B.24}$$

Alternatively these formulas may be replaced by

$$U_m = Q_m J_m(\Lambda\rho) + B_m H_m^{(2)}(\Lambda\rho) \tag{B.25}$$

$$V_m = P_m J_m(\Lambda\rho) + A_m H_m^{(2)}(\Lambda\rho) \tag{B.26}$$

where

$$\Lambda = \sqrt{k^2 + \Gamma^2}; \qquad -\pi < \arg \Lambda \leqslant 0 \tag{B.27}$$

We then have

$$E_{\rho m} = -\Gamma\Lambda \left[Q_m J'_m(\Lambda\rho) + B_m H_m^{(2)\prime}(\Lambda\rho) \right] - \frac{j\omega\mu_0 m}{\rho} \left[P_m J_m(\Lambda\rho) + A_m H_m^{(2)}(\Lambda\rho) \right] \tag{B.28}$$

$$E_{\phi m} = +\frac{\Gamma m}{\rho} \left[Q_m J_m(\Lambda\rho) + B_m H_m^{(2)}(\Lambda\rho) \right] + j\omega\mu_0 \Lambda \left[P_m J'_m(\Lambda\rho) + A_m H_m^{(2)\prime}(\Lambda\rho) \right] \tag{B.29}$$

$$E_{zm} = \Lambda^2 \left[Q_m J_m(\Lambda\rho) + B_m H_m^{(2)}(\Lambda\rho) \right] \tag{B.30}$$

$$H_{\rho m} = -\frac{j\omega\epsilon m}{\rho} \left[Q_m J_m(\Lambda\rho) + B_m H_m^{(2)}(\Lambda\rho) \right] - \Gamma\Lambda \left[P_m J'_m(\Lambda\rho) + A_m H_m^{(2)\prime}(\Lambda\rho) \right] \tag{B.31}$$

$$H_{\phi m} = -j\omega\epsilon\Lambda[Q_m J'_m(\Lambda\rho) + B_m H_m^{(2)'}(\Lambda\rho)] - \frac{\Gamma m}{\rho}[P_m J_m(\Lambda\rho) + A_m H_m^{(2)}(\Lambda\rho)]$$

$$\tag{B.32}$$

$$H_{zm} = \Lambda^2 [P_m J_m(\Lambda\rho) + A_m H_m^{(2)}(\Lambda\rho)] \tag{B.33}$$

Appendix C   Some useful formulas on Bessel functions

Unmodified Bessel's functions

The unmodified Bessel's equation

$$\left(\frac{d^2}{dz^2} + \frac{1}{z}\frac{d}{dz} + 1 - \frac{m^2}{z^2}\right)Z_m(z) = 0 \tag{C.1}$$

may be solved using any two of the linearly independent solutions $J_m(z)$, $N_m(z)$, $H_m^{(1)}(z)$ and $H_m^{(2)}(z)$. In the context of this book, m is an integer and z is a complex variable with $-\pi < \arg z \leqslant 0$. We use the fundamental set of solutions $J_m(z)$ and $H_m^{(2)}(z)$. The Wronskian of these functions is

$$J_m(z)H_m^{(2)'}(z) - J'_m(z)H_m^{(2)}(z) = \frac{-2j}{\pi z} \tag{C.2}$$

We have

$$J_m(z)H_{m-1}^{(2)}(z) - J_{m-1}(z)H_m^{(2)}(z) = \frac{-2j}{\pi z} \tag{C.3}$$

For any one of the functions $J_m(z)$ and $H_m^{(2)}(z)$:

$$\left(\frac{d}{dz} \pm \frac{m}{z}\right)Z_m(z) = \pm Z_{m\mp 1}(z) \tag{C.4}$$

Asymptotic values:

$$\left.\begin{array}{l} J_0(z) \simeq 1 \\[2mm] H_0^{(2)}(z) \simeq \frac{-2j}{\pi}\ln\left(\frac{Cz}{2}\right); \quad C = 1\cdot7810\ldots \end{array}\right\} \text{ for } |z| \ll 1 \tag{C.5}$$

$$\left.\begin{array}{l} J_m(z) \simeq \frac{1}{m!}\left(\frac{z}{2}\right)^m \\[3mm] H_m^{(2)}(z) \simeq j\frac{(m-1)!}{\pi}\left(\frac{2}{z}\right)^m \end{array}\right\} \text{ for } |z| \ll m, m \neq 0 \tag{C.6}$$

$$J_m(z) \simeq \sqrt{\frac{2}{\pi z}} \cos\left(z - \frac{2m+1}{4}\pi\right)$$

$$H_m^{(2)}(z) \simeq \sqrt{\frac{2}{\pi z}} \exp\left[-j\left(z - \frac{2m+1}{4}\pi\right)\right]$$

$\left.\vphantom{\begin{array}{c}a\\b\\c\end{array}}\right\}$ for $|z| \gg m$ and $|z| \gg 1$

(C.7)

Addition theorem: for a triangle with sides (r, r_0, R), where ϕ and ψ are the angles opposite to R and r, respectively, i.e. if

$$R = \sqrt{r^2 + r_0^2 - 2rr_0 \cos\phi}$$

and

$$r = \sqrt{R^2 + r_0^2 - 2Rr_0 \cos\psi}$$

we have for any Bessel function Z_m

$$e^{jn\psi} Z_n(kR) = \sum_{m=-\infty}^{\infty} e^{jm\phi} J_m(kr) Z_{m+n}(kr_0) \qquad (C.8)$$

In particular

$$Z_0(kR) = \sum_{m=-\infty}^{\infty} e^{jm\phi} J_m(kr) Z_m(kr_0) \qquad (C.9)$$

Green's function: the general solution of equation

$$\left(\frac{\partial^2}{\partial\rho^2} + \frac{1}{\rho}\frac{\partial}{\partial\rho} + \Lambda^2 - \frac{m^2}{\rho^2}\right) G_m(\rho, \rho_0) = \frac{-\delta(\rho - \rho_0)}{\rho_0} \qquad (C.10)$$

is

$$G_m(\rho, \rho_0) = \begin{cases} AJ_m(\Lambda\rho) + BH_m^{(2)}(\Lambda\rho) & \text{if } \rho < \rho_0 \\[2mm] AJ_m(\Lambda\rho) + BH_m^{(2)}(\Lambda\rho) \\[2mm] \qquad + \dfrac{j\pi}{2}\left[J_m(\Lambda\rho)H_m^{(2)}(\Lambda\rho_0) - J_m(\Lambda\rho_0)H_m^{(2)}(\Lambda\rho)\right] \\[2mm] \qquad\qquad\qquad\qquad \text{if } \rho > \rho_0 \end{cases}$$

(C.11)

Modified Bessel's functions

They are solutions of

$$\left(\frac{d^2}{dx} + \frac{1}{x}\frac{d}{dx} - 1 - \frac{m^2}{x^2}\right) X_m(x) = 0 \qquad (C.12)$$

In the context of this book, m is an integer and x is a complex variable with $-\pi/2 < \arg x \leqslant \pi/2$. We use the linearly independent solutions $I_m(x)$ and $K_m(x)$. They are related to the unmodified Bessel's functions by

$$I_m(x) = L_m(x) = j^m J_m(-jx) \qquad (C.13)$$

$$K_m(x) = K_{-m}(x) = \frac{\pi}{2}(-j)^{m+1} H_m^{(2)}(-jx) \qquad (C.14)$$

Consequently:

$$I_m(x)K'_m(x) - I'_m(x)K_m(x) \quad = \frac{-1}{x} \tag{C.15}$$

$$K_{m+1}(x)I_m(x) + K_m(x)I_{m+1}(x) = \frac{1}{x} \tag{C.16}$$

$$I'_m(x) \pm \frac{m}{x}I_m(x) \quad = I_{m \mp 1}(x) \tag{C.17}$$

$$K'_m(x) \pm \frac{m}{x}K_m(x) = -K_{m \mp 1}(x) \tag{C.18}$$

Asymptotic values:

$$\left. \begin{array}{l} I_0(x) \simeq 1 \\[2mm] K_0(x) \simeq -\ln\dfrac{Cx}{2}; \qquad C = 1{\cdot}7810\ldots \end{array} \right\} \text{ for } |x| \ll 1 \tag{C.19}$$

$$\left. \begin{array}{l} I_m(x) \simeq \dfrac{1}{m!}\left(\dfrac{x}{2}\right)^m \\[3mm] K_m(x) \simeq \dfrac{(m-1!)}{2}\left(\dfrac{2}{x}\right)^m \end{array} \right\} \text{ for } |x| \ll m, \; m \neq 0 \tag{C.20}$$

$$\left. \begin{array}{l} I_m(x) \simeq \dfrac{1}{\sqrt{2\pi x}}\, e^x \\[3mm] K_m(x) \simeq \sqrt{\dfrac{\pi}{2x}}\, e^{-x} \end{array} \right\} \text{ for } |x| \gg m \text{ and } |x| \gg 1 \tag{C.21}$$

Addition theorems: with r, r_0, R, ϕ, ψ defined as above

$$e^{jn\psi}I_n(uR) = \sum_{m=-\infty}^{\infty} (-1)^m e^{jm\phi}I_m(ur)I_{n+m}(ur_0) \tag{C.22}$$

$$e^{jn\psi}K_n(ur) = \sum_{m=-\infty}^{\infty} e^{jm\phi}I_m(ur)K_{n+m}(ur_0) \tag{C.23}$$

Green's function: the general solution of equation

$$\left(\frac{\partial^2}{\partial\rho^2} + \frac{1}{\rho}\frac{\partial}{\partial\rho} - u^2 - \frac{m^2}{\rho^2}\right)F_m(\rho,\rho_0) = \frac{-\delta(\rho-\rho_0)}{\rho_0} \tag{C.24}$$

is

$$F_m(\rho,\rho_0) = \begin{cases} AI_m(u\rho) + BK_m(u\rho) & \text{if } \rho < \rho_0 \\[2mm] AI_m(u\rho) + BK_m(u\rho) - [I_m(u\rho)K_m(u\rho_0) - I_m(u\rho_0)K_m(u\rho)] \\ \qquad\qquad\qquad\qquad \text{if } \rho > \rho_0 \end{cases} \tag{C.25}$$

where A and B are arbitrary constants.

For more complete information on Bessel's functions, see Erdéleyi *et al.* (1953).

Appendix D   One-sided Fourier transforms

Consider a function $f_+(z)$ which is defined for $z > 0$ and is such that

$$|f_+(z)| < e^{-\tau_+ z} \quad \text{if} \quad z \to \infty \tag{D.1}$$

The one-sided Fourier transform

$$F_N(h) = \int_0^\infty f_+(z) \exp(-jhz)\, dz \tag{D.2}$$

is defined (i.e. analytic) on the negative half-plane $\text{Im } h < \tau_+$ and we have

$$\frac{1}{2\pi} \int_{-\infty+j\beta}^{\infty+j\beta} F_N(h) e^{jhz}\, dh = \begin{cases} f_+(z) & \text{if} \quad z > 0 \\ 0 & \text{if} \quad z < 0 \end{cases} \quad \ldots \beta < \tau_+ \tag{D.3}$$

Similarly consider a function $f_-(z)$ which is defined for $z < 0$ and is such that

$$|f_-(z)| < e^{-\tau_- z} \quad \text{if} \quad z \to -\infty \tag{D.4}$$

The one-sided Fourier transform

$$F_P(h) = \int_{-\infty}^0 f_-(z) \exp(-jhz)\, dz \tag{D.5}$$

is defined (i.e. analytic) on the positive half-plane $\text{Im } h > \tau_-$ and we have

$$\frac{1}{2\pi} \int_{-\infty+j\beta}^{\infty+j\beta} F_P(h) e^{jhz}\, dh = \begin{cases} 0 & \text{if} \quad z > 0 \\ f_-(z) & \text{if} \quad z < 0 \end{cases} \quad \ldots \beta > \tau_- \tag{D.6}$$

The subscripts N and P are intended to recall the analytical regions of the one-sided transforms. If these regions overlap, i.e. if $\tau_- < \tau_+$, the two-sided Fourier transform of $(f_+ + f_-)$ is $(F_N + F_P)$ defined in the region $\tau_- < \text{Im } h < \tau_+$.

The theorem of the derivative is expressed as follows for one-sided Fourier transforms

$$f_+'(z) \to jhF_N(h) - f_+(+0) \tag{D.7}$$

$$f_-'(z) \to jhF_P(h) + f_-(-0) \tag{D.8}$$

This yields for second derivatives

$$f_+''(z) \to (jh)^2 F_N(h) - jhf_+(+0) - f'(+0) \tag{D.9}$$

$$f_-''(z) \to (jh)^2 F_P(h) + jhf_-(-0) + f_-'(-0) \tag{D.10}$$

Index